ADVANCES IN ENZYMOLOGY

AND RELATED AREAS OF
MOLECULAR BIOLOGY

Volume 72

LIST OF CONTRIBUTORS

JOHN S. BLANCHARD, Department of Biochemistry, Albert Einstein College of Medicine, Bronx, NY 10461

ARTHUR J.L. COOPER, Departments of Biochemistry and of Neurology and Neuroscience, Cornell University Medical College, New York, NY 10021

DANIEL L. PURICH, Department of Biochemistry and Molecular Biology, University of Florida College of Medicine, Gainesville, FL 32610-0245

DIETER HÄUSSINGER, Medizische Universitätsklinik, Heinrich-Heine-Universität Düsseldorf, Moorenstrasse 5, D-40225 Düsseldorf, Germany

KARI I. KIVIRIKKO, Collagen Research Unit, Biocenter and Department of Medical Biochemistry, University of Oulu, Kajaanintie 52 A, FIN-90220, Finland

YOSHITAKA IKEDA, Department of Biochemistry, Osaka University Medical School, 2-2 Yamadaoka, Suita, Osaka 565, Japan

TAINA PIHLAJANIEMI, Collagen Research Unit, Biocenter and Department of Medical Biochemistry, University of Oulu, Kajaanintie, 52 A, FIN-90220, Finland

NIGEL G.J. RICHARDS, Department of Chemistry, University of Florida, Gainesville, FL 32611

GIOVANNA SCAPIN, Department of Biochemistry, Albert Einstein College of Medicine, Bronx, NY 10461

SHELDON M. SCHUSTER, Department of Biochemistry and Molecular Biology, University of Florida, Gainesville, FL 32610

JANET L. SMITH, Department of Biological Sciences, Purdue University, West Lafayette, IN 47907

NAOYUKI TANIGUCHI, Department of Biochemistry, Osaka University Medical School, 2-2 Yamadaoka, Suita, Osaka 565, Japan

HOWARD ZALKIN, Department of Biochemistry, Purdue University, West Lafayette, IN 47907

ADVANCES IN ENZYMOLOGY
AND RELATED AREAS OF MOLECULAR BIOLOGY

F. F. Nord, Founding Editor

AMINO ACID METABOLISM, Part A

Volume 72

Edited by DANIEL L. PURICH
University of Florida College of Medicine
Gainesville, Florida

WILEY

1998

AN INTERSCIENCE® PUBLICATION

John Wiley & Sons, Inc.
New York • Chichester • Weinheim • Brisbane • Singapore • Toronto

QP601
.A1
A3
Vol. 72
1998
039115236

This book is printed on acid-free paper. ∞

Library of Congress Catalog Card Number: **41-9213**

ISBN 0-471-24643-3

Printed in the United States of America.

10 9 8 7 6 5 4 3 2 1

CONTENTS

MAY 07 1998

PUBLISHER'S FOREWORD

Many of the most notable achievements in the molecular life sciences were initiated by enzymologists. The assertion of Hopkins that cell life is "an ordered sequence of events governed by specific catalysts" has become a truism in the analytical investigation of all life processes. The field of enzymology has grown by leaps and bounds over the past half-century, and with explosive growth of molecular and cell biology, new enzymes and enzymatic activities are still routinely found. Terms such as "ribozyme," "abzyme" (or catalytic antibodies), "molecular motors" (e.g., dynein and kinesin), and "polyprotein proteinases" have entered the biochemical literature in just the past decade or so.

Professor F. F. Nord began *Advances in Enzymology* in 1941 after he had been appointed Professor of Biochemistry at Fordham University. *Advances* is the successor to *Ergebnisse der Enzymforschung*, a periodic review series started in 1930, while he was still a chemistry professor in Berlin. His idea was to identify areas of enzymology that had undergone significant recent growth. He then sought out expert opinions of others to find an appropriate author to communicate the nature of those strides and to illustrate how these findings could be of broader interest.

This tradition of excellence was maintained by Professor Alton Meister who, with the pubication of Volume 35, succeeded Dr. Nord as the series editor. A man of enormous intellect and awareness of the discipline, Dr. Meister produced 37 volumes of *Advances in Enzymology*. He was proud of the chapters in this series, not so much as a testimony to his good judgment (though certainly his instincts and insights about enzymology were always on the mark), but as contributions of intrinsic worth to other practitioners of a field that has sprung up around enzyme catalysis and regulation. His last effort appeared as Volume 71 about a year after his untimely death in 1995.

The series will now be continued by Daniel L. Purich, Professor of Biochemistry and Molecular Biology at the University of Florida College of Medicine.

New York, New York   John Wiley & Sons, Inc.
March 1998

PREFACE

Advances in Enzymology is now nearing its seventh decade as the leading periodic and authoritative review of the latest scientific achievements in enzymology. This field concerns itself with the multifaceted nature of enzymes—their reaction properties; their kinetic behavior; their catalytic mechanisms; their regulatory interations; their expression from genes; their zymogen and other storage forms; their mutant forms (both naturally occurring and man-made); their associated pathophysiology; and their virtually limitless use in agriculture, nutrition, biomedicine, and biotechnology. As the new editor of this series, I shall endeavor to maintain the standard of excellence so masterfully established by Dr. Nord and so skillfully pursued by the late Alton Meister.

The value of *Advances in Enzymology* ultimately will only endure if practicing enzymologists judge its chapters to add value to their own pursuits. Accordingly, this volume marks the advent of several changes that should appeal to its readers. This and many subsequent volumes will be thematically organized so that readers of one chapter will be more likely to be interested in several or all of the other contributions. This monograph-within-a-monograph approach places added burdens on the contributors and editor for the timely appearance of each volume, but the final product should provide broader and more integrated perspective on any particular topic. Another new feature will be the inclusion of an abstract describing the scope and content of each chapter (the abstracts can be found in the front matter of the volume). While this was previously an unnecessary feature of earlier volumes in this series, abstracts allow users of information retrieval services to identify material of interest. In the future we may also be able to provide such information in advance of the actual publication date, thereby minimizing the lag time between a volume's first appearance and wider public awareness of its contents.

In seeking to present readers with the latest and most accurate information about enzyme action, I invite other enzymologists to contact me about their interests, their criticisms, and above all their ideas for improving the impact of this series.

Gainesville, Florida   DANIEL L. PURICH
March 1998

ABSTRACTS

Advances in the Enzymology of Glutamine Synthesis

DANIEL L. PURICH

Meister's proposal of a γ-glutamyl-P intermediate in the glutamine synthetase reaction set the scene for understanding how the stepwise activation of the carboxyl group greatly increased its susceptibility toward nucleophilic attack and amide bond synthesis. Topics covered in this review include: the discovery of the enzymatic synthesis of glutamine; the role of glutamine synthetase in defining the thermodynamics of ATPases; early isotopic tracer studies of the synthetase reaction; the proposed intermediacy of γ-glutamyl-phosphate; the mechanism of methionine sulfoximine inhibition; stereochemical mapping of the enzyme's active site; detection of enzyme reaction cycle intermediates; borohydride trapping of γ-glutamyl-P; positional isotope exchanges catalyzed by glutamine synthetase; regulation of bacterial enzyme; and a brief account of how knowledge of the atomic structure of bacterial glutamine synthetase has clarified ligand binding interactions. Concluding remarks also address how the so-called "Protein Ligase Problem" may be solved by extending the catalytic versatility of carboxyl-group activating enzymes.

Hepatic Glutamine Transport and Metabolism

DIETER HÄUSSINGER

Although the liver was long known to play a major role in the uptake, synthesis, and disposition of glutamine, metabolite balance studies across the whole liver yielded apparently contradictory findings suggesting that little or no net turnover of glutamine occurred in this organ. Efforts to understand the unique regulatory properties of hepatic glutaminase culminated in the conceptual reformulation of the pathway for glutamine synthesis and turnover, especially as regards the role of sub-acinar distribution of glutamine synthetase and glutaminase. This chapter describes these processes as well as the role of glutamine in hepatocellular hydration, a process that is the consequence of cumulative, osmotically active uptake of glutamine into cells. This topic is also examined in terms of the effects of cell swelling on the selective stimulation or inhibition of other far-ranging cellular processes. The pathophysiology of the intercellular glutamine cycle in cirrhosis is also considered.

Enzymes Utilizing Glutamine as an Amide Donor

HOWARD ZALKIN AND JANET L. SMITH

Amide nitrogen from glutamine is a major source of nitrogen atoms incorporated biosynthetically into other amino acids, purine and pyrimidine bases, amino-sugars, and coenzymes. A family comprised of at least sixteen amidotransferases are known to catalyze amide nitrogen transfer from glutamine to their acceptor substrates. Recent fine structural advances, largely as a result of X-ray crystallography, now provide structure-based mechanisms that help to explain fundamental aspects of the catalytic and regulatory interactions of several of these aminotransferases. This chapter provides an overview of this recent progress made on the characterization of amidotransferase structure and mechanism.

Mechanistic Issues in Asparagine Synthetase Catalysis

NIGEL G. J. RICHARDS AND SHELDON M. SCHUSTER

The enzymatic synthesis of asparagine is an ATP-dependent process that utilizes the nitrogen atom derived from either glutamine or ammonia. Despite a long history of kinetic and mechanistic investigation, there is no universally accepted catalytic mechanism for this seemingly straightforward carboxyl group activating enzyme, especially as regards those steps immediately preceding amide bond formation. This chapter considers four issues dealing with the mechanism: (a) the structural organization of the active site(s) partaking in glutamine utilization and aspartate activation; (b) the relationship of asparagine synthetase to other amidotransferases; (c) the way in which ATP is used to activate the β-carboxyl group; and (d) the detailed mechanism by which nitrogen is transferred.

Mechanisms of Cysteine S-Conjugate β-Lyases

ARTHUR J. L. COOPER

Mercapturic acids are conjugates of S-(N-acetyl)-L-cysteine formed during the detoxification of xenobiotics and during the metabolism of such endogenous agents as estrogens and leukotrienes. Many mercaturates are formed from the corresponding glutathione S-conjugates. This chapter focuses on (a) the discovery of the cysteine S-conjugate β-lyases; (b) the involvement of pyridoxal-5-phosphate; (c) the influence of the electron-withdrawing properties of the group attached to the sulfur atom; and (d) the potential of cysteine S-conjugates as pro-drugs.

γ-Glutamyl Transpeptidase:
Catalytic Mechanism and Gene Expression

NAOYUKI TANIGUCHI AND YOSHITAKA IKEDA

The γ-glutamyl transpeptidases are key enzymes in the so-called γ-glutamyl cycle involving glutathione synthesis, the recovery of its constituents, and in the transport of amino acids. This membrane-bound ectoenzyme thus serves to regulate glutathione synthesis. This chapter deals with the active site chemistry of γ-glutamyl transpeptidase, including the role of side-chain groups on the light subunit as well as several serine residues in the catalytic process. Also considered are genomic studies indicating (a) the presence of a single gene in mouse and rat; (b) the occurrence of multiple genes in humans; (c) the involvement of multiple promoters for gene expression; and (d) how these multiple promoters may play a role in the tissue-specific expression of γ-glutamyl transpeptidases.

Enzymology of Bacterial Lysine Biosynthesis

GIOVANA SCAPIN AND JOHN S. BLANCHARD

Bacteria have evolved three strategies for the synthesis of lysine from aspartate via formation of the intermediate diaminopimelate (DAP), a metabolite that is also involved in peptidoglycan formation. The objectives of this chapter are descriptions of mechanistic studies on the reactions catalyzed by dihydrodipicolinate synthase, dihydrodopicolinate reductase, tetrahydrodipicolinate N-succinyl-transferase, N-succinyl-L,L-DAP aminotransferase, N-succinyl-L,L-DAP desuccinylase, L,L-DAP epimerase, L,L-DAP decarboxylase, and DAP dehydrogenase. These enzymes are discussed in terms of kinetic, isotopic, and X-ray crystallographic data that allow one to infer the nature of interactions of each of these enzymes with its substrate(s), coenzymes, and inhibitors.

Collagen Hydroxylases and the Protein Disulfide Isomerase Subunit of Prolyl 4-Hydroxylases

KARI I. KIVIRIKKO AND TAINA PIHLAJANIEMI

Prolyl 4-hydroxylases catalyze the formation of 4-hydroxyproline in collagens and other proteins with an appropriate collagen-like stretch of amino acid residues. The enzyme requires Fe(II), 2-oxo-glutarate, molecular oxygen, and ascorbate. This review concentrates on recent progress toward understanding the detailed mechanism of 4-hydroxylase action, including: (a) occurrence and function of the enzyme in animals; (b) general molecular properties; (c) intracellular sites of hydroxylation; (d) peptide substrates and mechanistic roles of the cosubstrates; (e) insights into the development of antifibrotic drugs; (f) studies of the enzyme's subunits and their catalytic function; and (g) mutations that lead to Ehlers-Danlos Syndrome. An account of the regulation of collagen hydroxylase activities is also provided.

ADVANCES IN ENZYMOLOGY

AND RELATED AREAS OF
MOLECULAR BIOLOGY

Amino Acid Metabolism, Part A

Volume 72

Edited by DANIEL L. PURICH

This volume is dedicated to the late Professor Alton Meister, the immediate past Editor of *Advances in Enzymology,* in recognition of his major contribuitons to amino acid metabolism. This is Part A in a subseries, entitled "Amino Acid Metabolism." Topics in Part A should be of immediate interest to those who are broadly concerned with amino acid assimilation and metabolism. Investigators interested in enzyme mechanism and regulation will also find this volume to be especially valuable.

Reprint of Meister's Classic Paper from Journal of Biological Chemistry, 1960, PC39 (by permission of American Society for Biochemistry and Molecular Biology):
ACTIVATED GLUTAMATE INTERMEDIATE IN THE ENZYMATIC SYNTHESIS OF GLUTAMINE

By P. R. KRISHNASWAMY, VAIRA PAMILJANS, and ALTON MEISTER, *Department of Biochemistry, Tufts University School of Medicine, Boston, Massachusetts*

Advances in Enzymology and Related Areas of Molecular Biology, Volume 72: Amino Acid Metabolism, Part A, Edited by Daniel L. Purich
ISBN 0-471-24643-3 © 1998 John Wiley & Sons, Inc.

Activated Glutamate Intermediate in the Enzymatic Synthesis of Glutamine*

P. R. Krishnaswamy, Vaira Pamiljans, and Alton Meister

From the Department of Biochemistry, Tufts University School of Medicine, Boston, Massachusetts

(Received for publication, May 31, 1960)

The enzyme that catalyzes glutamine synthesis (Reaction 1), also catalyzes a transfer reaction in which γ-glutamylhydroxamate is formed from glutamine and hydroxylamine in the presence of catalytic amounts of orthophosphate, or arsenate, and adenosine diphosphate; (see Meister (1) for review of literature).

$$\text{Glutamate} + \text{NH}_4^+ + \text{ATP} \xrightarrow{\text{Mg}^{++}} \text{glutamine} + \text{ADP} + \text{P}_i \quad (1)$$

No intermediates have been found in these reactions and partial reactions have not been demonstrated. Glutamate is not an obligatory intermediate in the transfer reaction (2), and studies with synthetic γ-glutamylphosphate did not provide evidence for intermediate participation of this compound, although an enzyme-bound anhydride is not excluded (3). Studies on the synthesis of D- and L-glutamines and γ-glutamylhydroxamates suggested a mechanism involving an initial activation of glutamate of relatively low optical specificity, followed by a more specific reaction of activated glutamate with NH_4^+ (4). The observation that pyrrolidone carboxylate was formed, when a mixture of L-glutamate, ATP, Mg^{++}, and enzyme was heated (5), provided a significant clue to an activated glutamate intermediate.

We now report evidence for formation, in the absence of NH_4^+, of enzyme-bound γ-carboxyl-activated L- and D-glutamates associated with cleavage of ATP. It is well known that γ-glutamyl derivatives cyclize more readily than glutamate when heated. We reasoned that if an enzyme-bound γ-activated glutamate were formed, it might cyclize *very* rapidly on heating. Therefore, we incubated large amounts of highly purified sheep brain enzyme with ATP, C^{14}-L-glutamate and Mg^{++}; when this mixture was heated at 100° for 5 minutes, pyrrolidone carboxylate was formed (Table I). Much less pyrrolidone carboxylate was

* The authors acknowledge the generous support of the National Heart Institute, National Institutes of Health, and the National Science Foundation. We thank Mr. Daniel Wellner for assistance in the ultracentrifugation studies.

3

TABLE I *Formation of pyrrolidone carboxylate*

Experiment No.	Reaction mixtures*	Pyrrolidone carboxylate formed	
		With L-glutamate†	With D-glutamate†
		mμmoles	*mμmoles*
1	Glutamate + ATP + Mg^{++} + enzyme	16.8	17.2
2	Glutamate + ATP + Mg^{++} + enzyme‡	0.7	0.5
3	Glutamate + Mg^{++} + enzyme	0.5	0.2
4	Glutamate + ATP + enzyme	0.7	0.5
5	Glutamate + ATP + Mg^{++}	0.7	0.5
6	Glutamate + ATP + Mg^{++} + enzyme + NH$_4^+$	0.6	4.3

* The reaction mixtures (0.5 ml) contained enzyme, 5 mg (see footnote†); glutamic acid-1-C^{14} (L: 0.17 μmole, 910,000 c.p.m.; D: 0.19 μmole, 92,000 c.p.m.); ATP, 5 μmoles; MgCl$_2$, 25 μmoles; NH$_4$Cl, 0.05 μmole; and 2-amino-2-(hydroxymethyl)-1,3-propane-diol-HCl buffer, pH 7.2, 25 μmoles. After 10 minutes (37°), they were heated at 100° for 5 minutes and then cooled; 0.5 ml of ethanol was added, and the precipitated protein was removed by centrifugation. The supernatant solution was analyzed for pyrrolidone carboxylate and glutamine by paper strip chromatography (n-butanol-acetic acid-H$_2$O; 4:1:1) and electrophoresis (pH 5.0), respectively. Sections of the strips were counted in a gas flow counter.

† The enzyme was about 70% pure as judged by ultracentrifugation studies. It was purified about 2000-fold from sheep brain by isoelectric precipitation, adsorption on Ca$_3$(PO$_4$)$_2$ gel, and differential heat inactivation. The latter step makes use of the interesting fact that ATP (+Mg^{++}) protects the enzyme against denaturation at 55°. Preparations of greater purity were obtained by (NH$_4$)$_2$SO$_4$ fractionation; however, this step was not employed for enzyme used in these studies, and precautions were taken to eliminate NH$_4^+$ from all solutions and reagents employed. The purified enzyme contained small amounts of material that absorbed at 260 mμ, the nature of which is being studied. Added ribonuclease did not inhibit the enzyme.

‡ The enzyme was inactivated before the experiment by heating at 100° for 5 minutes.

formed with heat-denatured enzyme, when enzyme, ATP, or Mg^{++} was separately omitted, when ATP was replaced by AMP, ADP, CTP, ITP, UTP, or GTP, or when enzyme was replaced by other proteins. When preincubation was reduced to 10 seconds, there was a 30% decrease in the pyrrolidone carboxylate formed. Heating at 100° for 1 to 20 minutes gave the same results, and about 70% as much pyrrolidone carboxylate was formed when the mixture was heated at 55° for five minutes. This suggests a highly reactive intermediate that undergoes cyclization about 600 times more rapidly than glutamine which does not cyclize appreciably in 5 minutes at 100° at pH 7.2. The amounts of pyrrolidone carboxylate formed were of the same order as that of the enzyme; however, exact stoichiometry has not yet been attempted. Only traces of, or no, glutamine were formed. When NH_4^+ was added to the system, glutamine was formed and pyrrolidone carboxylate formation was reduced to the blank level. Studies with D-glutamate gave similar results; however, pyrrolidone carboxylate formation in the presence of NH_4^+ was not reduced to blank values. This suggests that under these conditions orientation of the D-glutamate intermediate on the enzyme may be less favorable with respect to reaction with NH_4^+ than that of L-glutamate.[1] Although the α-amino group of glutamate may be considered to replace NH_4^+ in the cyclization reactions, the rates of cyclization (at 55°) do not differ markedly for the two isomers of glutamate. It seems probable that the cyclization reactions, like the reactions with hydroxylamine (*cf.* Levintow and Meister (4)), reflect presence of an activated carboxyl group. When a mixture equivalent to that of 1 in Table I was centrifuged in a separation cell (7), there was a loss of C^{14} corresponding to 17.1 mμmoles of L-glutamate from the upper portion of the cell, when the enzyme passed below the separation plate; there was no sedimentation of glutamate in the controls. In ultrafiltration studies of these reaction mixtures, with Schleicher and Schuell membranes, C^{14}-glutamate remained associated with the protein which did not pass the membrane, whereas this result was not observed in the controls. The amount of glutamate bound to protein in these studies was equal to the pyrrolidone carboxylate formed in the heating experiments. Experiments with ATP^{32} similar to those reported in Table I, except that less ATP was used, are described in Table II. The mixtures were treated with ethanol and centrifuged to remove protein. Only traces of phosphate were found in the absence of glutamate, whereas significant amounts of phosphate were formed with L- or D-glutamates. No glutamine was formed in controls with C^{14}-glutamates. Pyrrolidone carboxylate in amounts equivalent to phosphate was found in parallel heating experiments. We have found that the purified brain

---

[1] It is possible that activation of D-glutamate occurs in the intact animal; conversion to D-pyrrolidone carboxylate might occur in the presence of relatively low physiological concentrations of ammonia (*cf.* Ratner (6)).

P. R. *Krishnaswamy, V. Pamiljans, and A. Meister*

Table II *Formation of orthophosphate*

Reaction mixtures*	Orthophosphate formed	
	L-glutamate	D-glutamate
	mμmoles	*mμmoles*
Glutamate + ATP³² + Mg⁺⁺ + enzyme	6.6	5.6
ATP³² + Mg⁺⁺ + enzyme	0.2	0.1

* The reaction mixtures (0.2 ml) contained ATP, β, γ-P³², 60,000 c.p.m., 0.1 μmole; enzyme, 3.8 mg.; MgCl₂, 0.5 μmole; buffer, 10 μmoles; and glutamate, 0.068 μmole; incubated at 37° for 1 minute and then treated with 0.2 ml of ethanol. After centrifugation, the supernatant solution was chromatographed on paper (isopropyl ether-formic acid, 3:2) and the orthophosphate was determined.

enzyme does not catalyze exchange between orthophosphate and ATP in the presence of glutamate and Mg^{++} (NH_4^+ absent); the mild ethanol treatment may release phosphate from a labile bound form.

The data indicate that binding of glutamate to the enzyme requires ATP and Mg^{++}. Binding is associated with cleavage of ATP and the formation of γ-carboxyl-activated glutamate; these events take place in the absence of NH_4^+ and without formation of glutamine. These findings should be considered in relation to the suggestions that (a) glutamate and NH_4^+ react before reaction with ATP (8), (b) that ATP cleavage and reaction with NH_4^+ occur essentially simultaneously (9, 10), and (c) that ATP is cleaved to orthophosphate before reaction with glutamate (11). The present studies suggest binding of ATP and Mg^{++} before binding of glutamate and favor stepwise arrangement of reactants on the enzyme with NH_4^+ reacting last. When considered in the light of O¹⁸ data (12, 13), our results are consistent with but do not prove formation of enzyme-bound γ-glutamyl-phosphate. It seems probable that the same intermediate is formed in the synthesis, reversal of synthesis (14), and transfer reactions. The requirement for ADP in the transfer reaction suggests that ADP is also bound to the enzyme, and that ADP may be required for binding of glutamine when the reaction proceeds in the reverse direction as well as for the integrity of the activated glutamate intermediate. Further studies on the intermediate and on the role of ADP are being carried out.

6

REFERENCES

1. MEISTER, A., *Physiol. Revs.*, **36**, 103 (1956); in J. T. Edsall (Editor) *Amino acids, proteins, and cancer biochemistry*, Academic Press, Inc., New York, 1960, p. 85.
2. LEVINTOW, L., MEISTER, A., KUFF, E. L., AND HOGEBOOM, G. H., *J. Am. Chem. Soc.*, **77**, 5304 (1955).
3. LEVINTOW, L., AND MEISTER, A., *Federation Proc.*, **15**, 299 (1956).
4. LEVINTOW, L., AND MEISTER, A., *J. Am. Chem. Soc.*, **75**, 3039 (1953).
5. KRISHNASWAMY, P. R., PAMILJANS, V., AND MEISTER, A., *Federation Proc.*, **19**, 349 (1960).
6. RATNER, S., *J. Biol. Chem.*, **152**, 559 (1944).
7. YPHANTIS, D. A., AND WAUGH, D. F., *J. Phys. Chem.*, **60**, 630 (1956).
8. BOYER, P. D., MILLS, R. C., AND FROMM, H. J., *Arch. Biochem. Biophys.*, **81**, 249 (1959).
9. BUCHANAN, J. M., AND HARTMAN, S. C., *Advances in Enzymol.*, **21**, 199 (1959).
10. BUCHANAN, J. M., HARTMAN, S. C., HERRMANN, R. L., AND DAY, R. A., Symposium on Enzyme Reaction Mechanisms, (sponsored by the Biology Division, Oak Ridge National Laboratory, Oak Ridge, Tennessee); *J. Cellular Comp. Physiol.*, **54**, 139 (1959).
11. WIELAND, TH., PFLEIDERER, G., AND SANDMANN, B., *Biochem. Z.*, **330**, 198 (1958).
12. KOWALSKY, A., WYTTENBACH, C., LANGER, L., AND KOSHLAND, D. E., JR., *J. Biol. Chem.*, **219**, 719 (1956).
13. BOYER, P. D., KOEPPE, O. J., AND LUCHSINGER, W. W., *J. Am. Chem. Soc.*, **78**, 356 (1956).
14. LEVINTOW, L., AND MEISTER, A., *J. Biol. Chem.*, **209**, 265 (1954).

7

ADVANCES IN THE ENZYMOLOGY OF GLUTAMINE SYNTHESIS

By DANIEL L. PURICH, *Department of Biochemistry and Molecular Biology, University of Florida College of Medicine, Gainesville, Florida 32610*

CONTENTS

Advances in Enzymology and Related Areas of Molecular Biology, Volume 72: Amino Acid Metabolism, Part A, Edited by Daniel L. Purich
ISBN 0-471-24643-3   ©1998 John Wiley & Sons, Inc.

I. Introduction

Few would quarrel with the assertion that glutamine is a phenomenally versatile metabolite—a building block for proteins, an amide nitrogen donor in numerous biosynthetic reactions, and interorgan carrier of latent ammonia, a wound healing adjuvant, an effector in cellular hydration, and a glial cell metabolite with roles in neuronal nutrition and signal transmission. What may be less well recognized today is the extent to which glutamine synthetase itself has factored so centrally in the development of modern biochemistry, particularly mechanistic enzymology. While enzymologists had for years recognized that amide bond formation requires activation of glutamine's γ-carboxylate to generate a good leaving group at physiologic pH, notions of enzymic catalysis had for a long time centered on fully concerted mechanisms in which the bond-making and bond-breaking steps occurs virtually simultaneously. Indeed, even with Racker's earlier demonstration of a thiolester intermediate in the glyceraldehyde-3-P dehydrogenase reaction, there was lingering resistance to the general concept that bond breaking and bond making are among the most facile steps of enzymic catalysis. Meister's proposal of a γ-glutamyl-phosphate intermediate set the scene for understanding the stepwise activation of carboxyl groups, thereby increasing their susceptibility toward nucleophilic attack and amide bond synthesis. The goal of this chapter is to consider many of the milestones in characterizing the intermediates formed in the glutamine synthetase reaction, a field largely pioneered by the late Alton Meister. Those efforts helped establish fundamental organizing principles of contemporary enzymology, and they set the scene for the systematic investigation of other carboxyl group activating reactions.

II. Meister's Legacy to the Enzymology of Glutamine Synthesis

Meister authored upward of 500 scientific papers, and his enormous and sustained productivity, as well as the fundamental nature of the work itself, entreats one to consider the characteristics of his investigative style. As one familiar with his research career, I am reminded of the writings of the biophysicist Platt, who in the 1964 published an essay entitled "Strong Inference." Platt's goal was to consider the tenets of scientific investigation that had brought such great strides in such seemingly disparate fields as modern physics and molecular biology. He attributed this singular success to the ability to frame and solve problems concisely and decisively, and he asserted that these rapidly burgeoning fields shared a particularly effective method of doing scientific research, progressing systematically through a cumulative approach based on inductive inference, a method for which he coined the term strong inference (Platt, 1964). Tracing the elements of strong inference to the inductive approaches of Roger Bacon, Platt argued that the strategy involves the application of the following steps "to every problem in science, formally and explicitly and regularly: (a) devising alternative hypotheses; (b) devising a crucial experiment (or several of them), with alternative possible outcomes, each of which will, as nearly as possible, exclude one or more of the hypotheses; (c) carrying out the experiment so as to get a clean result; and (d) recycling the procedure, making sub-hypotheses or sequential hypotheses to refine the possibilities that remain; and so on." Platt suggested that many scientists get too wrapped up in the doing of experiments, rather than carefully considering the logic that maps the most effective path for solving problems, most often through recursive rounds of inference and critical experimentation. Enzymologists were among the first biochemists to recognize that systematic investigations of catalysis and regulation are also often greatly facilitated though the use of models. In principle, model building offers at least three advantages: first, because models are pictographic representations of a chemical process, such depictions often reveal gaps in one's understanding; second, the best models are minimalistic, quickly identifying essential mechanistic features; and third, models are generative, facilitating formulation of hypotheses about the nature of rival mechanistic schemes.

Anyone familiar with Meister's systematic attack on the enzymology of carboxyl group-activating enzymes, especially those of us who have sought to make our own contributions to this same endeavor, is well aware that he applied his own special brand of strong inference throughout his illustrious research career. Starting with his early work on transamination and extending throughout his exhaustive inquiry into the enzymology of glutamine and glutathione, he routinely created and tested models for enzyme mechanisms as well as developed entirely new metabolic pathways. Never one to adopt a single line of inquiry or a single test of a hypothesis, Meister's systematic and multiple attack on a problem, as well as his masterful talent of keeping a line of inquiry on course, allowed him to fulfill far more than the aspiration of making scientific discoveries of lasting value. The approach allowed him to pursue a life-long love affair with the biochemistry of amino acids.

III. Milestones in Glutamine Biochemistry

In seeking to develop better tools for discriminating the detailed catalytic and regulatory mechanisms for this enzyme, Meister and other notable enzymologists (among them Krebs, Koshland, Boyer, Stadtman, and Rose) substantially advanced our understanding of fundamental concepts of enzyme catalysis and metabolic control. This task was all the more challenging because there is no chromophoric cofactor that allows one to examine directly the formation and turnover of catalytic intermediates. While space does not permit a full consideration of Meister's achievements, much less the work of those mentioned above, a brief retrospective of several milestones in the biochemistry of glutamine synthetase appears to be well warranted.

A. DISCOVERY OF GLUTAMINE SYNTHETASE ACTIVITY

Krebs (1935) first reported that the amino acid glutamine could be synthesized from ammonia and glutamate in tissue slices of pig and rabbit kidney. He had conducted a series of experiments to show that glutamate, more than any other α-amino acid, increases oxygen uptake and that addition of glutamate also results in a reduction of ammonium ion. The compound formed from ammonia and glutamate was acid labile, and Krebs carried out additional studies with brain cortex slices and retina to demonstrate the synthesis of what he concluded to be L-glutamine. In fact, Krebs also remarked that the synthesis of glutamine depended on tissue respiration, an observation that presaged the involvement of adenosine triphosphate (ATP). Later work by Speck (1949) demonstrated that pigeon liver extracts form glutamine from glutamate and ammonia in the presence of phosphate, magnesium ion, and molecular oxygen, and he postulated the involvement of ATP formed during respiration. The stoichiometric reaction of ATP, glutamate, and ammonia was later directly demonstrated in aqueous extracts of pigeon liver and sheep brain acetone powders with the products identified as glutamine, adenosine diphosphate (ADP), and orthophosphate (P_i) (Elliott, 1951). Hydroxylamine can substitute for ammonia to form γ-glutamylhydroxamate, and glutamine synthetase also catalyzes the so-called γ-glutamyl transferase reaction even in the absence of added ATP:

$$\text{ADP, Me}^{2+}\text{, As}_i$$

$$\text{Glutamine} + NH_2OH \rightleftharpoons \gamma\text{-Glutamylhydroxamate} + NH_3$$

where the divalent metal ion (Me^{2+}) can be magnesium or manganese. It is noteworthy that arsenate (As_i) can substitute for orthophosphate in this reaction, and the γ-glutamyl transferase reaction probably offered the first hint that the glutamine synthetase might involve transient activation of the γ-carboxylate.

B. ATPASE THERMODYNAMICS

Because ATP hydrolysis is the universal driving force in metabolism, membrane transport, motility, and memory processes, the thermodynamics of the ATPase reaction is of enduring fundamental importance. The Gibbs free energy of ATP hydrolysis (and likewise the corresponding values for the hydrolysis of the other nucleoside-5'-triphosphates) depends on several variables. These variables consist mainly of the hydrogen ion concentration, absolute temperature, the concentration of uncomplexed magnesium, the presence of other monovalent (viz., sodium and potassium ions), and divalent cations (particularly calcium ion), as well as the solution ionic strength. Nonetheless, the ATPase reaction favors product formation to such a great extent that one cannot directly obtain sufficiently accurate experimental determinations for ATPase mass action ratio (i.e., $[ADP][P_i]/[ATP]$) by starting from ATP in the presence of an ATPase. In this regard, early investigators recognized that the glutamine synthetase and glutaminase reactions offered special advantages for those seeking accuracy and precision in determinations of the equilibrium constant for ATP hydrolysis. In this case, summation of these reactions yields the ATPase mass action ratio:

$$\text{Glutamate} + NH_3 + ATP = \text{Glutamine} + ADP + P_i$$

$$\underline{\text{Glutamine} + H_2O = NH_3 + \text{Glutamate}}$$

$$ATP + H_2O = ADP + P_i$$

Alberty (1968; 1969) and Phillips (1969) were among the first to recognize that the ability of magnesium ion to form a higher affinity complex with the ATP^{4-} form than with $HATP^{3-}$ explains why the ATPase equilibrium is so strongly influenced (a) by the concentration of "free" or uncomplexed magnesium ion, (b) by the concentration of protons that alter the protonation state of the nucleotide, (c) by the ionic strength of the supporting electrolyte and buffer, and (d) by the presence of other monovalent and divalent cations. More recently, Alberty and Goldberg (1992) presented a detailed consideration of the *transformed* Gibbs energy as the chief criterion for chemical equilibrium at specified temperature, pressure, pH, concentration of free magnesium ion, and ionic strength. One begins by writing the apparent equilibrium constant in terms of the total concentrations of reactants, like ATP_{total}, rather than in terms of the concentrations of individual molecular species, such as $MgATP^{2-}$, ATP^{4-}, or $HATP^{4-}$. The standard transformed Gibbs energy can then be calculated from the Gibbs energy by using the apparent equilibrium constant as well as the temperature dependence of the apparent equilibrium constant at specified pressure, pH, concentration of free magnesium ion, and ionic strength. From the apparent equilibrium constants and standard transformed enthalpies of reaction that have been measured in the ATP series and the dissociation constants of the weak acids and magnesium complexes involved, Alberty and Goldberg (1992) showed that one can calculate standard Gibbs energies of formation and standard enthalpies of formation of the species involved at zero ionic strength. They followed the convention that the standard Gibbs energy of formation as well as the standard enthalpy of formation for adenosine in dilute aqueous solutions be set equal to zero. On the basis of this convention, standard transformed Gibbs energies of formation and standard transformed enthalpies of formation of ATP, ADP, AMP, and adenosine at 298.15 K, 1 bar, pH 7, a concentration of free magnesium ions of 10^{-3} M, and an ionic strength of 0.25 M have been calculated. It is also worth noting that Alberty and Cornish-Bowden (1993) recently presented a cogent analysis of how this formalism allows one to understand the relationship between the apparent equilibrium constant "K" for a biochemical reaction and the equilibrium constant, K, for a reference reaction written as a chemical equation.

Finally, lest one consider the usage of Gibbs free energy (G) in biochemistry to be a cut and dried issue, Welch (1985) suggested that the symbol 'ΔG', as discussed in most biochemical textbook calculations of free energy change (e.g., freeze-clamp studies of steady-state levels of metabolites in a particular tissue), can lead to erroneous conclusions about biological processes. To describe the thermodynamic behavior of metabolic reactions within cells, he favors the use of the instantaneous change, symbolized by the expression ($\Delta G/\Delta \xi$), where ξ is the degree of advancement of the reaction. His article offers mathematical and graphical analyses of a sample reaction to demonstrate the fundamental difference between ΔG and $\Delta G/\Delta \xi$.

C. EARLY ISOTOPIC TRACER STUDIES OF THE GLUTAMINE SYNTHETASE REACTION

There was general recognition early on that the synthesis of glutamine might conceivably involve the transient formation of a covalent intermediate. Two rival hypotheses emerged: (a) that an amidophosphate intermediate could first form from ATP and ammonia, followed by attack of the γ-carboxylate on phosphorus with the expulsion of ammonia; and (b) that a γ-glutamyl-phosphate compound formed by attack of the γ-carboxylate on a MgATP complex, thereby expelling P_i and allowing NH_3 attack on the acyl-P to form glutamine. While a number of investigators sought to demonstrate the formation of a covalent compound directly, this was confounded by the reactivity of an unstable intermediate as well as the difficulty in preventing the presence of trace amounts of ammonia.

A major lead was provided by Boyer et al. (1956) who used [γ-^{18}O]-glutamate to trace the entry of oxygen atoms during the catalysis of the overall reaction from glutamate, ATP, and ammonia. These investigators discovered that ^{18}O-orthophosphate was synthesized, and Kowalsky et al. (1956) likewise showed the stoichiometric conversion of ^{18}O-orthophosphate from [γ-^{18}O]glutamate, with little or no incorporation of labeled oxygen into ADP. This finding was consistent with two possibilities: (1) a fully concerted mechanism whereby ammonia attack at the γ-carboxylate led to the direct transfer of one of the two carboxylate oxygen atoms to ATP to produce ADP and P_i; and (2) the stepwise synthesis of the γ-carboxyl-P from ADP and glutamate, followed by nucleophilic attack of NH_3 attended by release of P_i bearing an oxygen atom once held by the γ-carboxylate. These ideas also fit with the findings of Levintow and Meister (1954) who had examined the γ-glutamyl transferase reaction in the presence of radiocarbon-labeled glutamate. They concluded that the fact that no label appeared in either glutamine or the γ-glutamylhydroxamate argued that the amide and hydroxamate were sufficiently activated to form an unstable covalent intermediate, whereas glutamate was not.

While isotope exchange, especially that of Boyer et al. (1959) and Wedler and Boyer (1973), demonstrated that some of the steps proceeded at rates faster than the rate-determining steps, for some time equilibrium isotope exchange played no further role in determining the chemical processes involved in the catalytic mechanism of glutamine synthetase. On the other hand, equilibrium isotope exchange measurements did demonstrate promise for investigating how regulatory effectors alter the catalysis of enzymes (Wedler and Boyer, 1973; Purich and Allison, 1980). It remained for Rose to develop the positional isotope exchange protocol in the mid-1970s (see Section III.H).

D.   INTERMEDIACY OF γ-GLUTAMYL-PHOSPHATE IN GLUTAMINE SYNTHETASE REACTION

The next crucial step in elucidating the reaction mechanism was taken by Krishnaswamy et al. (1962) who reacted the enzyme with ATP and [^{14}C] glutamic acid in NH_3 depleted solutions. This procedure resulted in the synthesis of pyrrolidone carboxylate (also known as pyroglutamate and 5-oxoproline), a cyclic compound that can also be inefficiently formed in the nonenzymatic dehydration of glutamate or in the nonenzymatic displacement of the amide nitrogen of glutamine by nucleophilic attack of the α-amino group on the γ-amide. This suggested that exposure to the enzyme facilitated formation of a suitably activated covalent compound corresponding to the γ-acyl-P, which upon brief heating to 100°C would undergo intramolecular attack to form the cyclic amino acid. Both D- and L-glutamate were effective in the enzyme-catalyzed synthesis of the D and L forms of pyrrolidone carboxylate, a finding that was consistent with the generally recognized ability of the enzyme to use either enantiomer of glutamate. Well aware that γ-glutamyl derivatives are far more susceptible to nonenzymatic cyclization, Krishnaswamy et al. (1962) went on to demonstrate that glutamine formed the pyrrolidone-carboxylate some 600 times more slowly than in the corresponding enzymic step observed in the above experiment. This observation allowed Meister (1962) to offer a scheme involving the mandatory participation of an acyl-P intermediate in the glutamine synthetase reaction. Remarkably, that scheme (now modified modestly, as shown in Fig. 1) has withstood the test of time, and all experiments over the ensuing three decades have verified one or more of the central features of Meister's γ-glutamyl-P model.

Khedouri et al. (1964) chemically synthesized β-amino-glutaryl-P and demonstrated the enzymatic synthesis of β-amino-glutaramic acid (or β-glutamine) by ovine brain glutamine synthetase in the presence of ammonium ions. Likewise, they found that in the presence of hydroxylamine, β-amino-glutaryl-phosphate was enzymatically transformed into β-amino-glutaryl-hydroxamate. Moreover, when the synthetic acyl-P was incubated with ADP and enzyme, ATP was formed. None of these reactions occurred in the absence of enzyme, and β-aspartyl-P was completely ineffective in the above cited reactions. Later, studies by Allison et al. (1977) addressed the ter-reactant initial rate kinetics of β-glutamate, ATP, and hydroxylamine as substrates for this enzyme (see Section III.G). Together, these studies set the stage for rapid mix–quench experiments to measure the rates of converting the acyl-P intermediate to ATP in the presence of ADP or to glutamine in the presence of ammonium ion. This finding would allow one to identify the kinetic and thermodynamic barriers encountered by the synthetase.

E. MECHANISM OF METHIONINE SULFOXIMINE INHIBITION

Of the known enzymatic activities of glutamine synthetase, none is more revealing of the acyl-P intermediate than the inhibitory properties of methionine sulfoximine, or MSO (Meister, 1974). Of the four MSO stereomers, only L-methionine-S-sulfoximine irreversibly

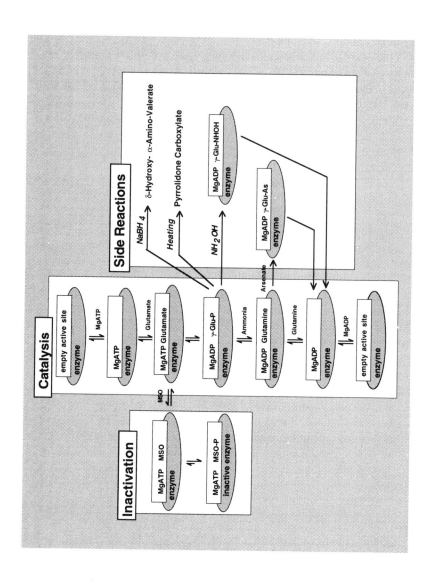

18

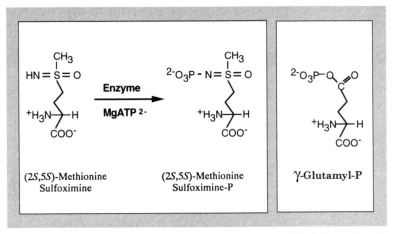

Figure 2.   Irreversible inhibition of glutamine synthetase by sulfoximine methionine. Only the (2S, 5S)-diasteriomer of MSO acts as a suicide inhibitor; the other three diasteriomers bind reversibly, yet fail to undergo phosphorylation. (*Note*: The phosphorylated inhibitor mimics the geometrical configuration of the proposed covalent intermediate formed in the synthetase reaction.)

inhibits glutamine synthetase, and the phosphorylated inhibitor appears to mimic the transition state of the synthetase reaction (Fig. 2). Methionine sulfoximine competes with both ammonia and glutamate for the ovine enzyme, suggesting that it occupies both pockets within the enzyme's active site. Computer-assisted mapping of ligand interactions indicates that the sulfoximine oxygen atom of L-methionine-(S)-sulfoximine binds in place of the acyl group's oxygen and that the methyl group occupies the ammonia binding site. In fact, such geometric considerations allow one to rationalize why

Figure 1.   Integrated mechanism for glutamine synthetase. The diagram shows the main route of catalysis as well as enzyme inactivation by methionine sulfoximine, and other side reactions that are now recognized as hallmarks of the formation of an acyl-P intermediate. (*Note*: Although the catalytic path implies an ordered binding mechanism with ATP as the leading substrate, later kinetic experiments have clearly established a random pathway for substrate addition.)

covalent phosphorylation takes place on the sulfoximine nitrogen of MSO. A similar steric orientation cannot be achieved with either of the D-isomers of MSO, and while L-methionine-(R)-sulfoximine does bind to the enzyme, the reversed positions of the sulfoximine nitrogen and oxygen atoms do not permit efficient phosphorylation. Meister (1974) also suggested (a) that the oxygen atom might be too acidic to attack ATP, and (b) that the observed binding of the methyl group in place of ammonia indicates that a hydrophobic interaction with un-ionized ammonia occurs, as opposed to ammonium ion. The latter would certainly favor nucleophilic attack on the acyl-P intermediary. Notably, inhibition by D- and L-methionine sulfone is reversible and can be easily reversed at higher glutamate concentrations; moreover, L- or D-methionine sulfone inhibition becomes more potent, but not irreversible in the presence of ATP.

Shrake et al. (1982) investigated the interactions of *Escherichia coli* glutamine synthetase with the resolved L-(S)- and L-(R)-diastereoisomers of the substrate analogue L-methionine-(SR)-sulfoximine. Reversible binding of the (S) isomer to unadenylylated manganese enzyme showed a stoichiometry of 1 equiv per subunit and negative cooperativity with a Hill coefficient of 0.7. The affinity of this enzyme complex was (a) highest for the (S) isomer; (b) lowest with the (R) isomer; and (c) intermediate for an equimolar (S) and (R)-isomer mixture. The affinity for the (S)-isomer was enhanced greater than 35-fold by ADP and was decreased approximately threefold by adenylylation of the enzyme. Earlier work by these investigators showed that UV spectral perturbations were markedly different for the binding of commercial L-methionine-(SR)-sulfoximine to unadenylylated and adenylylated manganese enzymes (Shrake et al., 1980). Yet, essentially the same protein difference spectrum is obtained for binding the resolved (S) and (R) diastereoisomers, and equimolar mixture of (S) and (R) isomers, and the commercial (S)- and (R)-isomeric mixture to a particular enzyme complex.

F. STEREOCHEMICAL MAPPING OF LIGAND INTERACTIONS AT THE ACTIVE SITE OF GLUTAMINE SYNTHETASE

Although X-ray crystallography provides the most definitive information on the stereochemical considerations governing substrate binding, atomic structures of enzyme complexes often fail to reveal the nature of ligand-induced conformational changes that strongly influence the kinetics and thermodynamics of enzyme–ligand interactions. Nonetheless, one can frequently infer stereochemical relationships by comparing observed binding constants, inhibition constants, and in some cases Michaelis constants, for interactions of inhibitors and substrates with enzymes (Gass and Meister, 1970; Purich et al., 1973). In the case of sheep brain glutamine synthetase, the goal was to discover why both D- and L-glutamate serve as substrates, and Gass and Meister (1970) studied these substrates along with 10 of their monomethyl derivatives. Of the latter, only three were active as substrates, and by making the assumption that amino acid substrates bind to glutamine synthetase in their most extended conformations, these investigators found that those hydrogens, whose methyl substituents yielded enzymatic activity, lie on one side of the molecule (i.e., the side that resides behind the plane for the structures of L-glutamate and D-glutamate as shown in Fig. 3). This finding led to the hypothesis that the enzyme approaches its amino acid substrate from above the plane of the paper, thereby permitting methyl groups of the three active methyl derivatives to project backward and away from the enzyme's surface. To test this proposal, these investigators prepared L-*cis*-1-amino-1,3-dicarboxy-cyclohexane as a conformationally restricted L-glutamate analogue wherein the two backward-extending hydrogens of D- and L-glutamate are connected by the three additional methylene groups within the cyclohexane ring (see Fig. 3). This analogue proved to be an effective substrate for the sheep brain enzyme, thereby fixing the stereochemistry of enzyme–substrate complexation without any atomic-level information on the enzyme's structure.

In retrospect, this conformationally restricted derivative would have also served as a valuable mechanistic probe of the mechanism of carboxyl group activation, because reaction of the analogue with ATP in an ammonia-free solution should form the corresponding acyl-P intermediate, and any intramolecular attack of the α-amino group to a pyroglutamate-like product (resembling a pyrrolidone carboxylate) should be far less likely. A similar argument applies to the use of β-glutamate, which should form an acyl-P compound that resists intramolecular attack by the amino group.

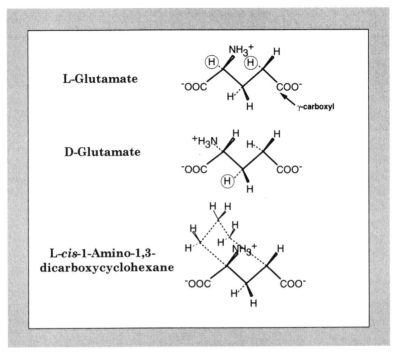

Figure 3.   Stereochemical representation of L-glutamate, D-glutamate, and L-*cis*-1-amino-1,3-dicarboxycyclohexane. Both D- and L-glutamate as well as aminoadipic acid, are substrates for the enzyme, but only 3 of 10 monomethyl derivatives of D- and L-glutamate are substrates. Assuming that the substrates are completely extended when bound to the enzyme, active substrates were obtained when the circled hydrogens are substituted by a methyl group. These hydrogens lie behind the plane of the paper, and they most probably face outward from the enzyme-bound substrate toward the bulk solvent. L-*cis*- 1-Amino-1,3-dicarboxycyclohexane is an alternative substrate for glutamate in the glutamine synthetase reaction. Note that the cyclohexane ring is connected by three successive methylenes that bridge the two replaceable (circled) hydrogen atoms in L-glutamate.

G.   THE KINETIC REACTION MECHANISM OF GLUTAMINE SYNTHETASE

The kinetic reaction mechanism of multisubstrate enzymes deals principally with the determination of the substrate binding order. Such information can often provide valuable evidence for the occurrence of ordered conformational changes that are evoked by substrate and/or cofactor binding to an enzyme. For this reason, enzyme kineticists have made a sustained effort to develop appropriate theory and experimental strategies to explain the rate behavior of multisubstrate enzymes. In the case of glutamine synthetase, although there is compelling evidence for the involvement of a γ-glutamyl-P intermediate in the catalytic process, neither the mammalian brain enzyme nor the bacterial enzyme catalyzes an ATP–ADP exchange reaction when glutamate is the only other substrate present (Meister, 1973; Stadtman and Ginsburg, 1973). While such observations have been often inappropriately used as support for concerted mechanisms (i.e., not involving a covalent intermediate), it is also possible that ADP might not easily dissociate from the enzyme–glutamine–P_i–ADP complex (Krishnaswamy et al., 1962).

Wedler and Boyer (1972) examined the equilibrium isotope exchange properties of *E. coli* glutamine synthetase. In so doing, they developed a straightforward and ingenious method for distinguishing ordered and random-order binding schemes. The obstacle in using equilibrium exchange methods with three-substrate reactions had been that there are so many potential pairwise exchanges to examine. For this reason, Wedler and Boyer (1972) chose to vary the absolute concentration of all substrates and products while maintaining their relative concentrations at thermodynamic equilibrium. For this purpose, one makes up a concentrated solution with the following products (P, Q, and R) and reactants (A, B, and C) at their equilibrium (eq) concentrations:

$$[P_{eq}][Q_{eq}][R_{eq}] / [A_{eq}][B_{eq}][C_{eq}] = K_{eq}$$

This solution is then successively diluted into assay buffer, and enzyme is added to each sample to assure that equilibrium is attained. At that point, one adds a radioactively labeled substrate (or product), and samples are quenched at various time points to assess the rate and extent of exchange into product. The crux of the matter is (a) that ordered binding mechanisms involve noncompetitive interactions that result in depressed rates of isotope exchange at high saturation of the active site and (b) that random mechanisms allow all components to competely interact with each other, such that there is no depression in plots of exchange rate versus the absolute concentration of all reactants (Purich and Allison, 1980). Wedler and Boyer

(1972) studied the glutamate–glutamine and P_i–ATP exchanges under the above conditions, and their data ruled out the ordered addition of substrates.

In a parallel series of experiments on the ovine enzyme, we determined initial rate kinetic mechanism using β-glutamate, NH_2OH, and ATP (Allison et al., 1977). We had planned to carry out later quench-flow studies on the formation of β-amino-glutaryl-P, which would not be susceptible to cyclization and expulsion of P_i. Our findings were fully consistent with a random order of substrate addition. We also applied the Wedler and Boyer (1972) technique to the ovine enzyme, and our isotope exchange results likewise agreed with the assignment of a random addition mechanism (Allison et al., 1977).

H. ISOTOPE TRAPPING METHOD FOR DETECTING ENZYME REACTION CYCLE INTERMEDIATES

As pointed out by Rose (1994) there are two basic questions typically asked in mechanistic investigations: "Are there covalent intermediates, and when do substrates and products leave the reaction cycle?" These questions can often be answered by using Rose's isotope trapping method, an approach that is based on Meister's invention of a clever pulse–chase experiment. In an effort to build the case for an acyl-P intermediate, Krishnaswamy et al. (1962) incubated glutamine synthetase with [^{14}C]glutamate (the "pulse") and ATP under conditions where they had scrupulously removed any ammonium ion (see Fig. 4); then, to trap the covalent intermediate, they added a solution containing hydroxylamine in the presence of a whopping excess of unlabeled glutamate (the "chase"). Krishnaswamy et al. (1962) reasoned that these conditions would favor the conversion of [^{14}C]glutamyl-P into radiolabeled γ-glutamyl-hydroxamate, and only a minor amount of labeled product would result from the steady-state turnover of reactants into products after combining the pulse and chase solutions. This would be the case because any unbound [^{14}C]glutamate would be diluted by the much larger pool of unlabeled glutamate, thereby assuring that the fraction of the diluted tracer γ-glutamyl-hydroxamate would be minimal, even after multiple catalytic rounds. An important advantage gained by utilizing hydroxylamine, as opposed to ammonium ion, was that the

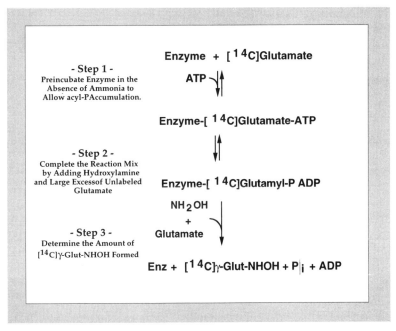

Figure 4.   Isotope trapping evidence for the formation of the acyl-P intermediate in the glutamine synthetase reaction.

investigators could fully prevent spurious results arising from any contaminating ammonium ion. These investigators assumed (a) that prior to any formation of the covalent intermediate, enzyme-bound [^{14}C]glutamate would exchange with the large pool of unlabeled glutamate, and this would result in a product with a low specific radioactivity; and (b) that after formation of the covalent intermediate, the enzyme-bound [^{14}C]glutamate would promptly react with hydroxylamine and yield the highly radioactive acyl-hydroxamate product. Because the latter was observed, they concluded that an acyl-P intermediate was the most likely explanation. As noted by Rose (1980), one may encounter cases wherein a covalent intermediate is formed but is not indicated by the results of this pulse–chase technique; likely explanation would be that the missing substrate is essential for maintaining the catalytically active configuration of the active site. Hence, the inability to trap such an intermediate cannot be regarded as definitive proof against the participation of a covalent intermediate.

I. BOROHYDRIDE TRAPPING OF GLUTAMINE SYNTHETASE'S ACYL-P INTERMEDIATE

Although the formation of an acyl-P intermediate in the glutamine synthetase reaction was amply supported by an impressive body of evidence, it was not until the mid-1970s that acyl-P intermediates were directly trapped using borohydride as the nucleophilic reductant. This approach was first used by Degani and Boyer (1973) to trap the β-aspartyl-P formed by an ion transport ATPase. They demonstrated that the presence of ATP was essential in order to observe any hydroxy-amino acid formation after borohydride reduction and acid hydrolysis to liberate the amino acids within the ATPases polypeptide chain. In the case of glutamine synthetase, Todhunter and Purich (1975) demonstrated that borohydride did not reduce glutamate, glutamine, or pyrrolidone-carboxylate in the absence of added enzyme; furthermore, glutamine, and pyrrolidone-carboxylate were unreactive toward borohydride in the presence of enzyme. Most significantly, we found that borohydride converted [^{14}C]glutamate into [^{14}C]δ-hydroxy-α-aminovalerate when the former was incubated in a solution containing 0.5-mM ATP, 1-mM magnesium ion, and 0.1-mM labeled glutamate. All samples were treated with Permutit to bind any ammonia present at trace levels. Escherichia coli glutamine synthetase was present at 20 μM, and under these conditions we demonstrated that 14-μM δ-hydroxy-α-aminovalerate was formed.

The amount of acyl-P formed in the above experiments corresponds to some 70% of the total enzyme concentration, and at first blush this appears to be rather large, especially when one considers that acyl-P compounds are substantially less stable than ATP. However, within the enzyme-bound state, the internal equilibrium Enz[ATPglutamate] = Enz[ADPglutamyl-P] may favor the acyl-P as a result of enzyme stabilization of the bound intermediate. Arguing in favor of this interpretation are the data of Krishnaswamy et al. (1960) on pyrrolidone carboxylate formation (discussed in Section II.C). Those investigators obtained 16.8 and 17.2 μM of the cyclic product from L- and D-glutamate, respectively, in the presence of 22-μM sheep brain glutamine synthetase, which as noted earlier utilizes either enantiomer. Their values correspond to 76 and 77% of the total available enzyme sites, and their values are remarkably similar to those reported for borohydride reduction experiments with the bacterial enzyme (Todhunter and Purich, 1975). The great stability of the enzyme-bound methionine sulfoximine-P complex is also quite compatible with this conclusion.

Borohydride reduction was the first direct trapping of the enzyme's acyl-P intermediate, and this author can well recall the admiration that Meister expressed regarding our finding. Only years later, while we were each smoking a cigar in his office, did Meister reveal that he was the reviewing editor for our paper. (Decades earlier, his spacious office at Cornell Medical College was the main editorial office for the *Journal of Biological Chemistry*.) Meister also commented that he had unsuccessfully attempted to reduce the glutamyl-P intermediate by using a hydrogen and palladium catalyst in the presence of the enzyme. In retrospect, borohydride is a nucleophilic reductant that closely resembles ammonia, and palladium is probably less well suited for the reduction reaction.

J. POSITIONAL ISOTOPE EXCHANGE EXPERIMENTS

While the observations cited immediately above indicate that a stabilized acyl-P intermediate is formed by the synthetase, the major unsettled question was whether such an intermediate is synthesized on a time scale that would be compatible with other kinetic properties of glutamine synthetase catalysis. Despite intensive efforts to develop a mix/quench protocol for estimating the rates of acyl-P formation, our group was unsuccessful in extending the borohydride method to obtain such kinetic data. Nevertheless, Midelfort and Rose (1976) made a major advance in mechanistic enzymology when they developed an isotope scrambling method for the detecting transient [Enz:ADP:P-X] formation from [^{18}O]ATP in ATP-coupled enzyme reactions. This ingenious method takes advantage of the torsional symmetry of the newly terminal phosphoryl oxygen atoms in the enzyme-bound ADP that is held near the newly formed acyl-P intermediate of glutamine synthetase. Adenosine triphosphate, specifically labeled with an ^{18}O atom in the β–γ bridge oxygen, was incubated with enzyme, and reversible cleavage of the $P_\beta O$-$P\gamma$ phosphoanhydride bond was detected by the appearance of ^{18}O in the β-nonbridge oxygen atoms of the ATP pool. This requires formation and reversal of the phosphoryl transfer steps leading to formation of enzyme-bound ADPacyl-P, such that the resynthesized ATP is free to depart from the enzyme and to mix with the pool of [β–γ bridge-labeled [^{18}O]ATP (see Fig. 5). The Midelfort and Rose (1976)

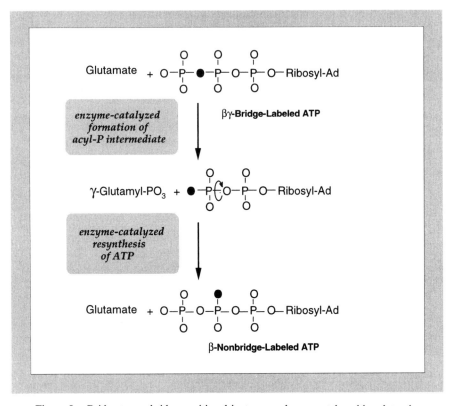

Figure 5. Bridge-to-nonbridge positional isotope exchange catalyzed by glutamine synthetase. Note that ATP (labeled with ^{18}O in the bridge position) is incubated with enzyme and glutamate. When formation of the acyl-P occurs, the oxygen atoms on the terminal phosphoryl of ADP become torsiosymmetric. With rotation about the αβ single bond and subsequent resynthesis of ATP, the ^{18}O atom scrambles into the nonbridge position, and can be detected in subsequent analysis of the ATP.

experiments with sheep brain and *E. coli* glutamine synthetases showed that cleavage of ATP to form enzyme-bound ADP and P–X requires glutamate. The exchange catalyzed by the *E. coli* enzyme with glutamate occurs in the absence of ammonia and is partially inhibited by added ammonium chloride, as would be expected if the exchange reflects the mechanistic pathway for glutamine synthesis. Their positional isotope exchange results provided compelling kinetic evidence for a two-step mechanism wherein phosphoryl transfer from ATP to glutamate to form the acyl-P intermediate precedes reaction with ammonia.

K. REGULATION OF BACTERIAL GLUTAMINE SYNTHETASE

The *de novo* synthesis of nitrogen-containing metabolites involves glutamine as an amide nitrogen donor. This includes the following end-products: GMP, CMP, *N*-acetyl-glucosamine, alanine, serine, tryptophan, and histidine (Fig. 6), and the formation of other nitrogenous metabolites are indirectly linked to glutamine synthesis. Because free ammonia can cross cellular membranes and can thereby collapse proton gradients and/or alter intracellular pH, multicellular organisms developed interorgan glutamine transport mechanisms to prevent ammonia toxicity. By contrast, bacteria can utilize ammonia and glutamine interchangeably as precursors in the biosynthesis of nitrogenous metabolites. Bacteria can thus take advantage of ambient foodstuffs that often can be rich in ammonium ions. Nonetheless, at neutral pH, ammonium ion preponderates over free ammonia, thereby requiring much higher total "ammonia" concentrations (i.e., the sum of $[NH_4^+]$ and $[NH_3]$) relative to the glutamine concentration. The synthesis of glutamine requires the stoichiometric hydrolysis of ATP, and the availability of ammonia serves as a signal to down-regulate glutamine synthetase activity, thereby conserving ATP. Any rise in available ammonia results in the rapid accumulation of glutamine with two regulatory consequences: (a) on a short time scale, a cascade of posttranslational modification results in the inactivation of glutamine synthetase (Stadtman, 1990); and (b) the sustained abundance of ammonia leads to a readjustment of nitrogenous metabolites in a manner that acts to repress the synthesis of glutamine synthetase (Magasanik and Bueno, 1985).

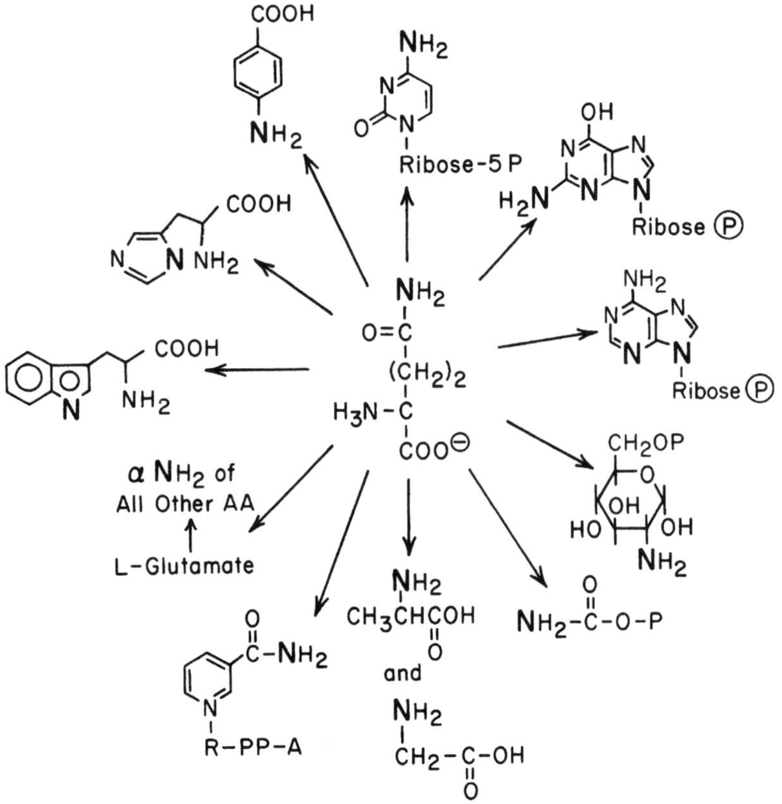

Figure 6. Metabolic end-products of glutamine. Those nitrogen atoms directly derived from the γ-amido group of glutamine are highlighted with boldface type. (*Note*: Although earlier kinetic and fluorescence experiments suggested that each end-product acted at its own allosteric regulatory site, NMR and crystallographic evidence indicates that these metabolites act at the active site.)

1. Cascade Control

Within gram-negative bacteria, glutamine synthetase itself serves as a central processing unit for interpreting many regulatory cues regarding the status of cellular nitrogen metabolism. Beyond the availability of substrates, the National Institutes of Health (NIH) group led by Stadtman identified that a hierarchy of metabolic control is at play, including (a) the covalent interconversion by a multi-step adenylylation/deadenylylation and uridylation/deuridylation cascade involving positive and negative effector control by metabolites whose concentrations depend on the availability of glutamine and ammonium ion; (b) the cumulative feedback inhibition by end-products of glutamine-dependent biosynthetic pathways; (c) the catalytic and structural sites for divalent metal ions; and (d) metabolite-regulated repression of glutamine synthetase gene expression.

With so much attention now accorded to the signal transduction protein kinases of eukaryotes, the fundamental nature of Stadtman's contributions to covalent interconverting enzymes often goes unappreciated by younger students of metabolic control. Stadtman (1990) traced the scaffolding of logic that brought bacterial nitrogen metabolism from its humble beginning with the recognition that bacterial culture conditions significantly influence the catalytic properties of glutamine synthetase to our present sophisticated level of understanding. Such gram-negative bacteria as E. coli and Salmonella typhimurium have evolved an elaborate mechanism for covalent interconversion of glutamine synthetase between active and inactive enzyme forms (Fig. 7). Unadenylylated glutamine synthetase uses magnesium ions for catalytic activity, but the adenylylated enzyme only has significant activity in the presence of manganese ion (Stadtman and Ginsburg, 1974). Activity in the presence of the latter metal ion is unlikely to be of physiologic significance. For this reason, the interconversion cascade allows these bacteria to avoid wasting glutamine (and hence ATP) whenever ammonium ion is available in sufficient quantitities. Any increase in the concentrations of metabolites such as glutamine and ammonium ion serve to activate adenylylation, which involves the ATP-dependent synthesis of a phosphodiester between AMP and tyrosine. On the other hand, any increase in the concentration of α-ketoglutarate signals that the bacteria have relatively low stores of nitrogenous metabolites, which would otherwise use transaminases to deplete the cell of its available pool of α-ketoglutarate (Adler et al., 1975).

The beauty of the adenylylation–deadenylylation cascade is that its interconverting enzymes have multiple sites for binding key indicators of the status of nitrogen metabolism in these cells. Multiple sites of action increase the information-processing capability of the cascade, and they give added potency to key metabolic effectors

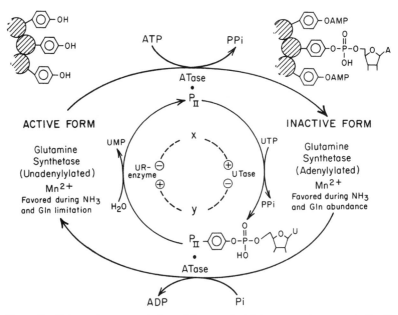

Figure 7. Schematic representation of the covalent interconversion cascade for *E. coli* glutamine synthetase through allosterically regulated nucleotidylation reactions. Adenylylation of glutamine synthetase is catalyzed by an adenylyl-transferase (ATase) under conditions that favor the uridylation of the P_{II} regulatory protein, and deadenylylation is also catalyzed by the ATase in the presence of deuridylated P_{II}. [Adapted from Adler et al., (1975)].

(Stadtman, 1990). Another significant property of such cascade cycles was recognized by Chock and Stadtman (1980)—namely, that these cascades lead to signal amplification and that the "gain" on the amplifier depends on the number of steps involved as well as the kinetic parameters of the interconverting enzymes.

2. Cumulative End-Product Inhibition

Another attractive proposal for the regulation of bacterial glutamine formation concerned the nature of inhibitory effects exerted by metabolic end-products. The basic hypothesis was that this enzyme, which itself catalyzes the first committed step in the multipathway synthesis of nitrogen-containing end-products, was subject to cumulative feedback inhibition (Woolfolk and Stadtman, 1967). Such regulatory behavior would allow feedback inhibitors to throttle the enzyme's catalytic activity without completely arresting glutamine biosynthesis, unless all of the end-products reached saturating concentrations. Each feedback inhibitor was believed to bind at its own topologically distinct regulatory site on the synthetase. This model has become a standard entry in general biochemistry textbooks (Metzler, 1976; Voet and Voet, 1995; Matthews and van Holde, 1996; Stryer, 1988).

Evidence for the cumulative inhibition model was adduced chiefly from initial rate studies using both the enzyme's biosynthetic and the γ-glutamyl-transferase activities (described in Section A.1). Before the adenylylation–deadenylylation cascade had been discovered, Woolfolk and Stadtman (1964; 1967) observed that raising the concentrations of any one of eight end-products to saturate the enzyme resulted in incomplete inhibition—a behavior that cannot be accounted for by classical competitive, noncompetitive, or uncompetitive inhibitor action. On the other hand, they found that the biosynthetic activity could be completely inhibited by a single end-product at saturating inhibitor levels (Ginsburg, 1969; Stadtman and Ginsburg, 1974).

How then can one explain the differences observed in measurements of the biosynthetic and transferase activities? The 1:1 stoichiometry of ATP, glutamate, and ammonium ion in the biosynthetic reaction renders its assay far more sensitive to substrate depletion and product inhibition. By contrast, the transferase reaction offers the clear advantage that one can maintain initial rate conditions. The transferase reaction almost certainly involves the transient formation of γ-glutamyl-arsenate, followed by nucleophilic attack by hydroxylamine to form the product γ-glutamyl-hydroxamate, which is conveniently assayed colorimetrically at 540 nm in the presence of ferric chloride. Adenosine diphosphate presumably maintains the active catalytic configuration of the nucleotide substrate pocket of the enzyme's active site, and kinetic studies suggest that ADP can persist on the enzyme through multiple rounds of the transferase reaction (Huang and Purich, unpublished findings). Indeed, ADP binding affinity depends on the divalent metal ion used as a cofactor in this reaction (Stadtman and Ginsburg, 1974).

Other initial rate studies of the γ-glutamyl-transferase activity in the absence and presence of guanosine triphosphate (GTP), a putative cumulative feedback inhibitor, indicated that commercial GTP contained sufficient trace amounts of guanosine diphosphate (GDP) to prevent the complete inhibition of transferase activity (Purich, unpublished findings). The basic idea is that if a competitive inhibitor (I) is contaminated by a substrate (S) (such that $\alpha = [I]/[S]$), then the total substrate is mathematically equivalent to $\{[S] + \alpha [I]\}$. The one-substrate initial rate equation then becomes

$$v = \{V_m [S] + \alpha [I]\}/\{[S] + K_m + (\alpha + K_m/K_i) [I]\}$$

Applying L'Hospital's rule allows one to obtain the limiting (or asymptotic) value of v at very high inhibitor concentrations:

$$v = \alpha V_m/\{\alpha + K_m/K_i)\}$$

Depending on the magnitudes of K_m and K_i, one can observe different plateaus of incomplete inhibition. For example, if α is about equal to K_m/K_i, then v will reach a stable plateau value around 0.5 V_m. Such plateau behavior can be obtained for GTP inhibition of the transferase activity, because GDP is an effective cofactor in this reaction. Thus, the observed incomplete inhibition of the enzyme by GTP and also ATP can be fully accounted for, and there is no need to invoke the far more complicated cumulative inhibition mechanism.

Because other cumulative feedback inhibitors also appeared to share structural features with the enzyme's substrates and/or products, Dahlquist and Purich (1975) reasoned that these inhibitors may well bind to the active site, rather than to topologically distinct sites whose occupancy must be elaborately linked to catalytic activity. They used resonance methods to examine the interaction of unadenylylated enzyme with several substrates and effectors. They confirmed that two Mn^{2+} are bound per enzyme subunit (Stadtman and Ginsburg, 1974; Hunt et al., 1975), and ESR titration experiments gave apparent dissociation constants of 20 and 120 nM. The dramatic line broadening observed in the nuclear magnetic resonance (NMR) spectra for alanine in the presence of both manganese and enzyme suggested alanine binds near the tightly bound manganese ion. Glutamate displaced enzyme-bound alanine, and the former likewise appeared to bind at a site that was about as distant as was alanine from the strongly bound manganese ion. Given this behavior, Dahlquist and Purich (1975) proposed that alanine and glutamine probably bind competitively within the enzyme's active site. The binding properties of alanine and ATP were also shown to be thermodynamically linked, such that the presence of one ligand increased the affinity of the enzyme for the other ligand. The presence of ATP dramatically sharpened the alanine methyl proton line width, when manganese and glutamine synthetase were present. Addition of ADP or phosphate alone had little effect on the alanine line width, but the addition of both ADP and phosphate showed the same dramatic sharpening as observed upon the addition of ATP alone, suggesting an induced-fit conformational change in the enzyme by ATP or by both ADP and phosphate. These findings led Dahlquist and Purich (1975) to propose that most, if not all, feedback inhibitors of glutamine synthetase bind at the active site in place of one or more substrates (see Section K on the atomic structure of glutamine synthetase).

3. Regulation of Glutamine Synthesis Expression

Before leaving the regulation of bacterial glutamine synthetase, some mention should be given to the fact that the expression of the synthetase gene (designated *glnA*) in *E. coli* is highly regulated. For example, Reitzer and Magasanik (1986) found that the *glnA* gene of the complex glnALG operon of E. coli is transcribed from tandem promoters. Expression from glnAp1, the upstream promoter, requires the catabolite activating protein and is repressed by nitrogen regulator I (NRI), the product of *glnG*. This produces a transcript with an untranslated leader of 187 nucleotides. Expression from glnAp2, the downstream promoter, also requires NRI as well as the *glnF* product. Full expression also requires growth in a nitrogen-limited environment. They also adduced evidence suggesting that the function of the *glnL* product is to mediate the interconversion of NRI between a form capable of activating glnAp2 and an inactive form in response to changes in the intracellular concentration of ammonia. Activation *glnA* gene transcription responds to nitrogen starvation and requires the phosphorylated form of the transcription factor NRI (Ninfa and Magasanik, 1986). Regulation of *glnA* transcription by nitrogen metabolites involves control of NRI phosphorylation through dephosphorylation (Kamberov et al. 1994). Because NRI catalyzes its self-phosphorylation using either acetyl phosphate (Feng et al., 1992) or the phosphorylated form of NRII, the bifunctional kinase/phosphatase (Weiss and Magasanik, 1988). Nitrogen regulator II can utilize ATP to undergo autophosphorylation at its histidine residue, and these phosphoryl groups serve as the substrate for NRI autophosphorylation. The NRIPs acyl phosphate moiety has a 4–8 min half-life at pH 7, and this "autophosphatase" activity of NRI-P is greatly stimulated by NRII in the presence of deuridylated P_{II} and ATP (Keener and Kustu, 1988). Kamberov et al. (1995) observed (a) that P_{II} is activated upon binding ATP and either α-ketoglutarate or glutamate, (b) that this liganded form of P_{II} binds much better to NRII, and (c) that the concentration of glutamine required to inhibit the uridylyltransferase activity is independent of the concentration of 2-ketoglutarate present. These significant findings led Kamberov et al. (1995) to propose that nitrogen sensation in *E. coli* involves the separate measurement of glutamine by the UTase/UR protein and α-ketoglutarate by the P_{II} protein.

L. ATOMIC STRUCTURE OF BACTERIAL GLUTAMINE SYNTHETASE

Unfortunately, detailed mechanistic studies, including the use of site-directed mutagenesis and modern spectroscopic probes, were hampered for many years by the lack of atomic-level structural information. Eisenberg's pursuit of an atomic structure was based on the clear recognition that X-ray crystallography offered the best opportunity to examine the most probable catalytic mechanism. Because there is every indication that the catalytic mechanisms of the ovine and bacterial glutamine synthetases are fundamentally congruent, he chose to pursue the bacterial enzyme's structure that would help to unravel the structural basis of this extraordinarily complex regulatory enzyme. X-ray crystallography of very large proteins, particularly those with numerous subunits, has remained a major challenge. The 600,000 molecular weight bacterial enzyme has dodecameric quaternary structure with two- and sixfold rotational axes of symmetry, appearing as a double layer of hexamers in negatively stained electron micrographs (Valentine et al., 1968). In an epic quest, which is now midway in its third decade, Eisenberg's research group has succeeded in defining fundamental features of glutamine synthetase's structure at the atomic level.

Although much of this structural work lies beyond the immediate scope of this chapter, the X-ray crystallographic findings on ligand-binding interactions with the synthetase help to place Stadtman's cumulative inhibitor model in clearer light. Liaw et al. (1993) examined the crystal structures of S. typhimurium glutamine synthetase (fully unadenylylated enzyme containing bound Mn^{2+}) complexed with the substrate L-glutamate or one of three feedback inhibitors (i.e., L-serine, L-alanine, and glycine). They also found that the inhibitors bind at the L-glutamate substrate binding site, with the "main chain" $^+NH_3$—CH—COO$^-$ occupying the same positions on the enzyme. Liaw et al. (1994) later investigated the structural basis of the feedback inhibition of the synthetase by AMP using crystal structures of unadenylylated glutamine synthetase (again containing two bound Mn^{2+} ions per subunit) complexed with AMP and the related nucleotides AMPPNP (a very weakly hydrolyzable ATP analogue), ADP, and GDP, as well as adenosine and adenine. The X-ray structures show that *all nucleotides bind at the same site,* the active site's pocket for ATP binding, as do adenosine and adenine. Thus from X-ray structures, adenosine monophosphate (AMP), adenosine, adenine, and GDP would be expected to inhibit Mn^{2+} enzyme by competing with the substrate ATP for the active site. The AMP thus competes with ATP, a conclusion that was fully consistent with their kinetic measurements using the biosynthetic assay.

These findings underscore the validity of the conclusion by Dahlquist and Purich (1975) that most, if not all, of the so-called cumulative inhibitors evoke their effects on glutamine synthetase by binding within one or more substrate adsorption pockets at the active site. An analogy can be made to the collective action of amino acids as substrates of the γ-glutamyl-transpeptidase reaction (Allison and Meister, 1981; Allison, 1985). In that case, one must consider how nonsaturating physiologic levels of the various substrates act together to affect transpeptidase activity. The collective action of competitive inhibitors X, Y, and Z on a one-substrate enzyme exhibiting Michaelis–Menten behavior is given by the following equation:

$$V_m v = 1 + (K_m/[S]) \{1 + ([X]/K_x) + ([Y]/K_y) + ([Z]/K_z)\}$$

where K_x, K_y, and K_z are inhibition constants. If the intracellular concentrations of each end-product (in this case, X, Y, and Z) cannot permit saturation of the enzyme, then these agents behave collectively as virtual cumulative inhibitors. This reviewer holds to the belief that such collective action can still afford the bacterium with effective feedback regulation by anabolic end-products of glutamine. An added dividend of this simpler mechanism of end-product action is that the enzyme's regulation is less susceptible to mutations that might have damaged one or more of the many binding sites required in the cumulative inhibition model. Instead, the same selective pressures exerted on maintaining catalytic activity also serve to keep the enzyme's responsiveness to end-products—a simple, but effective solution to an otherwise bewilderingly complicated problem.

IV. The "Protein Ligase" Problem: Extending the Versatility of Carboxyl Group Activating Enzymes

Biochemists have already begun to invent techniques for introducing a much wider spectrum of side-chain groups into proteins: (a) to alter ligand-binding specificity/strength, (b) to create novel biocatalysts, and (c) to modify conformational stability by increasing the lexicon of amino acids beyond those naturally occurring in proteins. While the ribosome is an exquisitely well-designed "enzyme" catalyzing carboxyl-group activation in repetitive cycles of peptide bond synthesis, its specificity and proofreading mechanisms limit the choice to those amino acids that naturally occur in proteins. This can be circumvented to some degree by using tRNA molecules that are charged with unnatural amino acids, but the technique is rather limited. One promising approach focuses on creating entirely new base pairs for DNA to increase the number of codons, and hence the lexicon of amino acids that can be introduced into polypeptide chains (Benner, 1994).

There is growing recognition of the need for additional approaches for introducing unnatural amino acids at defined positions, but the major obstacle is how to approximate and ligate synthetic peptides to obtain novel proteins. Complete chemical synthesis of polypeptides is still impractical, because one can rarely achieve yields approaching 100% in almost any synthetic organic reaction. Proteins typically contain anywhere from 400–1000 peptide bonds, and repetitive step-by-step solid-phase synthesis allows errors to accumulate. For this reason, one would desire to link preformed synthetic peptides (containing unnatural amino acids or other substances) to other peptides produced recombinantly, thereby forming large proteins with just a few ligations.

Dawson et al. (1994) described one potential route to the direct synthesis of native "all-peptide"-containing proteins. They favor using a chemoselective reaction that first forms a thiolester linkage between two unprotected peptide segments P_i and P_j, and this intermediate then spontaneously rearranges to yield a ligated polypeptide P_k (where $k = i + j$ residues) with a peptide bond at the junction of P_i and P_j. Dawson et al. (1994) demonstrated the feasibility of this strategy in a one-step synthesis of a cytokine containing multiple disulfides. This approach avoided the presence of reactive thiols, and after ligation, the polypeptide product was folded and oxidized to yield the native protein molecule.

The trick is to achieve specific activation of the terminal α-carboxyl group of one polypeptide under conditions only permitting attack by a second peptide's unprotected N-terminal amino group. Ideally, such a coupling reaction: (a) must lead to the formation of a true peptide bond; (b) should be as thermodynamically favorable as typical ATP-dependent amide synthetase reactions; (c) should not require protection/deprotection of otherwise reactive side-chain residue; and (d) should occur under mild reaction conditions. These are very restrictive requirements, and one may confidently predict that enzymes are most ideally suited to fulfill these criteria.

These considerations motivate the suggestion that greater attention should be devoted to transforming known carboxyl group activating enzymes into "polypeptide ligases." These engineered enzymes would then be suitable catalysts for semisynthesis of novel proteins from intermediate length polypeptides. Enzymes catalyzing peptide bond formation are especially attractive targets for such an enterprise, provided that one can remodel their substrate-binding pockets to permit binding of longer peptides. Glutathione synthetase and tubulin tyrosine ligase (ADP-forming) are among the leading candidate enzymes. These enzymes activate α-carboxyl groups in steps that are highly related to glutamine synthetase catalysis, and the tubulin shell tyrosine ligase already adds an amino acid to the C-terminus of the α-tubulin subunit (Deans et al., 1992). Another starting point might well be D-alanyl-D-alanine synthetase, which exhibits a mechanism quite similar to that of glutamine synthetase (Fan et al., 1997). While composed entirely of L-amino acids, this synthase binds two molecules of D-ala in its active site, one donating its carboxyl group and the other its amino group for peptide bond formation. Were one successful in converting this enzyme into a D-polypeptide ligase, then synthetic D-amino acid-containing peptides corresponding in sequence to the newly fashioned D-polypeptide ligase could be joined to form a protein that is its mirror image. Such an enzyme would then act as an L-polypeptide ligase, and one could create novel biologically active proteins by splicing L-polypeptides via ATP-driven peptide bond formation. Although there are obvious obstacles to surmount, this writer remains confident that the systematic remodeling of known ligases into polypeptide ligases will be fulfilled during my lifetime.

V. Concluding Remarks

ATP and other nucleoside-5'-triphosphates play a major role in facilitating catalysis by converting oxygen atoms into better leaving groups at physiologic pH. This can be accomplished by phosphorylation or nucleotidylation, followed by nucleophilic attack to release orthophosphate or a nucleoside-5'-monophosphate as the actual leaving group. The generality of mechanisms involving phosphorylation is amply illustrated by such divergent reactions as glutamine synthetase, carbamoyl-P synthase, citrate lyase, and adenylosuccinate synthase, which respectively form γ-glutamyl-P, carboxy-P, citryl-P, and 6-phospho-IMP intermediates. That these reaction mechanisms now seem so tame and predictable is largely attributable to the relentless effort and ingenuity of Alton Meister and his associates.

References

Adler, S. P., Purich, D. L., and Stadtman, E. R., *J. Biol. Chem.* **250**, 6264–6272 (1975).

Alberty, R. A., *J. Biol. Chem.*, **243**, 1337–1343 (1968).

Alberty, R. A., *J. Biol. Chem.*, **244**, 3290–3302 (1969).

Alberty, R. A. and Cornish-Bowden, A., **18**, 288–291 (1993).

Alberty, R. A. and Goldberg, R. N., *Biochemistry*, **31**, 10610–10615 (1992).

Allison, R. D., *Methods Enzymol.*, **113**, 419–437 (1985).

Allison, R. D., and Meister, A., *J. Biol. Chem.*, **256**, 2988–2992 (1981).

Allison, R. D., and Purich. D. L., *Methods Enzymol.*, **64**, 1–46 (1980).

Allison, R. D., Todhunter, J. A., and Purich, D. L., *J. Biol. Chem.*, **252**, 6046–6051 (1977).

Benner, S. A., *Trends Biotechnol.*, **12**, 158–163 (1994).

Boyer, P. D., Koeppe, O. J., and Luchsinger, W. W., *J Amer. Chem. Soc.*, **78**, 356–361 (1956).

Boyer, P. D., Mills, R. C., and Fromm, H. J., *Arch. Biochem. Biophys.*, **81**, 249–257 (1959).

Chock, P. B., and Stadtman, R. R., *Meth. Enzymol.*, **64**, 397–325 (1980).

Dahlquist, F. W., and Purich, D. L., *Biochemistry* (1975).

Dawson, P. E., Muir, T. W., Clark-Lewis, I., and Kent, S. B., *Science*, **266**, 776–779 (1994).

Deans, N. L., Allison, R. D., Purich, D. L., *Biochem. J.*, **286**, 243–251 (1992).

Degani, C., and Boyer, P. D., *J. Biol. Chem.*, **248**, 8222–8226 (1973).

Elliott, W. H., *Biochem. J.*, **49**, 106–112 (1951).

Fan, C., Park, I. S., Walsh, C. T., and Knox, J. R., *Biochemistry*, **36**, 2531–2538 (1997).

Feng, J., Atkinson, M. R., McCleary, W., Stock, J. B., Wanner, B. L., and Ninfa, A. J., *J. Bacteriol.*, **174**, 6061–6070 (1992).

Gass, J. D., and Meister, A., *Biochemistry*, **9**, 1380–1388.

Ginsburg, *Biochemistry*, **8**, 1726–1734 (1969).

Hunt, J. B., Smyrniotis, P. Z., Ginsburg, A, and Stadtman, E. R., *Arch. Biochem. Biophys.*, **166**, 102–124 (1975).

Kamberov, E. S., Atkinson, M. R., Feng, J., Chandran, P., and Ninfa, A. J., *Cell. Mol. Biol. Res.*, **40**, 175–191 (1994).

Kamberov, E. S., Atkinson, M. R., and Ninfa, A. J., *J. Biol. Chem.*, **270**, 17797–17807 (1995).

Keener, J., and Kustu, S., *Proc. Natl. Acad. Sci. USA*, **85**, 4976–4980 (1988).

Khedouri, E., Wellner, V., and Meister, A., *Biochemistry*, **3**, 824–828 (1964).

Kowalsky, A., Wyttenbach, C., Langer, L., and Koshland, D. E., Jr., *J. Biol. Chem.*, **219**, 719–725 (1956).

Krebs, H. A., *Biochem. J.*, **29**, 1951–1954 (1935).

Krishnaswamy, P. R., Pamiljans, V., and Meister, A., *J. Biol. Chem.*, **235**, PC39 (1960).

Krisnaswamy, P. R., Pamiljans, V., and Meister, A., *J. Biol. Chem.*, **237**, 2932–2940 (1962).

Levintow, L. and Meister, A., *J. Biol. Chem.*, **209**, 265–269 (1954).

Liaw, S. H., Kuo, I., and Eisenberg, D. S., *Protein Sci.*, **4**, 2358–2365 (1995).

Liaw, S. H., Pan, C., and Eisenberg, D. S., *Proc. Natl. Acad. Sci. USA*, **90**, 4996–5000 (1993).

Liaw, S. H., Jun, G., and Eisenberg, D., *Biochemistry*, **33**, 11184–11188 (1994).

Magasanik, B., and Bueno, R., *Curr. Top. Cell. Regul.*, **27**, 215–220 (1985).

Manning, J. M., Moore, S., Rowe, W. B., and Meister, A., *Biochemistry*, **8**, 2681–2685 (1969).

Mathews, C. K. and van Holde, K. E., *Biochemistry*, 2nd ed., Benjamin-Cummings, Menlo Park, CA, 1996, p. 707.

Meister, A. *The Enzymes*, 2nd ed., Boyer, P. D., Lardy, H., and Myrback, K. Eds., Vol. 6, 1962, pp. 443–468.

Meister, A., *Adv. Enzymol. Relat. Areas Mol. Biol.*, **31**, 183–218 (1968).

Meister, A., *The Enzymes*, Boyer, P. D., Ed., Vol. X, 1974, pp. 671–754.

Metzler, D. E., *Biochemistry: The Chemical Reactions of Living Cells*, Academic, New York, 1976, p. 813.

Midelfort, C. F. and Rose, I. A., *J. Biol. Chem.*, **251**, 5881–5887 (1976).

Ninfa, A. J. and Magasanik, B., *Proc. Natl. Acad. Sci. USA*, **83**, 5909–5913 (1986).

Phillips, R. C., *J. Biol. Chem.*, **244**, 3330–3342 (1969).

Platt, J. R., *Science*, **146**, 347–353 (1964).

Purich, D. L., and Allison, R. D. *Meth. Enzymol.*, **64**, 1–46 (1980).

Reitzer, L. J., and Magasanik, B., *Cell*, **45**, 785–792 (1986).

Ronzio, R. A. and Meister, A., *Proc. Natl. Acad. Sci. USA*, **59**, 164–170 (1968).

Rose, I. A., *Meth. Enzymol.*, **249**, 315–340 (1994).

Rowe, W. B., Ronzio, R. A., Meister, A., *Biochemistry*, **8**, 2674–2680 (1969).

Shrake, A., Ginsburg, A., Wedler, F. C., Sugiyama, Y. *J. Biol. Chem.*, **257**, 8238–8243 (1982).

Shrake, A., Whitley, E. J. Jr., and Ginsburg, A., *J. Biol. Chem.*, **255**, 581–589 (1980).

Speck, J. F., *J. Biol. Chem.*, **179**, 1405–1411 (1949).

Stadtman, E. R., and Ginsburg, A., *The Enzymes*, 3rd ed., IX, 755–807 (1974).

Stadtman, E. R. *Methods Enzymol.*, **182**: 793–809 (1990).

Stryer, L., *Biochemistry*, 3rd ed., Freeman, New York, 1988, p. 590.

Todhunter, J. A. and Purich, D. L., *J. Biol. Chem.*, **250**, 3505–3509 (1975).

Valentine, R. C., Shapiro, B. M., and Stadtman, E. R., *Biochemistry*, **7**, 2143–2152 (1968).

Voet, D., and Voet, J. G. *Biochemistry*, 2nd ed., Wiley, New York, 1995 p. 764.

Wedler, F. C., and Boyer, P. D., *J. Biol. Chem.*, **247**, 984–992 (1972).

Wedler, F. C. *Meth. Enzymol.*, **249**, 443–479 (1995).

Weiss, V. and Magasanik, B., *Proc. Natl. Acad. Sci. USA*, **85**, 8919–8923 (1988).

Welch, G. R., *J. Theor. Biol.*, **114**, 433–446 (1985).

Woolfolk, C. A. and Stadtman, E. R., *Biochem. Biophys. Res. Commun.*, **17**, 313–322 (1964).

Woolfolk, C. A. and Stadtman, E. R., *Arch. Biochem. Biophys.*, **118**, 736–755 (1967).

HEPATIC GLUTAMINE TRANSPORT AND METABOLISM

By DIETER HÄUSSINGER, *Medizinische Universitätsklinik, Heinrich-Heine-Universität Düsseldorf, D-40225 Düsseldorf, Germany*

CONTENTS

Advances in Enzymology and Related Areas of Molecular Biology, Volume 72: Amino Acid Metabolism, Part A, Edited by Daniel L. Purich
ISBN 0-471-24643-3 © 1998 John Wiley & Sons, Inc.

I. Introduction

Early in the work by Krebs in 1935, the liver was identified as having a particular role in glutamine metabolism, thereby contrasting other organs like brain or kidney in properties of glutaminase and glutamine synthetase. Much effort has been devoted since then to the characterization and regulation of the enzymes involved in glutamine metabolism (Curthoys et al., 1984; Curthoys and Watford, 1995; Huang and Knox, 1976; Horowitz and Knox, 1968; Katsunuma et al., 1968; McGivan et al., 1980; Meister and Anderson, 1983; Moldave and Meister, 1957; Ronzio and Meister, 1968; Cooper and Meister, 1972, 1974, 1977; Meister, 1974, 1984; Tate and Meister, 1972, 1974) and to the understanding of glutamine metabolism in different tissues including its physiological significance (for reviews see Atkinson and Bourke, 1984; Häussinger and Sies, 1984a; Häussinger, 1988, 1990, 1996; Kvamme, 1988; Meijer et al., 1990; Cooper, 1990). Balance studies across the whole liver gave either conflicting results or demonstrated no net glutamine turnover at all, favoring an early view in which the liver did not play a major role in glutamine metabolism, although the activities of glutamine metabolizing enzymes were found to be high. Later, a fundamental conceptional change in the field of hepatic glutamine metabolism was derived from the understanding of the unique regulatory properties of hepatic glutaminase, the discovery of hepatocyte heterogeneities in glutamine metabolism with metabolic interactions between differently localized subacinar hepatocyte populations, and the role of an intercellular cycling of glutamine in the liver acinus for the maintenance of whole body ammonia and bicarbonate homeostasis. More recently, another fascinating perspective arose when it became clear that glutamine exerts regulatory properties on hepatocellular function. These properties are not explained by the metabolism of the amino acid but which are mediated by alterations of the hepatocellular hydration state because of a cumulative, osmotically active uptake of glutamine into the cells. This chapter summarizes some basic concepts on hepatic glutamine transport and metabolism; for detailed analysis, the reader is referred to recent reviews and monographs (Häussinger and Sies 1984a; Häussinger, 1988, 1990a, 1996; Kovacevic and McGivan, 1983; Kvamme, 1988; Meijer, et al. 1990; Lang and Häussinger, 1993; Curthoys and Watford, 1995; Kilberg and Häussinger, 1992; Souba, 1991).

II. Glutamine Transport in Liver

A. PLASMA MEMBRANE TRANSPORT

Glutamine is transported across the plasma membrane and the inner-mitochondrial membrane by specific transport systems (for reviews see Kovacevic and McGivan, 1983, 1984). Two transport systems in the plasma membrane have been described: the Na^+-dependent transport system "N" (Kilberg et al., 1980) and a Na^+-independent system, which is probably involved in facilitated diffusion of glutamine from hepatocytes (Fafournoux et al., 1983). This latter export system has recently been designated "system n" (Pacitti et al., 1993); it is upregulated in the tumor-bearing rat and is down regulated by glutamine under these conditions (Inoue et al., 1995). Both hepatic glutamine uptake and release of newly synthesized glutamine from the liver are inhibited by histidine (Häussinger et al., 1985a).

Substrates for the Na^+-dependent system N are glutamine, histidine, and asparagine and system N activity is abolished following sulfhydryl specific protein modification by p-chloromercuribenzene sulfonate (Tamarappoo et al., 1994). The system N transporter has not yet been cloned, however, transport activity has been solubilized and reconstituted (Tamarappoo and Kilberg, 1991) and monoclonal antibodies were raised, which inhibit system N transport activity (Tamarappoo et al., 1992, 1994). These studies suggested that a 100-kDa plasma membrane protein mediates system N transport activity in rat hepatocytes (Tamarappoo et al., 1992). In studies on the initial rate of glutamine uptake into either whole cells (Kilberg et al., 1980, Fafournoux et al., 1983) or mitochondria (Kovacevic and Bajin 1982), glutamine influx and efflux rates far above the known rates of metabolic glutamine turnover were found. This led to the generally accepted view that glutamine transport across biological membranes in the liver is not controlling its metabolism. This view has changed completely, when it became clear that the glutamine transport systems in the plasma and mitochondrial membrane build up steady-state glutamine concentration gradients in the respective subcellular compartments (Häussinger et al., 1985b; Lenzen et al., 1987; Remesy et al., 1988), which determine the flux through glutamine-metabolizing enzymes in the liver. With a physiological extracellular glutamine concentration of 0.6 mM and a physiological extracellular pH 7.4, *in vivo* and in the isolated perfused rat liver, the cytosolic and mitochondrial glutamine concentrations are about 6 and 20 mM (Häussinger et al., 1985a). Raising the extracellular pH from 7.3 to 7.7, the mitochondrial glutamine concentration increases from 15 to 50 mM (Lenzen et al., 1987). It is evident that such pH-dependent fluctuations of the mitochondrial glutamine concentration, despite constancy of the extracellular concentration, are critical for flux through mitochondrial glutaminase with its K_m (glutamine) of 22–28 mM. This result also largely explains why flux through glutaminase is increased four- to fivefold upon raising the extracellular pH from 7.3 to 7.7. Thus, the control of glutamine metabolism by transport is exerted by the setting of the subcellular glutamine concentrations, rather than by the absolute velocity of initial glutamine import into the cells. However, the latter can also become limiting: stimulation of glutamine transamination by unphysiologically high concentrations of ketomethionine decreased flux through mitochondrial glutaminase by about 80%, because glutamine taken up across the plasma membrane is already consumed in the cytosolic compartment for the transamination (Häussinger et al., 1985b). Further indication for a role of glutamine transport as a determinant of its metabolism comes from recent observations on the metabolic adaptation to a high protein diet in rat liver (Remesy, et al., 1988) and the estimation of a flux control coefficient of 0.31 and -0.4 exerted by transport into (via system N) and out of the liver cell (via system L and/or n), respectively, on the overall pathway of glutamine degradation (Low et al., 1993).

Whereas glutamine uptake into hepatocytes is largely mediated by system N, uptake predominantly occurs via system ASC in hepatoma cells (Bode et al., 1995). Glutamine transport via system N is upregulated/stimulated in liver during endotoxinemia, by prostaglandins, tumor necrosis factor, interleukin-6 (Plumley et al., 1995; Watkins et al., 1994; Inoue et al., 1993; Bode et al., 1995a), in response to hepatocyte swelling (Häussinger et al., 1990a; Bode and Kilberg, 1991) and during amino acid free cell culture (Kovacevic and McGivan, 1984). Amiloride inhibits glutamine uptake by the liver and abolishes the alkalosis-induced increase of glutamine tissue levels (Lenzen et al., 1987). In contrast to amino acid transport system A, system N is largely unaffected by starvation (Hayes and McGivan 1982), insulin, or glucagon in freshly isolated rat hepatocytes. However, following 20-h cultivation of hepatocytes, system N activity decreases by about 70% and can be upregulated by insulin, glucagon, or dexamethasone in a cycloheximide-sensitive way (Gebhardt and Kleemann, 1987). In addition, hepatic glutamine uptake is inhibited at low pH (Lenzen et al., 1987; Christensen and Kilberg 1995).

B. MITOCHONDRIAL TRANSPORT

Glutamine transport across the mitochondrial membrane is rapid and is inhibited by mersalyl, but not by N-ethylmaleimide (Kovacevic and McGivan, 1983, 1984). L-Glutamine is transported faster than the D-glutamine. Nonaqueous fractionation analysis of livers perfused with a physiological glutamine concentration (0.6 mmol/L) revealed an apparent three- to fourfold accumulation of glutamine inside the mitochondria when compared to the cytosol (Häussinger et al., 1985a). This gradient and, consequently, the intramitochondrial glutamine concentration rose about threefold, when the extracellular pH was increased from 7.3 to 7.7, whereas the cytosolic glutamine concentration did not change significantly (Lenzen et al., 1987). These data indicated that pH control of flux through hepatic glutaminase is mediated by variations of the mitochondrial glutamine concentration and point to a regulatory role of the glutamine carrier in the mitochondrial membrane for hepatic glutamine breakdown. In isolated mitochondria, electroneutral glutamine uptake increases with the intramitochondrial pH and correlates with the pH gradient across the mitochondrial membrane (Soboll et al., 1991), however, the exact mechanism of glutamine transport across the mitochondrial membrane in liver remains to be established. When rat livers are perfused with a physiological portal glutamine concentration of 0.6 mM, the cytosolic and mitochondrial concentrations were 7 and 19 mM, respectively. These concentrations decreased when glutamine breakdown was stimulated by glucagon plus NH_4Cl to 4 and 8 mM, respectively, indicating again that transport across the mitochondrial membrane probably exerts control on flux through mitochondrial glutaminase (Häussinger et al., 1985a).

III. Glutamine and Liver Cell Hydration

Glutamine is, besides other substrates, hormones, and oxidative stress, a potent effector on hepatocellular hydration, that is, liver cell volume, which is now recognized as an independent and potent signal, which modifies hepatocellular metabolism and gene expression (for a review see Häussinger and Lang, 1993; Lang and Häussinger 1994; Häussinger, 1996). In principle, all amino acids that are taken up into hepatocytes by concentrative transport systems in the plasma membrane induce cell swelling, however, glutamine deserves special attention due to the high concentrative capacity of the Na^+-coupled glutamine transporting system N in liver and skeletal muscle, that is, due to its extraordinary swelling potency in these organs. Thus, glutamine supply to these organs and the activity of the Na^+-coupled transporter N are major determinants of cell hydration, and glutamine can trigger a variety of effects on hepatic metabolism and gene expression simply by increasing hepatocellular hydration. Consequently, the role of the Na^+-dependent transport system N is more than simply amino acid translocation across the plasma membrane. The transporter also acts as a transmembrane signaling system that modifies cell function in response to glutamine delivery by altering cell hydration (Häussinger and Lang, 1991).

A. MECHANISM

In liver, one of the most important challenges for cell volume homeostasis is the cumulative uptake of osmotically active substances, such as amino acids (Kristensen and Folke, 1984; Häussinger et al., 1990; Wettstein et al., 1990). The Na^+-dependent glutamine-transporting system N can build up intra or extracellular amino acid concentration gradients of up to 20-fold. The Na^+ on entering the hepatocyte together with the amino acid is extruded in exchange for K^+ by the electrogenic Na^+/K^+ ATPase. The accumulation of amino acids and K^+ into the cells leads to osmotic hepatocyte swelling, which in turn triggers volume regulatory K^+ efflux. This is illustrated in Figure 1: addition of glutamine to isolated perfused rat liver creates an intra- or extracellular glutamine concentration gradient of about 12-fold within about 12 min. During the first 2 min of glutamine accumulation into the hepatocytes, liver cell volume increases rapidly and hepatic net K^+ uptake during this phase is probably due to extrusion of Na^+ by the Na^+/K^+ ATPase. Thereafter, no further increase in cell volume is observed, despite continuing accumulation of glutamine inside the cell. This is achieved by a volume-regulatory K^+ efflux from the liver, which occurs until the steady-state intracellular glutamine concentration of about 35 mM is reached (Fig. 1). Most importantly, the hepatocytes remain in a swollen state as long as the amino acid is present. Here, volume regulatory ion fluxes in response to cumulative substrate uptake are apparently designed to blunt, but not to completely abolish the glutamine-induced cell swelling. During the rapid washout of glutamine following its withdrawal from the influent perfusate, the swollen hepatocytes shrink rapidly and a secondary volume-regulatory net K^+ uptake that lasts for more than 10 min restores cell volume to the starting level (Fig. 1). Glutamine-induced cell swelling is half-maximal at 0.6–0.8 mM, that is, at the physiological concentration in the portal vein and is maximal at 2 mM (Wettstein et al., 1990). Accordingly, physiological fluctuations of the portal glutamine concentration are accompanied by parallel alterations of liver cell volume. The degree of glutamine-induced cell swelling seems largely to be related to the steady-state intra- or extracellular amino acid concentration gradient. This gradient and accordingly the degree of cell swelling is modified by hormones, pH, and the nutritional state in a complex way due to effects on the expression of system N, the electrochemical Na^+ gradient as a driving force for Na^+ coupled transport, but also alterations of intracellular glutamine metabolism.

B. FUNCTIONAL CONSEQUENCES OF GLUTAMINE-INDUCED HEPATOCYTE SWELLING

Recent evidence suggests that the dynamic alterations of cellular hydration that occur physiologically within minutes under the influence of anisotonicity, hormones, and amino acids such as glutamine

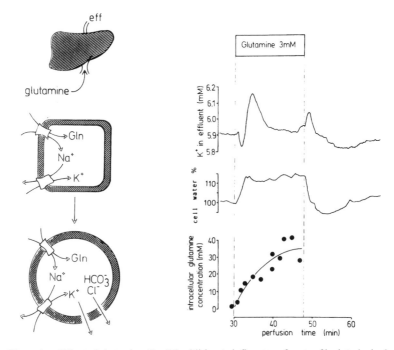

Figure 1.   Effect of glutamine (3 m*M*) addition to influent perfusate of isolated, single-pass perfused rat liver on intracellular glutamine accumulation, cell volume and volume-regulatory K^+ fluxes. As schematically shown on the left side, addition of glutamine to portal perfusate leads to rapid cell swelling due to cumulative, Na^+-dependent uptake of glutamine into liver cells. Cumulative glutamine uptake leads to cell swelling and activates a volume-regulatory K^+, Cl^-, and HCO_3^- efflux. The initial net K^+ uptake is explained by exchange of cotransported Na^+ against K^+ by Na^+/K^+ ATPase. Glutamine-induced cell swelling during the first 2 min of glutamine infusion activates volume regulatory K^+ (plus Cl^- and HCO_3^-) efflux. Cell water was continuously determined by monitoring liver mass, whose changes reflect under these conditions the intracellular water content as validated by isotope measurements of the intracellular water space. This volume-regulatory response prevents further cell swelling despite continuing glutamine accumulation inside the cell until a steady-state intracellular glutamine concentration of about 35 m*M* is reached. However, the liver cell remains in a swollen state as long as glutamine is infused. The extent of cell swelling modifies cellular function. [Adapted from Häussinger et al., 1990a.]

exert per se profound effects on cell function. These functional alterations in response to glutamine-induced swelling persist as long as the amino acid is present, are maintained even after completion of volume-regulatory ion fluxes, but are fully reversible upon restoration of the resting cell volume. Several long-known, but mechanistically poorly understood effects of glutamine, which could not be related to its metabolism, such as stimulation of glycogen synthesis (Katz et al., 1976; Lavoinne et al., 1987) or inhibition of proteolysis (for a review see Seglen and Gordon, 1984; Mortimore and Pösö 1987), can quantitatively be mimicked by swelling the cells in hypoosmotic media to the same extent as glutamine (Häussinger et al., 1990, 1991; Hallbrucker et al., 1991; Baquet et al., 1990, 1991a; vom Dahl and Häussinger 1996) and found their explanation in glutamine-induced hepatocyte swelling. Cell swelling in liver explains the inhibition of proteolysis (Häussinger et al., 1990a, 1991; Hallbrucker et al., 1991, vom Dahl and Häussinger, 1996) and the stimulation of protein synthesis (Stoll et al., 1992), glycogen synthesis (Baquet et al., 1990, 1991a), and flux through the pentose–phosphate shunt (Saha et al., 1992) and the activation of acetyl-CoA carboxylase (Baquet et al. 1991b, 1993) by glutamine. The protein-anabolic effect of glutamine-induced cell swelling is one basis for the understanding of severe clinical protein-catabolic states, which result from muscle glutamine depletion and muscle cell dehydration (Häussinger et al., 1993a; Roth et al., 1990). Glutamine-induced cell swelling also accounts for the alkalinization of endocytotic vesicles (Völkl et al., 1993; Schreiber et al., 1994; Busch et al., 1994), an increased polymerization state of β-actin (Theodoropoulos et al., 1992) and the stimulation of bile acid excretion by this amino acid (Häussinger et al., 1992a). Stimulation of taurocholate excretion into bile by glutamine-induced swelling involves an increase in canalicular transport capacity probably due to a microtubule-dependent insertion of intracellularly stored bile acid transporter molecules into the canalicular membrane (Boyer et al., 1992; Häussinger et al., 1992, 1993b). Like hypoosmotic exposure, addition of glutamine diminishes the excretion of oxidized glutathione (GSSG) into bile (Saha et al., 1992). In the endotoxinemic rat liver, glutamine-induced hepatocyte swelling increases the excretion of cysteinyl leukotrienes into bile (Wettstein et al., 1995).

The intracellular signaling mechanisms that couple glutamine-induced cell swelling to alterations of hepatocyte function are probably similar to those that are activated in response by hypoosmotic cell swelling. Here, a G-protein- and tyrosine kinase dependent activation of mitogen-activated protein (MAP-) kinases is triggered by cell swelling, which mediates the stimulation of bile acid excretion and the alkalinization of endocytotic vesicles (Schliess et al., 1995; Schreiber and Häussinger 1995; Noe et al., 1996). Also, glutamine activates MAP kinases (Schliess and Häussinger, unpublished result). Activation of this osmosignaling pathway toward MAP-kinases does not explain the inhibition of proteolysis (vom Dahl and Häussinger, unpublished result); however, the antiproteolytic effect of both glutamine- and hypoosmolarity-induced cell swelling requires intact microtubular structures (vom Dahl et al., 1995). In addition, hypoosmotic- and glutamine-induced hepatocyte swelling were shown to activate p70 ribosomal protein S6 kinase and phosphatidylinositol-3-kinase, which apparently mediate the activation of acetyl-CoA carboxylase and glycogen synthase (Krause et al., 1996). Furthermore, glutamine may interfere with protein phosphorylation via its degradation product glutamate, which causes activation of a glutamate and chloride-regulated protein phosphatase (Baquet et al., 1993; Gaussin et al., 1996). Thus, multiple signaling pathways are probably activated in response to glutamine-induced cell swelling.

Glutamine also affects gene expression. Glutamine increases the mRNA levels of phosphoenolpyruvate carboxykinase in rat liver (Newsome et al., 1994; Warskulat et al., 1996), glutamine synthetase (Warskulat et al., 1996), β-actin (Schulz et al., 1991; Theodoropoulos et al., 1992), argininosuccinate synthase (Quillard et al., 1996), ornithine decarboxylase (Tohyama et al., 1991), but represses the mRNA level for asparagine synthetase (Hutson and Kilberg, 1994). The mechanisms underlying these effects are not clear but may at least in part also relate to cell swelling.

Glutamine-induced hepatocyte swelling may also intere with kidney function. In the intact rat, infusion of glutamine into the portal vein leads within minutes to a strong decrease of glomerular filtration. The effect is not observed upon portal infusion of glutamate (which does not swell hepatocytes) or when the glutamine is infused into the jugular vein. As shown in denervation experiments and pharmacological studies, the effect of portal glutamine is due to the acti-

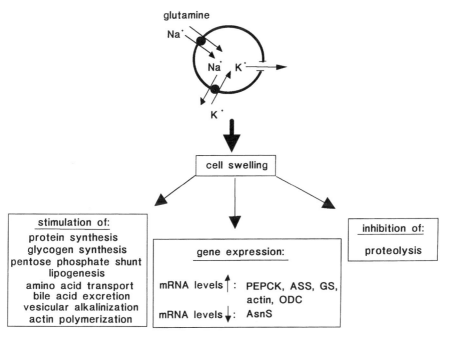

Figure 2. Functional consequences of glutamine-induced hepatocyte swelling. The intracellular signal transduction mechanisms that link glutamine-induced cell swelling to the functional consequences are not fully settled, but may involve an activation of mitogen-activated protein kinases, PI_3-kinase and ribosomal S6 kinase. Abbreviations: AsnS = asparagine synthetase, PEPCK = phosphoenolpyruvate carboxykinase, ODC = ornithine decarboxylase, ASS = argininosuccinate synthase, GS = glutamine synthetase. For further details see the text.

vation of a serotoninergic hepatorenal reflex, which is triggered most likely via hepatocyte swelling and/or an increase in sinusoidal resistance (Lang et al., 1991a, 1991b).

Figure 2 summarizes the effects of glutamine on hepatic function, which are thought to be due to hepatocyte swelling.

IV. Enzymes of Hepatic Glutamine Metabolism

Glutamine is an essential building block of proteins and occupies a central position in nitrogen metabolism (Meister, 1984). It is a storage and transport form for both glutamate and ammonia. Gluta-mine is a substrate of enzymes involved in purine and pyrimidine synthesis and conjugation reactions. Only hepatic glutaminase, glu-tamine synthetase, and glutamine aminotransferases will be consid-ered here.

A. LIVER GLUTAMINASE

Liver glutaminase (EC 3.5.1.2, L-glutamine amidohydrolase, phosphate-dependent glutaminase, formerly glutaminase I) is a mito-chondrial enzyme (Guha, 1962) with loose attachment to the inner-mitochondrial membrane (for review see McGivan et al., 1984). Solu-bilization of the enzyme increases the apparent K_m for glutamine from 6 to 21 mM and shifts glutamine dependence from hyperbolic to sigmoidal. Liver glutaminase is both immunologically and kinetically different from phosphate-dependent glutaminase from other organs. The enzyme is activated by phosphate (K_a = 5 mM), the K_m for glutamine is 28 mM (Huang and Knox 1976) and has a pH optimum between 7.8 and 8.2. Due to the concentrative properties of gluta-mine transporting systems in the plasma and mitochondrial mem-brane, the enzyme may operate despite physiologically low extracel-lular glutamine concentrations of about 0.6 mM at substrate concentrations close to its K_m. The enzyme appears to have a sub-unit size of 57–58 kDa (Heini et al., 1987; Smith and Watford, 1988), however, the actual molecular weight has not yet been determined exactly and estimations range between 170 and 320 kDa (Patel and McGivan, 1984; Heini et al., 1987; Huang and Knox, 1976; Smith and Watford, 1988). The liver glutaminase gene has been cloned (Chung-Bok and Watford 1994) and the mature mRNA has 2.2 kb. It is only expressed in neonatal and adult liver and is not detectable in fetal liver. In mitochondrial preparations, liver glutaminase is acti-vated by ammonia (Charles, 1968; Häussinger and Sies, 1975; Joseph and McGivan, 1978), with NH_3 being the activating species (McGi-van and Bradford, 1983), and by spermine (Kovacevic et al., 1995). In contrast to the kidney enzyme, liver glutaminase is not subject to inhibition by glutamate. Mitochondrial swelling activates glutami-nase as does an increase in pH (McGivan et al., 1984; Halestrap, 1989, 1993). In addition, citrate (Szweda and Atkinson, 1990) and N-acetylglutamate (Meijer and Verhoeven, 1986) activate glutaminase.

B. GLUTAMINE TRANSAMINASES

Glutamine transaminates with a number of α-ketoacids to yield the corresponding amino acids and α-ketoglutaramate; this ketoacid amide is hydrolyzed in a separate step by α-ketoacid-ω-amidase (EC 3.5.1.3.) to α-ketoglutarate and ammonia. ω-Amidase-catalyzed α-ketoglutaramate removal essentially drives the freely reversible transamination reaction toward glutamine consumption. There are two glutamine aminotransferases (EC 2.6.1.15) in the liver, the so-called L and K types, which are distinguished by their relative affinities toward different α-ketoacids and the use of albizziin instead of glutamine (Cooper and Meister, 1972, 1974, 1977; Meister, 1984). Both types are found in liver and about 80% of the enzyme activity is cytosolic (Livesy and Lund, 1980). Preferred ketoacid substrates for glutamine transamination are ketomethionine and phenylpyruvate, but there is little reactivity with ketoacids occurring physiologically at high concentrations such as pyruvate. Although enzyme activity is high in liver, flux through the enzyme is probably low due to low concentrations of suitable ketoacid substrates. The function of glutamine transaminases is seen in the preservation of the carbon skeletons of methionine and phenylalanine (Meister, 1984).

C. GLUTAMINE SYNTHETASE

Liver glutamine synthetase (EC 6.3.1.2.) is a cytosolic enzyme (Deuel et al., 1978) and the reaction mechanism has been studied in detail (Meister, 1974, 1984; Tate and Meister, 1972). The enzyme consists of eight subunits with a subunit molecular size of 44 kDa and resembles glutamine synthetase from bovine brain in its physical properties, amino acid composition, and substrate specificity. The enzyme is activated by α-ketoglutarate and is inhibited by methionine sulfoximine, glycine, and carbamyolphosphate (Tate and Meister, 1972).

The rat glutamine synthetase gene has been cloned (van de Zande et al., 1990). It is 9.5–10 kb long, has seven exons, and codes for two mRNA species of 1375 and 2787 nucleotides, respectively. The smaller mRNA is identical to the 5′ 1375 nucleotides of the long mRNA species and contains the entire protein coding region. The regulatory elements of the rat liver gene have been characterized and contain a putative AP-2 binding site, putative glucocorticoid-responsive elements and an HNF-3 binding site (Fahrner et al., 1993). Enhancer elements are found in the upstream region and the first intron, in which clusters of potential transcription factor binding sites (Sp1, HNF-3, and elements related to binding of members from the C/EBP family) are found (Gaunitz et al., 1997).

V. Hepatocyte Heterogeneity in Glutamine Metabolism

The liver acinus (Rappaport, 1976) or the metabolic lobulus (Lamers et al., 1989) is the functional unit of the liver. In the acinus, periportal (at the sinusoidal inflow), and perivenous (at the sinusoidal outflow) hepatocytes differ in their enzyme equipment and metabolic function. This functional hepatocyte heterogeneity or metabolic zonation (for reviews see Jungermann and Katz, 1989) is most markedly developed with respect to glutamine and nitrogen metabolism (for reviews see Traber et al., 1988; Häussinger, 1990; Meijer et al., 1990; Häussinger et al., 1992b; Gebhardt et al., 1994).

A. ENZYME DISTRIBUTION

Glutaminase and the enzymes of the urea cycle are present in periportal hepatocytes, whereas glutamine synthetase is found only in perivenous hepatocytes. This was shown in the intact perfused rat liver by comparing metabolic flux rates in antegrade and retrograde perfusions (Häussinger, 1983), in experiments on zonal cell damage (Häussinger and Gerok, 1984a; Gebhardt et al., 1988a), by immunohistochemistry (Gebhardt and Mecke, 1983; Saheki et al., 1983; Gaasbeek-Janzen et al., 1984; Smith and Campbell, 1988; Kuo et al., 1988), with hepatocyte preparations enriched in periportal and perivenous cells (Pösö et al., 1986; Watford and Smith, 1990), and more recently by *in situ* hybridization of the respective mRNA species (Moorman et al., 1988, 1990; Gebhardt et al., 1988b; Kuo et al., 1991). Ornithine aminotransferase mRNA colocalizes with glutamine synthetase mRNA in the liver acinus (Kuo et al., 1991). The borderline between the periportal urea-synthesizing compartment and the perivenous glutamine-synthesizing compartment is rather strict: glutamine synthetase is exclusively found in a small hepatocyte population (6–7% of all hepatocytes of an acinus) surrounding the terminal hepatic venules at the outflow of the sinusoidal bed (Fig. 3) and

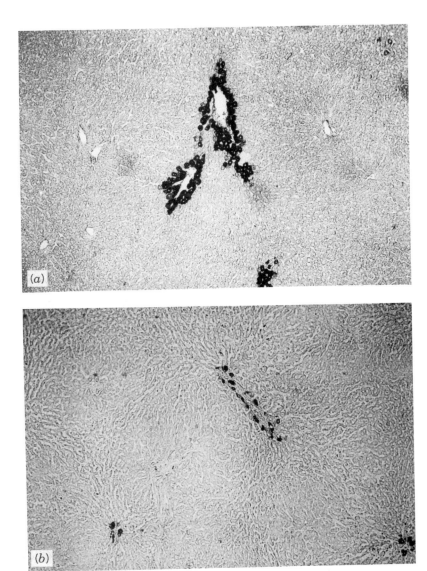

Figure 3.   Immunohistochemical localization of glutamine synthetase in normal rat liver (a) and 6 weeks after portocaval anastomosis (b). Glutamine synthetase is exclusively found in a small cell population (perivenous scavenger cells) at the perivenous outflow of the acinar bed, which sourrounds the terminal hepatic venules. [From Häussinger et al., 1992b.]

these cells are virtually free of urea cycle enzymes. In view of their function these cells have been named perivenous scavenger cells (Häussinger and Stehle, 1988). Label incorporation studies in the intact perfused rat liver have localized glutaminase in the periportal area of the liver acinus, that is, in a cell population that does not contain glutamine synthetase (Häussinger, 1983). This result was confirmed later in studies on the distribution of glutaminase mRNA in hepatocyte suspensions derived from periportal and perivenous areas of the acinus (Watford and Smith, 1990) and more recently by *in situ* hybridization (Moorman et al., 1994). According to the latter report, glutaminase is even restricted to a cell population at the periportal inflow, which is smaller than the compartment containing urea cycle enzymes. Thus, from the periportal to the perivenous end of the acinus three zones can be distinguished: zone 1 containing urea cycle enzymes and glutaminase, zone 2 containing only urea cycle enzymes, and zone 3 containing exclusively glutamine synthetase, but no glutaminase and urea cycle enzymes (Fig. 4). In adult human liver between zones 2 and 3 a further area is detectable, which neither contains urea cycle enzymes, glutaminase, or glutamine synthetase (Moorman et al., 1989). Whereas glutaminase and glutamine synthetase are heteogeneously distributed in the liver acinus, glutamine transaminases are present in the periportal as well as in the perivenous area of the acinus (Häussinger et al., 1985a), although slight activity gradients between the respective subacinar zones were not ruled out in these studies.

B. DYNAMICS AND REGULATION OF THE HETEROGENEOUS DISTRIBUTION OF GLUTAMINE SYNTHETASE

The glutamine synthetase gene is thought to be one of the oldest existing and functioning genes (Kumuda et al., 1993). However, its highly zonated expression is unique for mammalian liver and no zonation is found in avian and amphibian liver (Smith and Campbell, 1988). Slight differences in the size of the glutamine synthetase-positive zone can be noted between different rat strains and the zone is slightly smaller in livers from female compared to those from male rats (Sirma et al., 1996). In the early fetal rat all hepatocytes uniformly express glutamine synthetase and carbamoylphosphate synthetase mRNA and the heterogeneous and complementary distribu-

The Liver Acinus

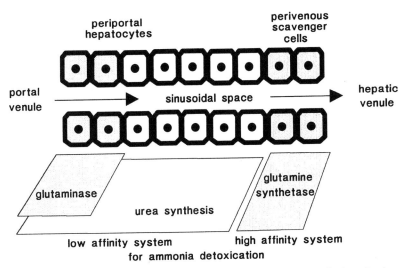

Figure 4. Structural and functional organization of hepatic ammonia detoxication. Urea and glutamine synthesis are anatomically switched behind each other and represent with respect to ammonia detoxication the sequence of a periportal "low-affinity, but high-capacity system" (ureogenesis) and a perivenous "high-affinity system" (glutamine synthesis). At a constant portal ammonia load flux through periportal urea synthesis determines the amount of ammonia reaching the perivenous compartment. Glutaminase is periportal and according to recent *in situ* hybridization experiments (Moorman et al. 1994) restricted to a small periportal cell population at the acinar inflow. In these cells, glutaminase acts as a pH and hormone modulated ammonia amplifier and is an important determinant for urea cycle flux. For details see text.

tion of these enzymes is fully developed 2 weeks after birth (Gaasbeek-Janzen et al., 1987). However, the first indications for a heterogenous distribution of glutamine synthetase, which is related to the vascular architecture, is detectable at the mRNA level after 18 embryonic days and 2 days later at the protein level (Moorman et al., 1990). The important role of liver architecture for the development of zonal heterogeneity is also suggested by comparative studies on the rat and spiny mouse liver (Lamers et al., 1987). The key observation was that in both species the development of compartments of gene expression for carbamoylphosphate synthetase and glutamine synthetase occurred independent of the process of birth, at a comparable stage of development, precisely when the adult type of architecture becomes established.

The adult pattern of glutamine synthetase expression in about two cell layers surrounding the terminal hepatic venules is remarkably stable. Glucocorticoids can upregulate glutamine synthetase, however, the acinar compartment containing the enzyme does not enlarge. Following destruction of the perivenous glutamine-synthetase containing hepatocytes by CCl_4, the scavenger cell compartment reappears within several days (Schöls et al., 1990). No *in vivo* conditions are known to induce glutamine synthase in the periportal compartment and an additional carbamoylphosphate synthetase expression in perivenous scavenger cells occurs only under extreme experimental conditions, such as administration of dexamethasone to diabetic or starved rats (de Groot et al., 1987; Moorman et al., 1990a). However, portocaval anastomosis diminishes the number of perivenous cells expressing glutamine synthetase (Fig. 3) and orchidectomy slightly diminishes the size of the glutamine synthetase-positive zone to levels found in female rats (Sirma et al., 1996). Glutamine synthetase- and carbamoylphosphate synthetase-positive hepatocytes are probably not the result of different hepatocyte lineages. This is suggested by explantation experiments, which have demonstrated that hepatocytes transplanted to the spleen preferentially express the carbamoylphosphate synthetase phenotype (Lamers et al., 1990), that transplantation to the fat pads supports the glutamine synthetase phenotype, provided the hepatocytes are present as single cells (Gebhardt et al., 1989), whereas the induction of hepatocyte in the pancreas results in preferential expression of both the carbamoylphosphate synthetase and the glutamine synthetase phenotype (Yeldani et al., 1990). Interestingly, as soon as hepatocyte transplants to the spleen have formed agglomerates with sinusoids that drain on venules, the hepatocytes surrounding the venules become strongly positive for glutamine synthetase and remain weekly positive for carbamoylphosphate synthetase (Lamers et al., 1990).

The mechanisms underlying the different subacinar gene expression are not yet settled, however, current evidence suggests that it is predominantly due to cell–cell/cell–matrix interactions and the development of hepatic architecture, whereas innervation, blood-borne humoral and metabolic factors probably play a minor role (for reviews, see Häussinger et al., 1992a; Gebhardt et al., 1994). The complementary distribution of urea cycle enzymes and glutamine synthetase in the liver acinus is found at both the mRNA and protein level, indicating regulation at a pretranslational level (Moorman et al., 1988). Studies on fetal liver fragment transplantation into testes indicated that the capacity for positional expression of glutamine synthetase is already established at fetal age, before expression of the enzyme can be detected and does not depend on portal blood supply (Shiojiri et al., 1995). A role of cell–cell interactions is also suggested from studies on the reappearance of glutamine synthetase positive cells following CCl₄-induced perivenous liver damage (Kuo and Darnell, 1991) and by the fact that cocultivation of periportal (glutamine-synthetase negative) hepatocytes with epithelial cells from rat liver or endothelial cells from human veins induces expression of glutamine synthetase in these hepatocytes (Schröde et al., 1990).

The mechanism underlying the sharp delineation of the expression of glutamine synthase is unclear, although the upstream regulatory region and the first intron of the glutamine synthetase gene were identified as major determinants for the pericentral expression pattern of the enzyme (Lie-Venema et al., 1995; Gaunitz et al., 1997). Steep enzyme gradients are to be expected if several effector molecules are involved in a cooperative binding to the same response element to bring out induction. A cooperative mechanism of cooperative response can result in an almost all-or-none type of induction. It would be attractive if such a mechanism would also *in vivo* display a reciprocity for carbamoylphosphate synthase and glutamine synthase gene expression.

C. TRANSPORT

So far, no firm evidence has been obtained for a zonation of the glutamine transporting systems N and n, although the spatial separation of glutaminase and glutamine synthetase in the liver acinus is suggestive for this. Studies with rat hepatocytes isolated from periportal and perivenous ends of the acinus showed no difference regarding Na^+-dependent transport of histidine, which like glutamine is a system N substrate, whereas the V_{max} of Na^+-independent histidine transport was slightly higher in perivenous than in periportal cells (Burger et al., 1989). However, these findings must be interpreted before the background that preparations of perivenous hepatocytes contain only up to 30% perivenous scavenger cells.

Most amino acids being delivered via the portal vein are predominantly taken up and processed in the much larger periportal compartment that contains high activities of gluconeogenic enzymes (Jungermann and Katz, 1989) in addition to urea cycle enzymes. One exception is the basolateral uptake of vascular glutamate, which occurs predominantly by the small perivenous hepatocyte population containing glutamine synthetase (Häussinger and Gerok, 1983; Häussinger et al., 1989; Stoll et al., 1991). Destruction of these cells by CCl_4 treatment strongly diminishes hepatic glutamate uptake (Häussinger and Gerok, 1983; Taylor and Rennie, 1987). The zonation of hepatic glutamate uptake was also shown by histoautoradiography (Fig. 5; Stoll et al., 1991) and studies on the incorporation of added [^{14}C-] glutamate into newly synthesized glutamine in perfused rat liver. Extrapolation to maximal rates of hepatic glutamine synthesis showed that about 90% of the labeled glutamate being taken up by the liver was released as labeled glutamine (Häussinger et al., 1989). This finding demonstrates that the small perivenous cell population containing glutamine synthetase accounts for at least 90% of total hepatic glutamate uptake. From the size of this cell population, that is, about 6–7% of all acinar hepatocytes, a more than 100-fold higher basolateral glutamate uptake activity is calculated for perivenous hepatocytes compared to periportal ones, which also demonstrates marked hepatocyte heterogeneities regarding plasma membrane transport systems. Basolateral glutamate uptake largely occurs via a Na^+-independent exchange mechanism (Häussinger et al., 1989) and glutamate release from the liver is markedly stimulated upon addition of benzoate, phenylpyruvate, and the ketoanalogues of branched-chain amino acids except ketovaline (Häussinger and Gerok 1984b). The Na^+-dependent glutamate transport is found exclusively in the canalicular membrane (Ballatori et al., 1986). This transporter exhibits less strict zonation, but shows a periportal–perivenous activity gradient. The loss of cell polarity upon isolation of hepatocytes may explain the conflicting data regarding Na^+-dependent glutamate uptake in the intact organ and isolated liver cells.

In addition to glutamate, other precursors for glutamine synthesis, that is, vascular oxoglutarate, malate, and related dicarboxylates are

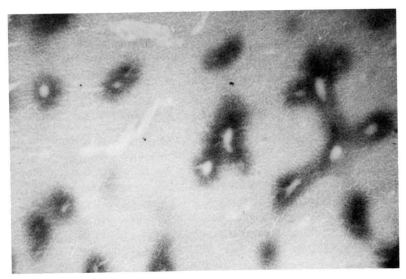

Figure 5.   Hepatocyte heterogeneity with regard to glutamate uptake. The histoautoradiography was obtained 15 s after bolus injection of [3H]labeled glutamate. Label
incorporation experiments demonstrated that vascular glutamate is taken up by
greater than 90% by perivenous scavenger cells. [From Stoll et al. 1991.]

almost exclusively taken up by these perivenous cells in a Na^+-
dependent way (Stoll et al., 1991; Stoll and Häussinger, 1989, 1991),
underlining again the specialization of these cells for glutamine synthesis. These dicarboxylates are present in micromolar concentrations in the plasma and are taken up by perivenous scavenger cells
with high affinity and serve as a carbon substrate for glutamine synthesis.

VI.   Hepatic Glutamine Metabolism

A.   REGULATION OF GLUTAMINE BREAKDOWN

Phosphate-dependent glutaminase is the major enzyme involved
in hepatic glutamine degradation and the regulatory properties of the
enzyme are the basis for its crucial role in hepatic nitrogen metabolism. Control on hepatic glutamine breakdown occurs at the transport and enzyme level. Although the *in vitro* activity of the enzyme

is fully dependent on the presence of phosphate, it is unlikely that the intact hepatocyte variations of glutaminase activity occur by changes in the phosphate concentration, since mitochondrial phosphate appears to be fairly well buffered (Akerboom et al., 1978). Glutaminase activity is subject to short- and long-term regulation.

1. Short-Term Regulation

Besides not being product inhibited by glutamate in contrast to the kidney enzyme (Krebs, 1935), the other unique property of liver glutaminase is its absolute requirement for its product ammonia as an essential activator (Charles, 1968; Häussinger and Sies, 1975, 1979; Häussinger et al. 1975, 1983; Joseph and McGivan, 1978; Verhoeven et al., 1983; McGivan and Bradford, 1983). Ammonia activation of glutamine breakdown in the perfused rat liver is maximal at NH_4^+ concentrations of 0.6 mM and is half-maximal at 0.2–0.3 mM, that is, at the physiological portal ammonia concentration (Häussinger and Sies, 1979). Because ammonia is generated in the intestine, an interorgan feed-forward system for hepatic glutamine metabolism has been proposed (Häussinger, 1979; Welbourne, 1987). When glutamine is infused at high concentrations of 5 mM in single pass perfused rat liver, glutamine breakdown occurs only after a lag phase of about 30 min, which is required for full self-activation of the enzyme by its reaction product ammonia. This lag phase can be shortened by a brief pulse of ammonia in order to activate the enzyme and to perpetuate activation by the ammonia formation during glutamine breakdown (Häussinger et al., 1975). Lowering the pH inhibits glutamine breakdown (Lueck and Miller, 1970; Häussinger et al., 1980; Häussinger et al., 1983, 1984), whereas alkalosis stimulates. Inhibition of flux through glutaminase at low pH is due to an increased concentration of ammonium ions required for activation of glutaminase (Verhoeven et al., 1983), but also due to an inhibition of glutamine transport across the plasma and mitochondrial membrane (Lenzen et al., 1987). Hepatic glutamine breakdown at glutaminase is also under hormonal control. The enzyme is activated by glucagon (Joseph and McGivan, 1978, Lacey et al., 1981; Häussinger et al., 1983; Häussinger and Sies, 1984a), aMP (Baverel and Lund, 1979), adrenaline, α-adrenergic stimulation, vasopressin, angiotensin II (Corvera and Garcia-Sainz, 1983; Häussinger and Sies, 1984b), and hypoosmotic hepatocyte swelling (Häussinger et al., 1990a). Glutaminase activation by glycine is in part due to cell swelling (Vincent et al., 1992). Activation of the enzyme by hormones and hypoosmolarity probably involves mitochondrial swelling *in situ*, which alters the attachment of the enzyme to the inner mitochondrial membrane (Halestrap, 1989, 1993). It is not yet clear whether glutamine-induced cell swelling will also activate mitochondrial glutaminase. If this were true, another interesting feed-forward regulation of enzyme activity by the substrate would ensue. In isolated hepatocytes, glutaminase is activated by okadaic acid and cell permeable protein kinase A activators (Brosnan et al., 1995). Glutaminase is also activated in endotoxinemia (Ewart et al., 1995).

2. Long-Term Regulation

When hepatocytes are cultivated, glutaminase activity and protein decrease, however, the activity is maintained when NH_4Cl is present (McGivan et al., 1991). This may indicate that ammonia is not only an important short-term regulator of hepatic glutaminase but also acts on glutaminase expression. Hepatic glutaminase activity is increased upon feeding high protein diets, during starvation, and uncontrolled diabetes, but is unaffected upon changes of the acid–base status (Watford et al., 1984; Watford, 1993). These alterations of glutaminase occur at the level of transcription and resemble the regulation of mRNA levels for phosphoenolpyruvate carboxykinase (PEPCK) (Zahn et al., 1994; Watford et al., 1994). However, glutaminase mRNA levels are only slightly affected by anisoosmolarity, whereas PEPCK mRNA levels strongly decrease or increase in response to hypoosmotic or hyperosmotic liver perfusion, respectively (Warskulat et al., 1996). In H4IIE rat hepatoma cells, glutaminase mRNA levels decrease in response to hyperosmolarity, whereas PEPCK mRNA levels increase (Warskulat et al., 1996). In isolated rat hepatocytes and hepatoma cells, glutaminase mRNA levels increase under the influence of cAMP and glucocorticoids (Curthoys and Watford, 1995), whereas in perfused rat liver exposure for 3 h to cAMP lowers glutaminase mRNA levels (Warskulat et al., 1996).

B.   REGULATION OF GLUTAMINE SYNTHESIS

In addition, glutamine synthetase is subject to short- and long-term regulation, however, under all conditions tested so far its unique subacinar distribution does practically not change.

1. Short-Term Regulation

Due to the exclusive downstream localization of glutamine synthetase in the liver acinus, studies on the short-term regulation of glutamine synthesis in the liver are difficult. Due to the metabolic activity of upstream periportal cells, the actual substrate concentrations in the vicinity of perivenous scavenger cells are not known. For example, with a constant ammonia supply to the portal vein, flux through the urea cycle in the periportal compartment will determine the amount of NH_4^+ reaching the periportal (scavenger) hepatocytes. Accordingly, all factors that regulate urea cycle flux will indirectly exert control on the more downstream located glutamine synthesis. The problem can be circumvented in part by studying glutamine synthesis in experiments with retrograde liver perfusion, that is, from the hepatic toward the portal vein. Using this approach, glutamine synthesis is half-maximal at NH_4^+ concentrations of about 0.1 mM and maximal at concentrations above 0.3 mM (Häussinger and Gerok, 1984a), whereas in the physiologically antegrade perfusion (from portal to hepatic vein) maximal rates require NH_4^+ concentrations above 1 mM (Häussinger et al., 1983). Isotope perfusion studies in the antegrade direction, which were allowed to assess flux separately through periportal glutaminase and glutamine synthetase, indicated that flux through glutamine synthetase is increased in alkalosis (Häussinger et al., 1983, 1984), decreased by phenylephrine and glucagon, and is modified by the portal ammonia, histidine, and glutamine concentration in a complex way (Häussinger et al., 1983, 1985a; Häussinger and Sies 1984b). In spite of a high V_{max} in vitro, the maximal flux through the enzyme in the intact hepatocyte is only a small fraction of the in vitro value (Lund, 1971). This discrepancy may be due to inhibitory Mn^{2+} ions (Joseph et al., 1979), product inhibition by glutamine (Häussinger et al., 1985a), or limiting glutamate availability. In line with the latter proposal, is the high K_m = 5 mM glutamate for isolated glutamine synthetase and the finding that the inhibition of glutamine synthesis in the intact organ by phenylephrine is accompanied by a stimulation of glutamate oxidation in perivenous scavenger cells (Häussinger and Sies, 1984b). Further, addition of α-ketoglutarate, which is exclusively taken up by perivenous scavenger cells, increases glutamine synthesis even in the presence of saturating NH_4^+ concentrations (Stoll et al., 1991), indicating that provision of the carbon skeleton for glutamine synthesis in situ can be rate controlling. The highest rates of glutamine synthesis in perfused rat liver are about 0.6 μmol/g/min (Stoll et al., 1991), which would correspond to 8–10 μmol/min/g perivenous cells. This value is about two-fold higher than the capacity of periportal cells to eliminate ammonia via the urea cycle.

2. Long-Term Regulation

Cultivation of hepatocytes for 20 h lowers the activity of glutamine synthetase by about 70% and can be increased again by growth hormone, dexamethasone, and triiodothyronine in a permissive way (Gebhardt and Mecke, 1979). Also, *in vivo* glutamine synthetase activity in liver is downregulated by hypophysectomy (Wong et al., 1980). Portocaval anastomosis lowers the activity of glutamine synthetase in liver by about 90% (Girard and Butterworth, 1992) and strongly reduces the number and immunochemical staining intensity of glutamine synthetase-positive cells (Häussinger et al., 1992b). In perfused rat liver, glutamine and hypoosmotic cell swelling increase the mRNA levels for glutamine synthetase (Warskulat et al., 1996), whereas dexamethasone application was reported to lower the mRNA levels in liver (Abcouwer and Souba, 1995). Insulin lowers the mRNA levels for glutamine synthetase in perfused rat liver (Warskulat et al., 1996). Induction of an NH_4Cl-acidosis transiently increases the mRNA levels for glutamine synthetase, glutaminase, and carbamoylphosphate synthetase in rat liver (Schoolwerth et al., 1994), however, it is not clear, whether this effect is related to acidosis or to the ammonium load.

VII. Integration of Glutamine Metabolism at the Acinar Level: Functional Relevance

A. GLUTAMINE SYNTHETASE AS PERIVENOUS AMMONIA SCAVENGER

In the intact liver lobule, the two major ammonia-detoxicating systems, urea and glutamine synthesis, are anatomically switched behind each other (Fig. 4). Accordingly, the portal blood will first get into contact with hepatocytes capable of urea synthesis, before glutamine-synthesizing (scavenger) cells just at the end of the acinar bed are reached (Häussinger, 1983). In functional terms, this organization represents the sequence of a periportal low affinity, but high-capacity system (ureogenesis) and a perivenous high-affinity system for ammonia detoxication (glutamine synthesis) (Häussinger, 1983; Häussinger and Gerok, 1984a). In isolated perfused rat liver, efficient ammonia extraction with physiologically low portal ammonia concentrations requires an intact glutamine synthetase activity (high-affinity system) and ammonia at physiological portal concentrations of 0.2–0.3 mM is converted by about two-thirds into urea and by about one-third into glutamine, although these pathways are structurally organized in sequence. A 7% [^{13}N]ammonia recovery in glutamine was also shown *in vivo* following a bolus injection of [^{13}N]ammonia (Cooper et al., 1987); this value may underestimate the real contribution of perivenous hepatocytes to hepatic ammonia clearance due to label dilution by endogenously formed ammonia in periportal cells (Cooper 1990). Thus, *in vitro* and *in vivo* a considerable fraction of the ammonia delivered via the portal vein from the intestine reaches the perivenous end of the liver acinus. Here, perivenous glutamine synthetase acts as a high-affinity scavenger for the ammonia, which escaped periportal detoxication by urea synthesis (Häussinger, 1983, 1990). This also holds for ammonia produced during amino acid breakdown in the much larger periportal compartment: Under these conditions ammonia is released from periportal cells into the sinusoidal space, despite the high urea cycle enzyme activity in this compartment, and is eliminated by perivenous hepatocytes via glutamine synthesis. This not only underlines the comparatively low affinity of periportal urea synthesis for ammonia and the importance of ammonia scavenging by perivenous hepatocytes for efficient hepatic ammonia detoxication, but also demonstrated for the first time metabolic interactions between different subacinar cell populations (Häussinger, 1983). The fact that ammonia is formed from amino acids in periportal cells, but is simultaneously consumed for glutamine synthesis in perivenous cells, points to different directions of flux through glutamate dehydrogenase in periportal and perivenous cells, respectively (Häussinger and Sies, 1984a). The difference in ammonia affinity between urea and glutamine synthesis results from the approximately 10-fold difference in K_m values for NH_4^+ of carbamoyl phosphate synthetase, the rate-controlling enzyme of the urea cycle (Lusty, 1978; Meijer et al., 1985) and glutamine synthetase (Deuel et al., 1978), respectively. This difference is also found for the $K_{0.5}$ values for urea and glutamine synthesis from NH_4Cl in isolated perfused rat liver (Häussinger and Gerok, 1984a; Häussinger et al., 1975). The difference is even more pronounced in human liver [$K_{0.5}(NH_4^+)$ for urea and glutamine synthesis are 3.6 and 0.11 mM, respectively (Kaiser et al., 1988)].

The important scavenger role of perivenous glutamine synthesis for the maintenance of physiologically low ammonia concentrations in the hepatic vein becomes rapidly evident after inhibition of glutamine synthetase by methionine sulfoximine (Häussinger, 1983) or after destruction of perivenous cells by CCl_4 treatment (Häussinger and Gerok 1984a). In the latter case, hyperammonemia ensues due to an almost complete scavenger cell failure to synthesize glutamine, although periportal urea synthesis is not affected.

B. GLUTAMINASE: A PH AND HORMONE-CONTROLLED AMMONIA AMPLIFYING SYSTEM

Whereas glutamine synthetase is perivenous, glutaminase is found in periportal hepatocytes (Häussinger, 1983), and has a joint mitochondrial localization together with carbamoylphosphate synthetase. In view of the potent activation of liver glutaminase by its product ammonia in the physiological concentration range (Häussinger and Sies, 1979), glutaminase function is seen to amplify ammonia inside the mitochondria in parallel to that delivered via the portal vein or arising during intrahepatic amino acid breakdown (Häussinger, 1983). Note that direct measurements of the actual ammonia concentration inside the mitochondria are still lacking (for a review see Häussinger, 1990). Because urea synthesis is normally controlled by flux through carbamoylphosphate synthetase (Meijer et al., 1985), which largely depends on the actual ammonia concentration inside the mitochondria, the extent of amplification of ammonia via glutaminase becomes an important determinant of urea cycle flux in view of the physiologically low ammonia concentrations. These concentrations are about one order of magnitude below the K_m (ammonia) of carbamoyl phosphate synthase. Evidence for glutamine-amide nitrogen as a major source for urea synthesis has also been obtained in later *in vivo* studies (Nissim et al., 1992; Welbourne, 1986; Cooper, 1990). It has been suggested that glutamine-derived ammonia is even a preferred substrate for urea synthesis, because glutamine-nitrogen incorporation into urea is, in contrast to the incorporation of infused NH_4^+, almost insensitive to carbonic anhydrase inhibition (Häussinger, 1986) and there is some evidence of glutamine-nitrogen channeling into carbamoylphosphate synthetase (Meijer, 1985). In view of the recent suggestion that glutaminase is restricted to a subpopulation of urea cycle enzyme-containing periportal hepatocytes (see Fig. 4), the possibility has to be considered that glutaminase-mediated ammonia amplification may also increase the ammonia delivery to more downstream hepatocytes containing urea cycle enzymes. If this were true, the small periportal cell population containing both glutaminase and urea cycle enzymes would function as an ammonia amplifying compartment, which determines the ammonia supply for ureogenic hepatocytes being devoid of glutaminase.

As anticipated from a control of urea cycle flux by glutaminase activity, factors known to affect the urea cycle flux are indeed associated with parallel activity changes of the "ammonia amplifier" glutaminase. Apart from the portal ammonia concentration, this includes the effects of glucagon, α-adrenergic agonists, vasopressin, acidosis/alkalosis, and feeding of a high protein diet (for a review, see Häussinger, 1990). Interestingly, liver glutaminase is activated by liver cell swelling, as it occurs during concentrative, mainly Na^+-dependent, uptake of amino acids into the hepatocyte (Häussinger et al., 1990). Thus, an increased portal amino acid load will increase hepatocyte volume and the accompanying activation of glutaminase is expected to favor amino acid–nitrogen disposal via urea synthesis. Inhibition of flux through liver glutaminase in acidosis (Häussinger et al., 1983, 1984) inhibits urea cycle flux and irreversible hepatic bicarbonate consumption associated with ureogenesis.

C. INTERCELLULAR GLUTAMINE CYCLING

In the intact liver acinus, periportal glutaminase and perivenous glutamine synthetase are simultaneously active, resulting in a periportal breakdown and perivenous resynthesis of glutamine (Häussinger and Sies, 1979; Häussinger, 1983). This energy-consuming cycling of glutamine was termed the intercellular glutamine cycle (Häussinger, 1983). Glutamine cycling has been confirmed *in vivo* and *in vitro* (Welbourne, 1986; Cooper et al., 1987, 1988; Vincent et al., 1989; Cooper, 1990). In the presence of physiological glutamine and ammonia concentrations, flux through the glutamine cycle is 0.1–0.2 μmol/g/min and is under complex metabolic, hormonal, and pH control (Häussinger et al., 1983, 1984; Häussinger, 1983). In addition, the hepatocellular hydration state affects glutamine cycling. In rat liver being perfused with glutamine and ammonia at near-physiological concentrations, hypoosmotic cell swelling switches net glutamine release from the liver to net uptake, due to both an increased flux through periportal glutaminase and a decreased flux through glutamine synthetase (Häussinger et al., 1990a). In the special case of a well-balanced acid–base situation, intercellular glutamine cycling allows to maintain a high urea flux, despite the low affinity of carbamoyl phosphate synthetase for ammonia and the presence of physiologically low ammonia concentrations: Glutamine consumed during ammonia amplification in periportal hepatocytes is resynthesized in the perivenous compartment from the ammonia that escaped upstream urea synthesis (Fig. 6). Intercellular glutamine cycling, however, also provides an effective means for adjusting ammonia flux into urea or glutamine according the needs of the acid–base situation. In acidosis, flux through glutaminase decreases and flux through glutamine synthetase increases, resulting in a net production of glutamine at the expense of urea (Häussinger et al., 1984) (Fig. 7). Thus, intercellular glutamine cycling provides an effective means by which the liver can switch from net glutamine consumption to net glutamine output depending on the experimental conditions (Häussinger et al., 1983, 1984; Welbourne, 1986).

D. ROLE IN ACID–BASE HOMEOSTASIS

Urea synthesis must be viewed as a major pathway for irreversible HCO_3^- disposal. By this means, the liver is considered as an important organ that is involved in the regulation of the systemic acid–base homeostasis (Atkinson and Camien, 1982; Atkinson and Bourke, 1984; Häussinger et al., 1984; Bourke and Häussinger, 1992). The structural and functional organization of ammonia and glutamine metabolizing pathways in the liver lobule is one prerequi-

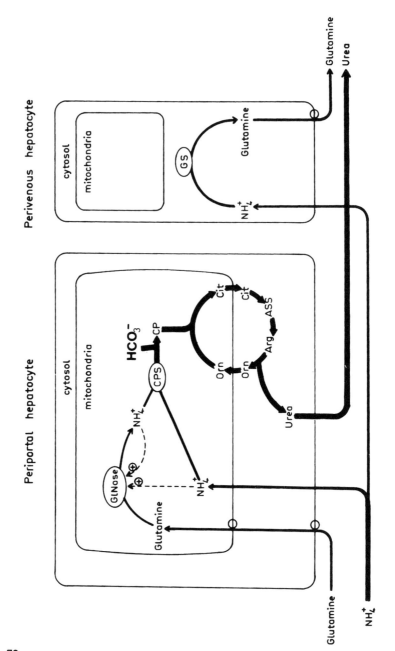

site for such a role of the liver in acid–base homeostasis (Häussinger et al., 1984; Häussinger, 1988, 1990). Due to this organization, periportal urea cycle flux can be adjusted to the needs of systemic acid–base balance without threat of hyperammonemia. The reason is that glutamine synthesis in perivenous scavenger cells acts as a "back-up system" for ammonia detoxication, guaranteeing nontoxic ammonia levels in effluent hepatic venous blood even when urea cycle flux is downregulated in order to diminish hepatic bicarbonate consumption in acidosis. In general, pH control of urea cycle flux occurs at the level of provision of substrates for carbamoylphosphate synthetase, but not within the urea cycle itself (Häussinger, 1990). The remarkable pH sensitivities of glutamine transport and glutaminase play an important role among the various factors that adjust bicarbonate-consuming urea cycle flux to the needs of pH homeostasis. As lined out above, mitochondrial glutaminase acts as a pH-regulated ammonia amplifier: lowering the extracellular pH from 7.4 to 7.3 already inhibits the enzyme by 70%. Inhibition of flux through glutaminase in acidosis diminshes urea synthesis and therefore irreversible bicarbonate consumption in the liver. The liver becomes a net producer of glutamine due to diminished glutamine consumption

←───

Figure 6.   Structural and functional organization of hepatic ammonia and glutamine metabolism. Periportal hepatocytes contain urea cycle enzymes and glutaminase, but no glutamine synthetase. The latter enzyme is restricted to a small perivenous hepatocyte population near the outflow of the sinusoidal bed, which is virtually free of urea cycle enzymes (perivenous scavenger cells). Periportal glutaminase is activated by its product ammonia and is highly sensitive to changes of the extracellular pH. Glutaminase acts as a pH modulated amplifier of the ammonia concentration and is an important determinant of urea cycle flux. Ammonia that escaped periportal urea synthesis is disposed by the perivenous scavenger glutamine synthetase. At normal extracellular pH, glutaminase flux equals glutamine synthetase flux (so-called intercellular glutamine cycle): Portal ammonia is completely converted into urea. *In situ* hybridization experiments on the distribution of glutaminase mRNA suggest that the location in a subpopulation of periportal, urea synthesizing hepatocytes at the very inflow of the acinus (Moorman et al., 1994). If this also holds for the glutaminase protein, a third cell population has to be introduced into the scheme, which is located between the two representative cell types depicted here and which contains urea cycle enzymes only, but neither enzyme of glutamine metabolism. In such a case, glutaminase may fulfill a role as ammonia amplifier also for the more downstream urea synthesis in this intermediate compartment.

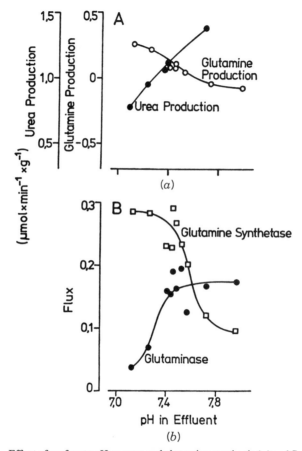

Figure 7. Effect of perfusate pH on urea and glutamine synthesis (*a*) and flux through glutaminase and glutamine synthetase (*b*). Livers were perfused with near-physiological ammonium and glutamine concentrations. The conversion of portal ammonium into urea without accompanying net glutamine turnover (i.e., flux through glutaminase equals flux through glutamine synthetase) characterizes the well-balanced pH situation. [From Häussinger et al., 1984.]

in periportal cells and an increased synthesis in perivenous scavenger cells (Fig. 7) (Häussinger et al., 1983, 1984; Welbourne, 1986). In acidosis, the glutamine is hydrolyzed by renal glutaminase, which is activated in acidosis, and surplus ammonium ions are excreted as such into urine. Thus, an interorgan team effort between liver and kidney maintains both bicarbonate and ammonium homeostasis. For further considerations see Häussinger (1988, 1990) and Bourke and Häussinger (1992).

VIII. Pathophysiological Aspects

Metabolic alkalosis is frequently found in patients with liver cirrhosis (Record et al., 1975; Oster, 1983), even in the absence of alkalosis-precipitating factors, such as vomiting, diuretic, or antacid therapy (Häussinger et al., 1990b). A simple explanation is that metabolic alkalosis results from an impaired HCO_3^- disposal via urea synthesis in the diseased liver. In line with this, an inverse correlation between the *in vitro* determined capacity to synthesize urea and the *in vivo* determined plasma bicarbonate level was demonstrated in humans (Häussinger et al., 1990c). The pathobiochemical events linking the disturbance of acid–base to that of ammonia metabolism in cirrhosis, however, are more complex (Fig. 8). The capacity of the cirrhotic liver to synthesize urea is diminished by about 80%, giving rise to metabolic alkalosis (Häussinger et al., 1990, 1992c). This alkalosis in turn is seen as one signal for activation of liver glutaminase. Indeed, flux through glutaminase was shown to be increased about five-fold in human liver cirrhosis (Kaiser et al., 1988; Häussinger et al., 1990c, 1992c; Matsuno and Goto 1992). This compensatory response augments—via increased mitochondrial ammonia amplifying (i.e., glutaminase flux)—urea synthesis and in the compensated cirrhotic patient near-normal rates of urea synthesis can be maintained. Thus, the cirrhotic patient approaches a new, albeit more alkaline steady state, which allows him to maintain a life-compatible urea cycle flux despite a marked reduction of the capacity for urea synthesis. Indeed, compensated cirrhotic patients excrete near-normal amounts of urea, although the ureogenic capacity of cirrhotic liver tissue *in vitro* is decreased by 80%. This compensation, however, requires the presence of metabolic alkalosis in order to keep the ammonia amplifying system glutaminase active. When acidosis develops (e.g., during infection, sepsis, or cardiac insufficiency), this compensatory mechanism is shut off and severe hyperammonemia develops.

Human-liver cirrhosis is characterized by a defect of perivenous scavenger cells: The capacity to synthesize glutamine is decreased

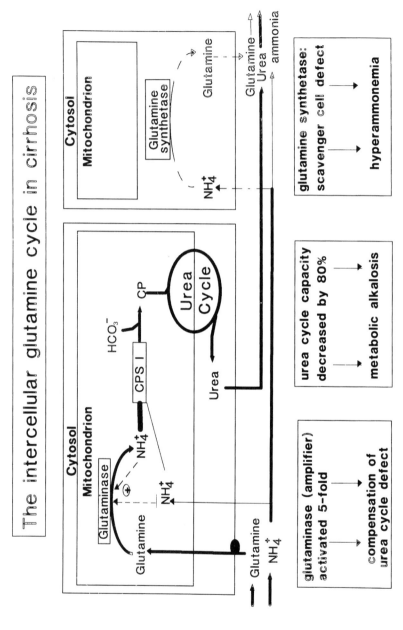

Figure 8. The intercellular glutamine cycle in liver cirrhosis. For details see the text.

by about 80% in cirrhosis (Kaiser et al., 1988; Häussinger et al., 1990c, 1992c; Racine-Samson et al., 1996). This scavenger cell defect may be related to portosystemic shunting (Häussinger et al., 1992a, Girard and Butterworth, 1992) and is a major factor for the development of hyperammonemia in liver cirrhosis. These finding made on human liver were later on also confirmed in animal models of cirrhosis (Gebhardt and Reichen, 1994). Hyperammonemia can also be induced by drugs and toxins, which primarily impair the function of perivenous hepatocytes, whereas the periportal urea-synthesizing hepatocytes need not to be affected (Häussinger and Gerok, 1984a). In addition, reversal of sinusoidal blood flow, as it is sometimes observed in cirrhosis, can produce hyperammonemia, because under these conditions the perivenous scavenger cells (high-affinity ammonia detoxication) are switched before the periportal urea synthesis (low-affinity detoxication) (Häussinger, 1983). A scavenger cell defect in cirrhosis may also be responsible for diuretic-induced hyperammonemia. Today, several frequently used diuretics are potent inhibitors of hepatic mitochondrial carbonic anhydrase V as well as urea synthesis (Häussinger et al., 1986). A 5–10% inhibition of urea synthesis by mefruside, xipamide, and thiazides in near-therapeutic concentrations will not impair hepatic ammonia detoxication in healthy individuals with intact perivenous scavenger cells. However, in cirrhosis, the scavenger cells are defective and even a slight inhibition of urea synthesis will lead to systemic hyperammonemia. This explains why the above-mentioned loop diuretics are well tolerated by normal individuals, but can precipitate hyperammonemia in the cirrhotic patient. Albeit to a lesser extent, alterations in glutaminase and glutamine synthetase activity typical for cirrhosis are already found in liver fibrosis (Kaiser et al., 1988; Häussinger et al., 1990c, 1992c). The metabolic zonation in the human hepatic lobulus is preserved in fibrotic lesions, but is lost in cirrhotic nodules (Racine-Samson et al. 1996).

The age-related increase in oxidized protein in rat liver and brain is accompanied by a loss of glutamine synthetase activity (Stadtman et al., 1992). Also, acetaminophen was shown to bind to glutamine synthetase and to inhibit its catalytic activity (Bulera et al., 1995). It is an attractive speculation to link such chemical modifications of glutamine synthetase to the occurrence of hyperammonemic states in clinical settings.

A. PERSPECTIVE

Although our current knowledge suggests that glutamine plays an important role in liver biochemistry and pathophysiology, many questions remain open. These questions relate to the molecular mechanisms that are responsible for the zone-specific expression patterns of glutaminase and glutamine synthetase, the intercellular communication between the different hepatocyte populations, the zonal distribution of glutamine transport, and the intracellular signaling mechanisms that are activated by glutamine. It is hoped that this chapter will stimulate future research.

References

Abcouwer, S. F. and Souba W. W. *J. Surg. Res.*, **59**, 59–65 (1995).

Akerboom, T. P. M., Bookelman, H., Zuurendonk, P. F., van der Meer, R., and Tager, J. M., *Eur. J. Biochem.*, **84**, 413–420 (1978).

Atkinson, D. E. and Bourke, E., *Biochem Sci Sci.*, **9**, 297–300 (1984).

Atkinson, D. E. and Camien, M., *Biochem. Sci.*, *9*, 297–300 (1982).

Ballatori, N., Moseley, R. H., and Boyer, J. L., *J. Biol. Chem.*, **261**, 6216–6221 (1986).

Baquet, A., Gaussin, V., Bollen, M., Stalmans, W., and Hue, L., *Eur. J. Biochem.*, **217**, 1083–1089 (1993).

Baquet, A., Hue, L., Meijer, A. J., van Woerkom, G. M., and Plomp, P. J. A. M. *J. Biol. Chem.*, **265**, 955–959 (1990).

Baquet, A., Lavoinne, A., and Hue, L., *Biochem. J.*, **273**, 57–62 (1991a).

Baquet, A., Maisin, L., and Hue, L., *Biochem. J.*, **278**, 887–890 (1991b).

Baverel, G. and Lund, P., *Biochem. J.*, **184**, 599–606 (1979).

Bode, B. P., Abcouwer, S. F., and Souba, W. W., In *Modulation of hepatic glutamine metabolism in cancer*, Cynober, L. A., Ed. CRC Press, Boca Raton, FL, 1995a, pp. 275–286.

Bode, B. P., Kaminski, D. L., Souba, W. W. and Li, A. P., *Hepatology*, **21**, 511–520 (1995).

Bode, B. P. and Kilberg, M.S., *J. Biol. Chem.*, **266**, 7376–7381 (1991).

Bourke, E. and Häussinger, D., *Contrib. Nephrol.*, **100**, 58–88 (1992).

Boyer, J. L., Graf, J., and Meier, P. J., *Ann. Rev. Physiol.*, **54**, 415–438 (1992).

Brosnan, J. T., Ewart, H. S., and Squires, S. A. *Adv. Enzymol. Reg.*, **35**, 131–146 (1995).

Bulera, S. J., Birge, R. B., Coen, S. D., and Khairallah, E. A., *Toxicol. Appl. Pharmacol.*, **134**, 313–320 (1995).

Busch, G. L., Schreiber, R., Dartsch, P. C., Völkl, H., vom Dahl, S., Häussinger, D., and Lang, F., *Proc. Natl. Acad. Sci. USA*, **91**, 9165–9169 (1994).

Burger, H. J., Gebhardt, R., Mayer, C., and Mecke, D., *Hepatology*, **9**, 22–28 (1989).

Charles, R., Ph. D. Thesis, Mitochondirale citrulline synthese: een ammoniak fixerend en ATP verbruikend process. Rototype, Amsterdam, The Netherlands (1968).

Christensen, H. V. and Kilberg, M. S., *Nutr. Rev.*, **53**, 74–76 (1995).

Chung-Bok, M. I. and Watford, M., *FASEB J*, **8**, A1442 (1994).

Colombo, J. P., Berüter, J., Bachmann, C., Peheim, E., *Enzyme*, **22**, 391–398 (1977).

Cooper, A. L., In *Ammonia metabolism in normal and portocaval-shunted rats*, Grisolia, S. and Felipo, V. Eds., Plenum New York, (1990), pp. 23–46.

Cooper, A. J. L. and Meister, A., *Biochemistry*, **11**, 661–671 (1972).

Cooper, A. J. L. and Meister, A., *J. Biol. Chem.*, **249**, 2554–2561 (1974).

Cooper, A. J. L. and Meister, A., *Crit. Rev. Biochem.*, **4**, 281–303 (1977).

Cooper, A. J. L., Nieves, E., Coleman, A. E., Filc-DeRicco, S., and Gelbard, A. S., *J. Biol. Chem.*, **262**, 1073–1080 (1987).

Cooper, A. J., Nieves, E., Rosenspire, K. C., Filc-DeRicco, S., Gelbard, A. S., and Brusilow, S. W., *J. Biol. Chem.*, **263**, 12268–12273 (1988).

Corvera, S. and Garcia-Sainz J. A., *Biochem. J.*, **210**, 957–960 (1983).

Curthoys, N. P., Shapiro, R. A., and Haser, W. G., In *Enzymes of renal glutamine metabolism*, Häussinger, D. and Sies, H., Eds., Springer-Verlag, Berlin, (1984) pp. 16–31.

Curthoys, N. P. and Watford, M., *Ann. Rev. Nutr.*, **15**, 133–159 (1995).

de Groot, C. J., ten Voorde, C. H. J., van Andel, R. E., te Kortschot, A., Gaasbeek Janzen, J. W., Wilson, R. H., Moorman, A. F. M., Charles, R., and Lamers, W. H., *Biochim. Biophys. Acta*, **908**, 231–240 (1987).

Deuel, T. F., Louie, M., and Lerner, A., *J. Biol. Chem.*, **253**, 6111–6118 (1978).

Ewart, H. S., Quian, D., and Brosnan, J. T., *J. Surg. Res.*, **59**, 245–249 (1995).

Fafournoux, P., Demigne, C., Remesy, C., and LeCam, A., *Biochem. J.*, **216**, 401–408 (1983).

Fahrner, J., Labruyere, W. T., Gaunitz, C., Moorman, A. F. M., Gebhardt, R., and Lamers, W. H., *Eur. J. Biochem.*, **213**, 1067–1073 (1993).

Gaasbeek Janzen, J. W., Gebhardt, R., ten Voorde, C. H. J., Lamers, W. H., Charles, R., and Moorman, A. F. M., *J. Histochem. Cytochem.*, **35**, 49–54 (1987).

Gaasbeek Janzen, J. W., Lamers, W. H., Moorman, A. F. M., de Graaf, A., Los, J. A., and Charles, R., *J. Histochem. Cytochem.*, **32**, 557–564 (1984).

Gaunitz, F., Gaunitz, G., Papke, M., and Gebhardt, R., *Biol. Chem. Hoppe-Seyler*, **378**, (1997).

Gaussin, V., Hue, L., Stalmans, W., and Bollen, M., *Biochem. J.*, **316**, 217–224 (1996).

Gebhardt, R. and Kleemann, E., *Eur. J. Biochem.*, **166**, 339–344 (1987).

Gebhardt, R., and Mecke, D., *Eur. J. Biochem.*, **100**, 519–525 (1979).

Gebhardt, R. and Mecke, D., *EMBO J.*, **2**, 567–570 (1983).

Gebhardt, R. and Reichen, J., *Hepatology*, **20**, 684–691 (1994).

Gebhardt, R., Burger, H. J., Heini, H., Schreiber, K. L., and Mecke, D., *Hepatology*, **8**, 822–830 (1988a).

Gebhardt, R., Ebert, A., and Bauer, G., *FEBS Lett.*, **241**, 89–93 (1988b).

Gebhardt, R., Gaunitz, F., and Mecke, D., *Adv. Enz. Reg.*, **34**, 27–56 (1994).

Gebhardt, R., Jirtle, R., Moorman, A. F. M., Lamers, W. H., and Michalopoulos, G., *Histochem.*, **92**, 337–342 (1989).

Girard, G. and Butterworth, R. F., *Dig. Dis. Surg.*, **37**, 1121–1126 (1992).

Guha, S. R., *Enzymologia*, **24**, 310–326 (1962).

Häussinger, D., *Eur. J. Biochem.*, **133**, 269–274 (1983).

Häussinger, D., *Biol. Chem. Hoppe-Seyler*, **367**, 741–750 (1986).

Häussinger, D., *pH Homeostasis—Mechanisms and Control*, Academic, London (1988).

Häussinger, D., *Biochem. J.*, **267**, 281–290 (1990).

Häussinger, D., *Biochem. J.*, **313**, 697–710 (1996).

Häussinger, D. and Gerok, W., *Eur. J. Biochem.*, **136**, 421–425 (1983).

Häussinger, D. and Gerok, W., *Chem. Biol. Interact.*, **48**, 191–194 (1984).

Häussinger, D. and Gerok, W., *Eur. J. Biochem.*, **143**, 491–497 (1984).

Häussinger, D. and Lang, F., *Biochem. Cell Biol.*, **69**, 1–4 (1991).

Häussinger, D. and Lang, F., *Trends Pharmacol. Sci.*, **13**, 371–373 (1992).

Häussinger, D. and Sies, H., *Abstr. Commun. 9th FEBS Meeting*, No. 1497 (1975).

Häussinger, D. and Sies, H., *Eur. J. Biochem.*, **101**, 179–184 (1979).

Häussinger, D. and Sies, H., *Glutamine Metabolism in Mammalian Tissues*, Springer-Verlag, Heidelberg, Germany (1984a).

Häussinger, D. and Sies, H., *Biochem. J.*, **221**, 651–658 (1984b).

Häussinger, D. and Stehle, T., *Eur. J. Biochem.*, **175**, 395–403 (1988).

Häussinger, D., Weiss, L., and Sies, H., *Eur. J. Biochem.*, **52**, 421–431 (1975).

Häussinger, D., Akerboom, T. P. M., and Sies, H., *Hoppe-Seylers Z. Physiol. Chem.*, **361**, 995–1001 (1980).

Häussinger, D., Gerok, W., and Sies, H., *Biochim. Biophys. Acta*, **755**, 272–278 (1983).

Häussinger, D., Gerok, W., and Sies, H., *Trends Biochem. Sci.*, **9**, 300–302 (1984).

Häussinger, D., Lamers, W. H., Moorman, A. F. M., *Enzyme*, **46**, 72–93 (1992b).

Häussinger, D., Steeb, W., and Gerok, W., *Klin. Wschr.*, **68**, 175–182 (1990c).

Häussinger, D., Steeb, R., and Gerok, W., *Clin. Investig.*, **70**, 411–415 (1992c).

Häussinger, D., Stehle, T., and Gerok, W., *Biol. Chem. Hoppe-Seyler*, **366**, 527–536 (1985b).

Häussinger, D., Kaiser, S., Stehle, T., and Gerok, W., *Biochem. Pharmacol.*, **35**, 3317–3322 (1986).

Häussinger, D., Lang, F., Bauers, K., and Gerok, W., *Eur. J. Biochem.*, **188**, 689–695 (1990a).

Häussinger, D., Roth, E., Lang, F., and Gerok, W., *Lancet*, **341**, 1330–1332 (1993a).

Häussinger, D., Stoll, B., Stehle, T., and Gerok, W., *Eur. J. Biochem.*, **185**, 189–195 (1989).

Häussinger, D., Hallbrucker, C., Saha, N., Lang, F., and Gerok, W., *Biochem. J.*, **288**, 681–689 (1992a).

Häussinger, D., Saha, N., Hallbrucker, C., Lang, F., and Gerok, W., *Biochem. J.*, **291**, 355–360 (1993b).

Häussinger, D., Hallbrucker, C., vom Dahl S., Lang, F., and Gerok, W., *Biochem. J.*, **272**, 239–242 (1990b).

Häussinger, D., Soboll, S., Meijer, A. J., Gerok, W., Tager, J. M., and Sies, H., *Eur. J. Biochem.*, **152**, 597–603 (1985a).

Häussinger, D., Hallbrucker, C., vom Dahl, S., Decker, S., Schweizer, U., Lang, F., and Gerok, W., *FEBS Lett.*, **283**, 70–72 (1991).

Halestrap, A. P., *Biochim. Biophys. Acta*, **973**, 355–382 (1989).

Halestrap, A. P., In: Lang, F., and Häussinger, D., Eds., In *Interactions of cell volume and cell function*, Springer-Verlag, Heidelberg, Germany, pp. 279–307 (1993).

Hallbrucker, C., vom Dahl, S., Lang, F., Gerok, W., and Häussinger, D., *Eur. J. Biochem.*, **197**, 717–724 (1991).

Hayes, M. R. and McGivan, J. D., *Biochem. J.*, **204**, 365–368 (19xx).

Heini, H. G., Gebhardt, R., and Mecke, D., *Eur. J. Biochem.*, **162**, 541–546 (1987).

Horowitz, M. L. and Knox, W. E., *Enzymol. Biol. Clin.*, **9**, 241–255 (1968).

Huang, Y. Z. and Knox, W. E., *Enzyme*, **21**, 408–426 (1976).

Hutson, R. G. and Kilberg, M. S., *Biochem. J.*, **304**, 745–750 (1994).

Inoue, Y., Bode, B. P., and Souba, W. W., *Am. J. Surg.*, **169**, 173–178 (1995).

Inoue, Y., Pacitti, A. J., and Souba, W. W., *J. Surg. Res.*, **54**, 393–400 (1993).

Joseph, S. K. and McGivan, J. D., *Biochem. J.*, **176**, 837–844 (1978).

Joseph, S. K., Bradford, N. M., and McGivan, J. D., *Biochem. J.* **184**, 477–480 (1979).

Jungermann, K. and Katz, N. *Physiol. Rev.*, **69**, 708–764 (1989).

Kaiser, S., Gerok, W., and Häussinger, D., *Eur. J. Clin. Invest.*, **18**, 535–542 (1988).

Katsunuma, T., Temma, M., and Katunuma, N., *Biochem. Biophys. Res. Commun.*, **32**, 433–437 (1968).

Katz, J., Golden, S., and Wals, P. A., *Proc. Natl. Acad. Sci.*, **73**, 3433–3437 (1976).

Kilberg, M. S., Handlogten, M. E., and Christensen, H. N., *J. Biol. Chem.*, **255**, 4011–4019 (1980).

Kilberg, M. S. and Häussinger, D. *Amino Acid Transport—Mechanisms and Control*, Plenum, New York, (1992).

Kovacevic, Z. and Bajin, K. *Biochim. Biophys. Acta*, **687**, 291–295 (1982).

Kovacevic, Z., Day, S. H., Collett, V., Brosnan, J. T., and Brosnan, M. E., *Biochem. J.*, **305**, 837–841 (1995).

Kovacevic, Z. and McGivan, J. D., *Physiol. Rev.*, **63**, 547–605 (1983).

Kovacevic, Z. and McGivan, J. D., In *Glutamine transport across biological membranes*. Hässinger, D. and Sies, H., Eds., Springer-Verlag, Heidelberg, Germany, (1984) pp. 47–58.

Krause, U., Rider, M. H., and Hue, L., *J. Biol. Chem.*, **271**, 16668–16673 (1996).

Krebs, H. A., *Biochem. J.*, **29**, 1951–1959 (1935).

Kristensen, L. O. and Folke, M., *Biochem. J.*, **221**, 265–268 (1984).

Kumada, Y., Benson, D. R., Hillemann, D., Hosted, T. J., Rochefort, D. A., Thompson, C. J., Wohlleben, W., and Tateno, Y., *Proc. Natl. Acad. Sci. USA*, **90**, 3009–3013 (1993).

Kuo, F. C. and Darnell, J. E., *Mol. Cell Biol.*, **11**, 6050–6058 (1991).

Kuo, C. F., Hwu, W. L., Valle, D., and Darnell, J. E., *Proc. Natl. Acad. Sci. USA*, **88**, 9468–9472 (1991).

Kuo, C. F., Paulson, K. E., and Darnell, J. E., *Mol. Cell Biol.*, **8**, 4966–4971 (1988).

Kvamme, E. *Glutamine and Glutamate in Mammals*, CRC Press, Boca Raton, FL (1988).

Lacey, J. H., Bradford, N. M., Joseph, S. K., and McGivan, J. D., *Biochem. J.*, **194**, 29–33 (1981).

Lamers, W. H., Been, W., Charles, R., Moorman, and A. F. M., *Hepatology*, **12**, 701–709 (1990).

Lamers, W. H., Gaasbeek-Janzen J. W., te Kortschot, A., Charles, R., and Moorman, A. F. M., *Differentiation*, **35**, 228–235 (1987).

Lamers, W. H., Hilberts, A., Furt, E., Smith, J., Jones, C. N., van Noorden, C. J. F., Gaasbeek Janzen, J. W., Charles, R., and Moorman, A. F. M., *Hepatology*, **10**, 72–76 (1989).

Lang, F. and Häussinger, D., *Interaction of cell volume and cell function*, Springer-Verlag, Heidelberg, Germany, (1993).

Lang, F., Öttl, I., Freudenschuss, K., Honeder, M., Tschernko, E., and Häussinger, D., *Pflügers Arch*, **419**, 111–113 (1991).

Lang, F., Tschernko, E., Schulze, E., Öttl, I., Ritter, M., Völkl, H., Hallbrucker, C., and Häussinger, D., *Hepatology*, **14**, 590–594 (1991).

Lavoinne, A., Baquet, A., and Hue, L., *Biochem. J.*, **248**, 429–437 (1987).

Lenzen, C., Soboll, S., Sies, H., Häussinger, D., *Eur. J. Biochem.*, **166**, 483–488 (1987).

Lie-Venema, H., Labruyere, W. T., van Roon, M. A., de Boer, P. A., Moorman, A. F. M., Berns, A. J., and Lamers, W. H., *J. Biol. Chem.*, **270**, 28251–28256 (1995).

Livesey, G. and Lund, P., *Biochem. Soc. Trans.*, **8**, 540–541 (1980).

Low, S., Salter, M., Knowles, R. G., Pogson, C. I., and Rennie, M. J., *Biochem. J.*, **295**, 617–624 (1993).

Lueck, J. D. and Miller, L. L., *J. Biol. Chem.*, **245**, 5491–5497 (1970).

Lund, P., *Biochem. J.*, **124**, 653–660 (1971).

Lusty C., *Eur. J. Biochem.*, **85**, 373–383 (1978).

Matsuno, T. and Goto, I., *Cancer Res.*, **52**, 1192–1194 (1992).

McGivan, J. D., Boon, K., and Doyle, F. A., *Biochem. J.*, **274**, 103–108 (1991).

McGivan, J. D. and Bradford, N. M., *Biochim. Biophys. Acta*, **759**, 296–302 (1983).

McGivan, J. D., Bradforf, N. M., Verhoeven, A. J., and Meijer, A. J., Hässinger, D. and Sies, H., Eds., *Glutamine Metabolism in Mammalian Tissues* Springer-Verlag, Berlin, Germany, (1984) pp. 122–137.

McGivan, J. D., Lacey, J. H., and Joseph, S. K., *Biochem. J.*, **192**, 537–542 (1980).

Meijer, A. J., *FEBS Lett.*, **191**, 249–251 (1985).

Meijer, A. J., Lamers, W. H., and Chamaleau, R. A. F. M., *Physiol. Rev.*, **70**, 701–748 (1990).

Meijer, A. J., Lof, C., Ramos, I., and Verhoeven, A. J., *Eur. J. Biochem.*, **148**, 189–196 (1985).

Meijer, A. J. and Verhoeven, A. J., *Biochem. Soc. Trans.*, **14**, 1001–1004 (1986).

Meister, A., *Enzymes*, **10**, 699–754 (1974).

Meister, A., Hässinger, D. and Sies, H., Eds., In *Glutamine Metabolism in Mammalian Tissues*, Springer-Verlag, Heidelberg, Germany, (1984) pp. 3–15.

Meister, A. and Anderson, M. E., *Annu. Rev. Biochem.*, **52**, 711–760 (1983).

Moldave, K. and Meister, A., *J. Biol. Chem.*, **229**, 463 (1957).

Moorman, A. F. M., de Boer, P. A. J., Charles, R., and Lamers, W. H., *FEBS Lett.*, **276**, 9–13 (1990b).

Moorman, A. F. M., de Boer, P. A., Das, A. T., Labruyre, W. T., Charles, R., and Lamers, W. H., *Histochem. J.*, **22**, 457–468 (1990a).

Moorman, A. F. M., de Boer, P. A. J., Geerts, W. J. C., van de Zande, L. P. W. G., Charles, R., and Lamers, W. H., *J. Histochem. Cytochem.*, **36**, 751–755 (1988).

Moorman, A. F. M., Vermeulen, J. L. M., Charles, R., and Lamers, W. H., *Hepatology*, **9**, 367–372 (1989).

Moorman, A. F. M., de Boer, P. A., Watford, M., Dingemanse, M. A., and Lamers, W. H., *FEBS Lett.*, **356**, 76–80 (1994).

Mortimore, G. E. and Pösö, A. R., *Ann. Rev. Nutr.*, **7**, 539–564 (1987).

Newsome, W. P., Warskulat, U., Noe, B., Wettstein, M., Stoll, B., Gerok, W. and Häussinger, D., *Biochem. J.*, **304**, 555–560 (1994).

Nissim, I., Cattano, C., Nissim, I., and Yudkoff, M., *Arch. Biochem. Biophys.*, **292**, 393–401 (1992).

Noe, B., Schliess, F., Wettstein, M., Heinrich, S., and Häusinger, D., *Gastroenterology*, **110**, 858–865 (1996).

Oster, J. R., In *The kidney in liver disease* Epstein, M., Ed., Elsevier New York, (1983) pp. 147–182.

Pacitti, A. J., Inoue, Y., and Souba, W. W., *Am. J. Physiol.*, **265**, G90–98 (1993).

Patel, M. and McGivan, J. D., *Biochem. J.*, **220**, 583–590 (1984).

Plumley, D. A., Watkins, K., Bode, B. P., Pacitti, A. J., and Souba, W. W., *JPEN*, **19**, 9–14 (1995).

Pösö, A. R., Penttilä, K. E., Suolinna, E. M., and Lindros, K. O., *Biochem. J.*, **239**, 263–267 (1986).

Quillard, M., Husson, A., and Lavoinne, A., *Eur. J. Biochem.*, **236**, 56–59 (1996).

Racine-Samson, L., Scoadec, J. Y., D. Errico, A., Fiorentino, M., Christa, L., Moreau, A., Grigioni, W. F., and Feldmann, G., *Hepatology*, **24**, 104–113 (1996).

Rappaport, A. M., *Beitr. Pathol.*, **157**, 215–243 (1976).

Record, C. O., Iles, R. A., Cohen, R. D., and Williams, R., *Gut*, **16**, 144–149 (1975).

Remesy, C., Morand, C., Demigne, C., and Fafournoux, P., *J. Nutr.*, **118**, 569–578 (1988).

Ronzio, R. and Meister, A., *Proc. Natl. Acad. Sci. USA*, **59**, 164–170 (1968).

Roth, E., Karner, J., and Ollenschläger, G., *JPEN*, **14**, 130S–136S (1990).

Saha, N., Stoll, B., Lang, F., and Häussinger, D., *Eur. J. Biochem*, **209**, 437–444 (1992).

Saheki, T., Yagi, Y., Sase, M., Nakano, K., and Sato, E., *Biomed. Res.*, **4**, 235–238 (1983).

Schliess, F., Schreiber, R., and Häusinger, D., *Biochem. J.*, **309**, 13–17 (1995).

Schöls, L., Mecke, D., and Gebhardt, R., *Histochemistry*, **94**, 49–54 (1990).

Schoolwerth, A. C., deBoer, P. A. J., Moorman, A. F. M., Lamers, W. H., *Am. J. Physiol.*, **267**, F400–F406 (1994).

Schreiber, R. and Häussinger, D., *Biochem. J.*, **309**, 19–24 (1995).

Schreiber, R., Stoll, B., Lang F., and Häussinger, D., *Biochem. J.*, **303**, 113–120 (1994).

Schröde, W., Mecke, D., and Gebhardt, R., *Eur. J. Cell Biol.*, **53**, 35–41 (1990).

Schulz, W. A., Eickelmann, P., Hallbrucker, C., Sies, H., and Häussinger, D., *FEBS Lett.*, **292**, 264–266 (1991).

Seglen, P. O. and Gordon, P. B., *J. Cell. Biol.*, **99**, 435–444 (1984).

Shiojiri, N., Wada, J. I., Tanaka, T., Noguchi, M., Ito, M., and Gebhardt, R., *Lab. Invest*, **72**, 740–747 (1995).

Sirma, H., Williams, G. M., and Gebhardt, R., *Liver*, **16**, 166–173 (1996).

Smith, D. D. and Campbell, J. W., *Proc. Natl. Acad. Sci.*, **85**, 160–164 (1988).

Smith, E. M. and Watford, M., *Arch. Biochem. Biophys.*, **260**, 740–751 (1988).

Smith, E. M. and Watford, M., *J. Biol. Chem.*, **265**, 10631–10636 (1990).

Soboll, S., Lenzen, C., Rettich, D., Gründel, S., and Ziegler, B., *Eur. J. Biochem.*, **197**, 113–117 (1991).

Souba, W. W., *Ann. Rev. Nutr.*, **11**, 285–308 (1991).

Stadtman, E. R., Starke-Reed, P. E., Oliver, C. N., Carney, J. M., and Floyd, R. A., *EXS*, **62**, 64–72 (1992).

Stoll B., Gerok, W., Lang, F., and Häussinger, D., *Biochem. J.*, **287**, 217–222 (1992).

Stoll, B. and Häussinger, D., *Eur. J. Biochem.*, **181**, 709–716 (1989).

Stoll, B. and Häussinger, D., *Eur. J. Biochem.*, **195**, 121–129 (1991).

Stoll, B., McNelly, S., Buscher, H. P., and Häussinger, D., *Hepatology*, **13**, 247–253 (1991).

Stoll, L. I. and Atkinson, D. E., *J. Biol. Chem.*, **287**, 217–222 (1992).

Szweda L. I. and Atkinson, D. E., *J. Biol. Chem.*, **265**, 20869–20873 (1990).

Tamarappoo, B. K., Handlogten, M. E., Laine, R. O., Serrano, M. A., Dugan, J., and Kilberg, M. S., *J. Biol. Chem.*, **267**, 2370–2374 (1992).

Tamarappoo, B. K. and Kilberg, M. S., *Biochem. J.*, **274**, 97–101 (1991).

Tamarappoo, B. K., Singh, H. P., and Kilberg, M. S., *J. Nutr.*, **124**, 1493S–1498S (1994).

Tate, S. S. and Meister, A., *J. Biol. Chem.*, **247**, 5312–5321 (1972).

Tate, S. S. and Meister, A., *Acad. Sci. USA*, **74**, 3329–333 (1974).

Taylor, P. M. and Rennie, M. J., *FEBS. Lett.*, **221**, 370–374 (1987).

Theodoropoulos, T., Stournaras, C., Stoll, B., Markogiannakis, E., Lang, F., Gravani, A., and Häussinger, D., *FEBS Lett.*, **311**, 241–245 (1992).

Tohyama, Y., Kameji, T., and Hayashi, S., *Eur. J. Biochem.*, **202**, 1327–1331 (1991).

Traber, P. G., Chianale, J., and Gumucio, J. J., *Gastroenterology*, **95**, 30–43 (1988).

van de Zande, L. P. G. W., Labruyere, W. T., Arnberg, A. C., Wilson, R. H., van den Bogaert, A. J. W., Das, A. T., Frijters, C., Charles, R., Moorman, A. F. M., and Lamers, W. H., *Gene*, **87**, 225–232 (1990).

Verhoeven, A. J., van Iwaarden, J. F., Joseph, S. K., and Meijer, A. J., *Eur. J. Biochem.*, **133**, 241–244 (1983).

Vincent, N., Martin, G., and Baverel, G., *Biochim. Biophys. Acta*, **1014**, 184–188 (1989).

Vincent, N., Martin, G., and Baverel, G., *Biochim. Biophys. Acta*, **1175**, 13–20 (1992).

Völkl, H., Friedrich, F., Häussinger, D., and Lang, F., *Biochem. J.*, **295**, 11–14 (1993).

vom Dahl, S., Stoll, B., Gerok, W., and Häussinger, D., *Biochem. J.*, **308**, 529–536 (1995).

vom Dahl, S. and Häussinger, D., *J. Nutr.*, **126**, 395–402 (1996).

Warskulat, U., Newsome, W. P., Noe, B., Stoll, B., and Häussinger, D., *Biol. Chem. Hoppe-Seyler*, **377**, 57–65 (1996).

Watford, M., Berdanier, C., and Hargrove, J. L., Eds., In *Nutrition and gene expression*, CRC Press, Boca Raton, FL (1993) pp. 335–352.

Watford, M. and Smith, E. M., *Biochem. J.*, **267**, 265–267 (1990).

Watford, M., Smith, E. M., and Erbelding, E. J., *Biochem. J.*, **224**, 207–221 (1984).

Watford, M., Vincent, N., Zhan, Z., Fanelli, J., Kowalski, T. J., and Kovacevic, Z., *J. Nutr.*, **124**, 493–499 (1994).

Watkins, K. T., Dudrick, P. S., Copeland, E. M., and Souba, W. W., *J. Trauma.*, **36**, 523–528 (1994).

Welbourne, T. C., *Biol. Chem. Hoppe-Seyler*, **367**, 301–305 (1986).

Welbourne, T. C., *Am. J. Physiol.*, **253**, F1069–F1076 (1987).

Wettstein, M., Noe, B., and Häussinger, D., *Hepatology*, **22**, 235–240 (1995).

Wettstein, M., vom Dahl, S., Lang, F., Gerok, W., and Häussinger, D., *Biol. Chem. Hoppe-Seyler*, **371**, 493–501 (1990).

Wong, B. S., Chenoweth, M. E., and Dunn, A., *Endocrinology*, **106**, 268–274 (1980).

Yeldandi, A. V., Tan, X., Dwivedi, R. S., Subbarao, V., Smith, D. D., Scarpelli, D. G., Rao, M. S., and Reddy, J. K., *Proc. Natl. Acad. Sci.*, **87**, 881–885 (1990).

Zhan, Z., Vincent, N., and Watford, M., *Int. J. Biochem.*, **26**, 263–268 (1994).

ENZYMES UTILIZING GLUTAMINE AS AN AMIDE DONOR

By HOWARD ZALKIN, *Professor of Biochemistry, Purdue University* and JANET L. SMITH, *Professor of Biological Sciences, Purdue University*

CONTENTS

Advances in Enzymology and Related Areas of Molecular Biology, Volume 72: Amino Acid Metabolism, Part A, Edited by Daniel L. Purich
ISBN 0-471-24643-3   ©1998 John Wiley & Sons, Inc.

I. Introduction

The amide of glutamine is the major source of nitrogen for the biosynthesis of amino acids, purine and pyrimidine nucleotides, amino sugars, and coenzymes. A family of at least 16 glutamine amidotransferases functions to catalyze amide nitrogen transfer from glutamine to acceptor substrate molecules in different biosynthetic pathways. Some of the amidotransferases, those involved in the biosynthesis of glutamate, histidine, and tryptophan, for example, are restricted to microorganisms. Other amidotransferases, such as those involved in *de novo* nucleotide and amino sugar biosynthesis, are ubiquitous in all cells except for parasites.

The amidotransferases were initially reviewed in this series in 1973 (Buchanan, 1973). The emphasis of this paper was on enzymology and biochemical characterization. A comprehensive update, written in 1993 (Zalkin, 1993), described advances to our understanding of the amidotransferase family resulting primarily from analyses of cloned genes, high-level production of recombinant enzymes in *Escherichia coli*, and application of site directed mutagenesis to probe enzyme function and mechanism. Advances since 1993 have resulted from determinations of several amidotransferase structures by X-ray crystallography. This has led to specific models for structure-based mechanisms for catalysis and allosteric regulation that have guided biochemical and molecular biological experiments. The intent of this chapter is to provide an overview of the progress made in understanding amidotransferase structure and mechanism since the previous article. Our emphasis is on new insights into mechanisms of catalysis and regulation, which results from structural work largely from one of our laboratories. The reader is referred to the 1993 article on *The Amidotransferases* (Zalkin, 1993) for properties of enzymes not covered here and for additional background information.

II. Structure–Function Relationship and Enzyme Classification

Most of the amidotransferases catalyze three reactions: (*a*) an amide transfer from glutamine to a specific acceptor substrate, R, Eq. 1; (*b*) an amination reaction in which NH_3 replaces glutamine, Eq. 2; and (*c*) a glutaminase, Eq. 3.

$$Gln + R \rightarrow RNH_2 + Glu \tag{1}$$

$$NH_3 + R \rightarrow RNH_2 \tag{2}$$

$$Gln + H_2O \rightarrow Glu + NH_3 \tag{3}$$

Although Eq. 1 is considered to represent the physiological reaction, in several cases the NH_3-dependent reaction has been shown to function in cells (Zalkin, 1993). The reactions shown by Eqs. 2 and 3 are important for three reasons. First, Reaction 1, the glutamine-dependent reaction, is the sum of Reactions 2 and 3. Therefore, a two-step mechanism for amide transfer, glutamine hydrolysis to yield glutamate plus enzyme-sequestered NH_3, followed by NH_3-dependent synthesis of the product is one of the proposed routes for N transfer. Second, Reactions 2 and 3 are catalyzed by specific amidotransferase domains or subunits. In some enzymes, the domains for the two half-reactions are fused into a single protein chain. In other amidotransferases, these domains are parts of separate but interacting subunits (Zalkin, 1993). The domain–subunit that catalyzes the reaction with NH_3 (Eq. 2) has been called synthase, synthetase, or acceptor. The domain–subunit that hydrolyzes glutamine (Eq. 3) has been referred to as glutamine amide transfer (GAT), glutamine, or glutaminase. We will use these terms interchangeably to conform with current usage for each enzyme. There is a third consequence of the partial reactions given by Eqs. 2 and 3. The occurrence of distinct domains and subunits for NH_3-dependent synthesis and for glutamine hydrolysis has led to the proposal that amidotransferases may have evolved from NH_3-dependent enzymes by recruitment of a glutamine domain derived by gene duplication from an unknown ancestor (Nagano et al., 1970; Li and Buchanan, 1971).

Two different ancestral glutamine domains are inferred from analyses of sequences derived from cloned amidotransferase genes.[1] Each of the two enzyme classes thus identified is defined by a conserved glutamine domain. The key features are an NH_2-terminal nucleophile (Zalkin, 1993; Smith et al., 1994) for the Ntn amidotransferases and a catalytic triad (Amuro et al., 1985; Tesmer et al., 1996) for the Triad amidotransferases. Glutamine domain nomenclature has evolved during the past several years. Initially, these domains–subunits were designated F- and G-type based on *E. coli* gene nomenclature for *purF*-encoded glutamine PRPP amidotransferase and *trpG*-encoded anthranilate synthase component II, the glutamine subunit for anthranilate synthase (Zalkin, 1993). Glutamine PRPP amidotransferase and anthranilate synthase were studied in one of our laboratories and were two of the first amidotransferase sequences obtained. A more straightforward nomenclature of Classes I and II for G- and F-type amidotransferases, respectively, was adopted by one of us (Smith, 1995) and has appeared in the literature. With the recent availability of X-ray structures, the nomenclature has been revised (Kim et al., 1996) to reflect the defining structural features of the glutamine domains that are required for catalysis. Accordingly, the G-type (Class I) enzymes are in the Triad class and F-type (Class II) enzymes are in the Ntn class. Enzymes within each class have a distinctive fingerprint of invariant residues [Fig. 1(*a*)], which allow easy recognition of the enzymes, and a distinct three-dimensional structure [(Fig. 1(*b* and *c*)].

---

[1] After completion of this article a third glutamine domain, related to a family of amidases, was reported (Curnow et al.). An update is included in Section 4.F.

→

Figure 1.   Structural features of glutamine domains of amidotransferases. (*a*) Invariant residues that define the glutamine domains of Ntn and Triad amidotransferases. The numbering is for *E. coli* glutamine PRPP amidotransferase (Ntn) and *E. coli* GMP synthetase (Triad), enzymes of known structure. The number of residues for these two glutamine domains is given in parentheses. (*b*) Glutamine domain of Ntn amidotransferases. The glutamine domain of the active conformation of *E. coli* glutamine PRPP amidotransferase (Krahn et al., 1997) is shown as a stereoribbon drawing. Residues from the Ntn sequence fingerprint are shown explicitly. The catalytic residue Cys-1, alkylated by the glutamine analogue DON, is drawn in darker bonds. (*c*) Glutamine domain of Triad amidotransferases. The glutamine domain of *E. coli* GMP synthetase (Tesmer et al., 1996) is shown as in (*b*). The catalytic residue Cys-86 is shown without modification.

Ntn: C^1-R^{26}-G^{27}-G^{32}-R^{73}-P^{86}-N^{101}-G^{102}-D^{127} (236)

Triad: G^{59}-G^{84}-C^{86}-G^{88}-Q^{90}-H^{143}-H^{181}-P^{182}-E^{183} (206)

(a)

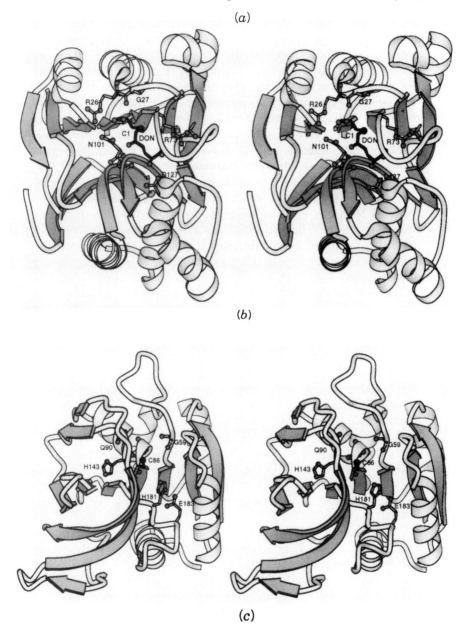

(b)

(c)

There are nine invariant amino acids in the 236-residue Ntn domain (*E. coli* glutamine PRPP amidotransferase numbering) and nine
invariant amino acids in the 206-residue Triad domain (*E. coli* GMP
synthetase numbering). The roles of these conserved residues have
been deduced from X-ray structural models and mutagenesis experiments and are described in sections III.A and IV.A, which are
devoted to these enzymes. The distinctive features of the Ntn
subfamily that permit easy visual recognition are three catalytic residues: Cys-1, the NH_2-terminal nucleophile and basic α-amino terminus, and Asn-101-Gly-102 (*E. coli* glutamine PRPP amidotransferase
numbering), residues that stabilize an oxyanion in the catalytic tetrahedral intermediate. The Triad enzymes are defined by a Cys/His/
Glu catalytic triad. The Cys nucleophile is in a Gly-Xaa-Cys-Xaa-
Gly sequence (residues 84–88, *E. coli* GMP synthetase numbering)
followed by a His-Pro-Glu sequence (residues 181–183 in *E. coli*
GMP synthetase). The Ntn amidotransferases belong to a larger homologous family of Ntn hydrolases (Brannigan et al., 1995). The
Triad amidotransferases are not known to be homologous to any
other protein.

The Ntn amidotransferases have an invariant organization, an
NH_2-terminal glutamine domain followed by an acceptor domain.
The domain arrangement is fixed by the requirement for an NH_2-
terminal nucleophile but it is not clear why glutamine and acceptor
functions could not be on interacting subunits. The Triad amidotransferases have more varied structures. The glutamine and acceptor functions are fused domains in some enzymes and interacting
subunits in others. In either case, the enzymes may be monofunctional or multifunctional depending on the organism. The Triad enzyme carbamyl-P synthetase has interacting glutaminase and synthetase subunits in prokaryotes and fused domains in multifunctional
enzymes in eukaryotes. In the multifunctional carbamyl-P synthetases the glutaminase domain is NH_2-terminal to the contiguous synthetase domain, but this arrangement is reversed in other Triad enzymes.

Three lines of evidence support the dissection of the catalytic
transformation (Eq. 1) into two partial reactions (Eqs. 2 and 3): (a)
in vitro biochemical assays, (b) the organization of primary structures, and (c) the three-dimensional structures. We therefore describe the amidotransferases as "complex enzymes" because they

employ two active sites to catalyze one biochemical reaction. The complexity arises in coordinating catalysis at the two active sites. Biochemical data show that the active sites can function independently, but *in vivo* they are normally coupled. To the usual task of understanding the function of an enzyme active site must be added the challenge to understand how the coupling takes place.

III.   Ntn Amidotransferases

A.   GLUTAMINE PRPP AMIDOTRANSFERASE   .

Glutamine PRPP amidotransferase catalyzes the first reaction in *de novo* purine nucleotide synthesis (Eq. 4)

Glutamine + PRPP → Phosphoribosylamine + Glutamate + PP_i

$$(4)$$

Sequences have been derived from cloned genes from more than 20 organisms including bacteria, eukarya, and archea. However, only the enzymes from *E. coli* and *Bacillus subtilis* have been purified to homogeneity and extensively characterized. In addition, X-ray structures have been determined for the *E. coli* (Muchmore et al., 1998; Krahn et al., 1997) and *B. subtilis* (Smith et al., 1994) enzymes.

1.   Two Enzyme Classes

The *E. coli* and *B. subtilis* enzymes are representative of distinctive glutamine PRPP amidotransferase classes (Fig. 2). The classes are distinguished by an NH_2-terminal propeptide and an [Fe–S] cluster. Enzymes of the *E. coli* class have neither a propeptide nor an [Fe–S] cluster. Enzymes of the *B. subtilis* class are synthesized with a propeptide and they likely all contain an [Fe–S] cluster. Propeptide cleavage is required to generate the mature active enzyme having an NH_2-terminal catalytic cysteine. This type of processing is a hallmark for enzymes in the Ntn hydrolase family (Brannigan et al., 1995). Four conserved cysteines in the acceptor domain of glutamine PRPP amidotransferase are ligands for a [4Fe–4S] cluster. The [4Fe–4S] cluster in *B. subtilis* glutamine PRPP amidotransferase has been rigorously characterized (Vollmer et al., 1983; Smith et al., 1994). Direct evidence for an [Fe–S] cluster in enzymes from other

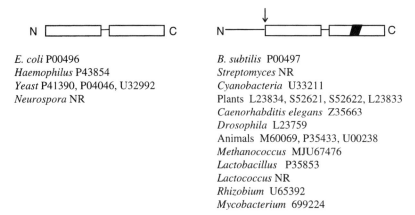

E. coli P00496
Haemophilus P43854
Yeast P41390, P04046, U32992
Neurospora NR

B. subtilis P00497
Streptomyces NR
Cyanobacteria U33211
Plants L23834, S52621, S52622, L23833
Caenorhabditis elegans Z35663
Drosophila L23759
Animals M60069, P35433, U00238
Methanococcus MJU67476
Lactobacillus P35853
Lactococcus NR
Rhizobium U65392
Mycobacterium 699224

Figure 2.   Glutamine PRPP amidotransferase classes. The prototype enzymes from
E. coli and B. subtilis are diagrammed at the top and show N- and C-domains (open
boxes), N-terminal propeptide (thin line), processing site (vertical arrow), and an
[Fe–S] cluster (dark region). Accession or NCBI ID numbers are given below for
each of the enzymes. The abbreviation NR is not reported.

organisms is limited to Fe analyses in purified, but not necessarily
homogeneous, preparations (Hartman, 1963; Rowe and Wyngaarden, 1968; Leff et al., 1984). Nevertheless, conservation of the cysteine ligands is used to infer the presence of an [Fe–S] cluster. The
role of the [4Fe–4S] cluster in the B. subtilis amidotransferase is to
sense nutrient availability and regulate O_2-mediated enzyme inactivation and turnover (Switzer, 1989). Nutrient limitations that lead
to sporulation in stationary-phase growth promote oxidative decomposition of the [Fe–S] cluster, enzyme aggregation, and proteolytic
degradation. We can speculate that the [Fe–S] cluster in the glutamine PRPP amidotransferase from other organisms also serves to
regulate enzyme turnover. Although evidence bearing on the role
of the propeptide is minimal, there is a good correlation between
propeptide and [Fe–S] cluster. Our working hypothesis is that the
propeptide restricts Cys-1 from folding into its final interior position
until cotranslational [Fe–S] ligation to the cysteinyl ligands has occurred. In this way, assembly of an aberrant [Fe–S] center with
an incorrect cysteine ligand may be avoided. However, errors in

maturation of the *B. subtilis* enzyme appear to be rare in the absence of a propeptide. This result is based on the observation that the recombinant *B. subtilis* glutamine PRPP amidotransferase made in *E. coli* with or without the propeptide has an indistinguishable specific activity. Based on structures of the *E. coli* and *B. subtilis* enzymes and multiple alignment of amino acid sequences from 23 sources, all of the enzymes are homologous. Pairs of enzymes within a class have 36–92% identical sequences. Between classes identity is 29–39%.

2. *Regulation by Nucleotides and Allosteric Control*

Glutamine PRPP amidotransferase catalyzes the committed step of purine biosynthesis and is thus positioned to control the entire pathway. Although gene regulation contributes to regulating the pathway in bacteria, in eukaryotes most if not all of the purine-specific control is at the protein level in the form of feedback inhibition. The effector molecules are nucleoside mono- and diphosphate end products of purine synthesis, and all are negative regulators. Feedback regulation of the enzyme is in many respects a classic example of allosteric control. For example, nucleotide inhibition is strongly cooperative. However, in other respects feedback control is more complicated. Some nucleotide inhibitors display partially competitive kinetics with respect to the substrate PRPP. Inhibition by certain pairs of nucleotides is strongly synergistic, implying multiple, interdependent sites for nucleotide binding.

The structural expectation for an allosteric enzyme is that it will be a symmetric oligomer existing in two very different conformations, one more active by far than the other (Perutz, 1989; Mattevi et al., 1996). Effector molecules exert their control by binding preferentially to one state. This binding alters the equilibrium between states and is the basis for cooperativity. In all cases of allosteric control studied to date, the most dramatic difference between active and inactive states is a shift of subunits within the symmetric oligomer. Effector molecules directly affect the conformational equilibrium by binding between subunits. Other more subtle structural changes may be induced within the domain of the "active" or "inactive" conformation in response to effector or substrate, but the structural change between these gross conformations is profound.

3. Crystal Structures

The first two glutamine PRPP amidotransferase structures to be determined were feedback-inhibited forms of the *B. subtilis* (Smith et al., 1994; Tomchick et al., 1995) and *E. coli* (Muchmore et al., 1998) enzymes. These structures proved that the catalytic glutamine and acceptor domains are also structural domains, and that enzymes with and without an [Fe–S] cluster are structural homologs. The binding mode of feedback inhibitors was also found to be consistent with the biochemical data, but somewhat unusual for allosteric enzymes. For the *E. coli* enzyme, there are also structures of an inactive conformer having the glutamine analogue DON covalently bound to the glutamine site (Kim et al., 1996), and of an active conformer containing both DON and a stable carbocyclic PRPP analogue (Krahn et al., 1997). These structures have confirmed the hypotheses about the mode of substrate binding and the basis for substrate specificity, as explained in Section III.A.4. The structure of the active conformation has begun to provide answers about how the catalytic activities of the two domains are coordinated. Structure-based ideas about enzyme function and regulation have, in many cases, been tested by site-directed mutagenesis.

The catalytic domains of glutamine PRPP amidotransferase are organized as structural modules. The glutamine domain comprises the N-terminal half of the polypeptide and the acceptor domain the C-terminal half. Active sites of the two catalytic domains face one another across the domain boundary. In both domains, conserved residues cluster around the active sites as expected. The catalytic domains will be discussed separately. The interaction of feedback inhibitors with the enzyme will then be discussed, followed by the allosteric transition between inactive and active forms and the coordination of activities of the two catalytic domains.

4. The Glutamine Domain

The glutamine domain of the Ntn amidotransferases is a four-layer structure with two internal antiparallel β-sheets surrounded by helical layers [Fig. 1(*b*)]. The catalytic Cys-1 residue occurs on a central strand of one of the β-sheets. A pocket for glutamine binding is between the β-sheets at one edge of the domain. Residues that are conserved among the Ntn family of amidotransferases are clustered around the nucleophilic residue and line this pocket.

The arrangement of invariant residues in the glutamine domain of glutamine PRPP amidotransferase suggested to us roles for these residues in the glutaminase reaction (Fig. 3). The catalytic elements for glutamine hydrolysis are the nucleophilic side chain of Cys-1, an oxyanion hole provided by $N_{\delta 2}$ of Asn-101 and NH of Gly-102, and the α-amino group of Cys-1, which functions as a proton donor to the leaving group. Specificity for the α-amino and α-carboxyl groups, respectively, of substrate glutamine is provided by the side chains of Asp-127 and Arg-73. Other invariant residues play supporting roles by maintaining the structure of the active site. The Arg-26-Gly-27 dipeptide is an "anchor" for both the backbone of Cys-1 and the side chain of Asn-101 through specific hydrogen bonds. The cis conformation adopted by invariant Pro-87 forms a wall of the glutamine-binding pocket. The DON complexes of the inactive and active conformers of the enzyme have confirmed the location of the oxyanion hole and also of Asp-127 and Arg-73 in conferring glutamine specificity. The complexes are consistent with the mechanism described above and shown in Figure 3.

In earlier work, Mei and Zalkin (1989) proposed that conserved His and Asp residues function in acid–base catalysis for steps of glutamine amide transfer. More recent evidence, however, indicates that the proposed His-101 and Asp-29 side chains are too far removed from Cys-1 to participate in catalysis (Smith et al., 1994). Furthermore, these residues are not conserved using improved alignments with the larger group of more recently acquired sequences of Ntn amidotransferases (Zalkin, 1993).

The glutamine binding pocket in all crystal structures of glutamine PRPP amidotransferase is a cavity. The walls of this cavity are formed by conserved residues from the glutamine domain, and by two nonconserved peptides from the acceptor domain, which do not contribute to the specificity of the site. These are the loop of residues 409–412 and the C-terminal structural element, which is a long α-helix in the *E. coli* enzyme and a short loop in the *B. subtilis* enzyme. The cavity is elliptical, with catalytic residues surrounding Cys-1 at one end and glutamine-recognition residues at the other. Cavities in proteins are by definition inaccessible to molecules from the bulk solvent, including water. Based on the first crystal structure of the

(a)

(b)

98

inactive conformation of the *B. subtilis* enzyme, the low level of glutaminase activity in the absence of substrate PRPP was attributed to inaccessibility of the glutaminase active site (Smith et al., 1994). The expectation was that this site would be more accessible in the active form of the enzyme. However, it is now apparent that the active site in all crystallographic models is "inaccessible" in both active and inactive forms of the enzyme, with and without a substrate analogue. We therefore conclude that normal conformational breathing of the protein in solution is of large enough amplitude and sufficient frequency for glutamine access to the active site, and that accessibility to the site is not a major factor in control of glutaminase activity. In support of this idea is the observation that atomic thermal parameters for peptides that line the glutamine-recognition end of the cavity, particularly in structures of unbound forms of the enzyme, are large relative to other parts of the protein.

5. *The Acceptor Domain*

The acceptor domain of glutamine PRPP amidotransferase is a representative phosphoribosyltransferase (PRTase). Phosphoribosyltransferases are enzymes that have PRPP as a substrate or product. Most of these substrates have a sequence fingerprint that was proposed as a PRPP binding peptide (Hove-Jensen et al., 1986) and has been demonstrated as such in several recent structural studies (Scapin et al., 1994; Eads et al., 1994; Scapin et al., 1995; Schumacher et al., 1996). The core fold of the domain is a parallel β-sheet with α-helices on either side, similar to the folds of many nucleotide-binding enzymes. Like the "P-loops" of these nucleotide-binding

---

Figure 3.   The glutaminase active site of Ntn amidotransferases. (*a*) Interactions of glutamine in the glutaminase active site of Ntn amidotransferases. All residues shown are among the Ntn fingerprint residues. Numbering of amino acids here and in (*b*) is for *E. coli* glutamine PRPP amidotransferase. (*b*) Proposed mechanism for the glutaminase reaction of Ntn amidotransferases. The α-amino group of Cys-1 (C1) increases the nucleophilicity of the Cys-1 side chain and donates a proton to the NH_2 group leaving the tetrahedral oxyanion intermediate. Both Asn-101 and Gly-102 stabilize the glutaminyl and the glutamyl oxyanions. Enzyme-bound NH_3 is released (glutaminase) or, in the presence of the acceptor substrate, reacts to form the aminated product.

enzymes, the PRPP-binding peptide or "PRPP loop" occurs in a loop between the third β-strand and α-helix at a cross-over point in the topology of the β-sheet. The substrate PRPP binds to the PRPP loop with extensive contacts between the enzyme and the ribose-5-phosphate moiety of PRPP. The monophosphate sits at the N-terminus of an α-helix, as in many nucleotide binding proteins. A Mg^{2+} ion is bound to the pyrophosphate of PRPP, which also contacts basic side chains of the protein. The entire PRPP site is highly solvent exposed in the inactive forms of the amidotransferase, but is sequestered from bulk solvent in the cPRPP-bound active form.

A common feature of PRTase structures is a long, flexible loop between the second and third β-strands, including conserved residues for each PRTase. The flexible loop has been proposed to play an essential role during catalysis, when it presumably becomes more ordered. In glutamine PRPP amidotransferase, this loop encompasses residues 323–353 (numbering for the *E. coli* enzyme) and includes 11 conserved residues. The loop exhibits a variety of poorly ordered conformations in the inactive forms of the *E. coli* enzyme, while in the inactive forms of the *B. subtilis* enzyme, it has a single conformation resembling a flag. In none of these structures are the five invariant residues of the loop in contact with conserved residues elsewhere in the protein. The role of the invariant residues becomes apparent in the active form of the enzyme, where the loop adopts a single, well-ordered conformation. The loop folds over the PRTase active site, has extensive contacts with the substrate analogue cPRPP, and sequesters the active site from solvent.

The PRTase "flexible loop" has been designated the "allosteric loop" in glutamine PRPP amidotransferase because of contacts of residues at the base of the loop with bound nucleotide feedback inhibitors. These residues are conservatively substituted among glutamine PRPP amidotransferases, indicating subtle, species-dependent differences in allosteric control of the enzyme.

6. Feedback Inhibitor Binding Sites

Nucleotides bind to two distinct sites in glutamine PRPP amidotransferase (Fig. 4). One site is the PRPP substrate-binding site, with ribose-5-phosphate of the nucleotide binding as the same moiety of PRPP binds. This explains the partially competitive kinetics of nu-

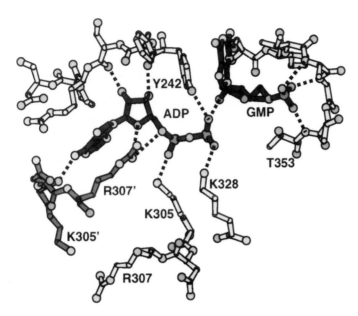

Figure 4. Feedback inhibitor binding to glutamine PRPP amidotransferase. Shown in dark bonds are the feedback inhibitors guanosine monophosphate (GMP), bound in the PRTase catalytic site, and adenosine diphosphate (ADP), bound in the allosteric site of the *B. subtilis* enzyme. Synergistic inhibition by this nucleotide pair is explained in part by a hydrogen bond between the β-phosphate of ADP and the O2′ hydroxyl of GMP. The allosteric site is formed by peptides from two subunits of the tetrameric enzyme, depicted here in light and dark gray bonds with unprimed and primed residue labels, respectively. Hydrogen bonds are shown as dashed lines.

cleotide inhibition. The second site is between subunits of the enzyme, like classic allosteric effector sites, but is unusual in being adjacent to the active site. The intersubunit binding of nucleotides to this allosteric site explains the cooperativity of nucleotide inhibition, and the proximity of allosteric and catalytic sites explains the strong synergism of some nucleotide pairs. The catalytic site strongly prefers nucleoside monophosphates, but has only weak base specificity. Conversely, the allosteric site is strongly base specific, but can accommodate both nucleoside mono- and diphosphates. The allosteric loop, described above, figures prominently in substrate binding in the active enzyme, in subunit contacts at the tight twofold interface of the inactive enzyme, and in binding of feedback inhibitors to two allosteric sites related by this molecular symmetry.

Biochemical and structural studies have established the specificity of the allosteric sites for guanine in the *E. coli* enzyme (Zhou et al., 1994) and for adenine in the *B. subtilis* enzyme (Tomchick et al., 1995). Surprisingly, the specificity arises from hydrogen bonds between the nucleotide base and the protein backbone. This specificity is accomplished by slightly different orientations of the nucleotides in the allosteric sites of the two enzymes. There is also weak base specificity in the catalytic site, which may arise from the ability of guanine nucleotides in the syn conformation to form a hydrogen bond between N2 and the hydroxyl group of Ser-347 in the PRPP loop (*B. subtilis* numbering). The basis for adenine specificity in this site of the *E. coli* enzyme is not known.

Synergism between the appropriate nucleoside diphosphate in the allosteric site and a nucleoside monophosphate in the catalytic site is explained in part by a hydrogen bond between the β-phosphate of the nucleoside diphosphate and the 2′-hydroxyl of the nucleoside monophosphate (Chen et al., 1997) (Fig. 4).

7. The Allosteric Transition and Coordination of Two Catalytic Activities

The separation of the two catalytic domains of glutamine PRPP amidotransferase by a 15-Å, solvent-filled space is an important feature of the inactive conformation of the enzyme. This spatial arrangement of active sites is incompatible with many features of catalytic activity. For example, NH_3 does not exchange with solvent in the overall reaction, and PRPP is easily hydrolyzed, suggesting that the two active sites must be accessible to one another but not to bulk solvent during catalysis. However, in the inactive conformation of the enzyme, it would not be possible for NH_3 produced by glutamine hydrolysis to reach PRPP without exchanging with solvent, nor could direct nitrogen transfer from glutamine to PRPP take place. Thus the separation of active sites, the sequestration of one, and the over-accessibility of the other seem to be designed for inactivity.

The structural change in the enzyme upon transition to the active form is expected to result in an open connection between the active sites, and a general closing of the overall site to bulk solvent. This can be imagined to occur in one of two ways. The two active sites could merge into one complex active site by hinge bending between domains. Alternatively, the two active sites could remain separate and channel NH_3 between them. In either case, the active sites should be closed to bulk solvent and open to one another.

The structure of the active form of *E. coli* glutamine PRPP amido-transferase reveals that the two catalytic domains remain physically separated and do not merge. Rather, there is a substantial restructuring of the allosteric loop so that it closes over the space between the active sites, effectively sequestering both sites from bulk solvent. A narrow tunnel forms between the glutamine active site and the acceptor active site. The pathway between the active sites is hydrophobic and is lined with conserved residues, some from the reordered allosteric loop. The simplest and most obvious explanation of this structure is that the enzyme has created a channel for NH_3 to pass from the glutamine active site to the acceptor active site. The hydrophobic nature of the channel is probably important for efficient transfer of NH_3 between active sites (no sticking points) as well as for elimination of water, which would rapidly hydrolyze PRPP.

The structure of the active form of glutamine PRPP amidotransferase provides strong support for catalytic mechanisms for nitrogen transfer by Ntn amidotransferases in which free NH_3 is produced by the enzyme and transferred to the acceptor substrate. This structure does not support models in which there is direct transfer of nitrogen to the acceptor substrate, either from glutamine itself or from the tetrahedral intermediate in glutamine hydrolysis.

Coordination of the activities of the glutamine and acceptor domains is thus achieved by signaling between the active sites. Specifically, PRPP binding in the acceptor domain stimulates glutaminase activity in the glutamine domain by 10- to 20-fold. The reactivity of Cys-1 with analogues of glutamine is comparably stimulated (Messenger and Zalkin, 1979; Kim et al., 1996). The catalytic deficiency in PRPP-independent glutamine hydrolysis is in the K_m for glutamine, which appears to be controlled by the Ntn-invariant residue Arg-73. The Arg-73 residue is a glutamine-specificity element of the Ntn amidotransferase fingerprint and interacts with the α-carboxylate of glutamine in the active but not the inactive form of the enzyme. Substitutions at Arg-73 increase the K_m for glutamine (Kim et al., 1996). The signal between catalytic domains that PRPP is bound may arrive through the adjacent residue Tyr-74. Mutations at Tyr-74 decouple glutamine hydrolysis from formation of the product phosphoribosylamine.

B. GLUCOSAMINE 6-PHOSPHATE SYNTHASE

Glucosamine 6-phosphate synthase catalyzes the first reaction in hexosamine biosynthesis (Eq. 5). The product glucosamine 6-phosphate

$$\text{Glutamine} + \text{D-Fructose 6-phosphate} \rightarrow$$

$$\text{D-Glucosamine 6-phosphate} + \text{Glutamate} \qquad (5)$$

is used for synthesis of UDP-N-acetylglucosamine (UDP = uridine diphosphate), a precursor in the biosynthesis of all macromolecules containing amino sugars. Interest in the human enzyme stems from the requirement for metabolism of glucose to a hexosamine in the desensitization of glucose transport to insulin, and the role of this process in non-insulin-dependent diabetes mellitus and obesity (Marshall et al., 1991). Metabolism of glucose to glucosamine is also necessary for the transcriptional stimulation by glucose of transforming growth factor-α (TGF-α) in vascular smooth muscle cells (Sayeski and Kudlow, 1996).

A *glmS* gene encoding glucosamine 6-phosphate synthase has recently been cloned from the extreme thermophile *Thermus thermophilus* (Fernández-Herrero et al., 1995). Enzymes from thermophiles may facilitate crystallization and can help to identify amino acids important for catalysis as a result of differences in amino acid composition compared to enzymes from mesophiles. Specifically relevant to glucosamine 6-phosphate synthase is the decreased content of cysteine, an amino acid that is rare in proteins from thermophiles. The *Thermus* enzyme contains only two cysteines at positions 1 and 295, positions that are invariant in glucosamine 6-P synthase from other organisms. The Cys-1 residue is the N-terminal nucleophile for catalysis, whereas a role for Cys-295 remains to be identified.

1. Glutamine and Acceptor Domains

Glutaminase and synthase domains of 240 and 368 amino acids, respectively, have been isolated from *E. coli* glucosamine 6-phosphate synthase following limited proteolysis by chymotrypsin (Denisot et al., 1991). A unique property of this amidotransferase is the absence of NH_3-dependent synthesis of glucosamine 6-phosphate. In terms of a tunnel mechanism for glutamine PRPP amidotransferase, this can be explained by assuming that external NH_3 does not have access to a tunnel that connects the glutamine and fructose 6-phosphate acceptor sites. Truncated genes encoding the glutamine and acceptor domains were overexpressed, purified, and crystallized (Obmolova et al., 1994). The high-resolution structure of the glutamine domain has been reported (Isupov et al., 1996), and the structure of the acceptor domain has been determined very recently (Teplyakov, personal communication). A description of the structure of the glutamine domain and mechanistic implications follow.

2. Glutamine Domain Structure

The overall fold of the glutamine domain of glucosamine 6-phosphate synthase is identical to that of the glutamine domain of glutamine PRPP amidotransferase, as expected for homologous enzymes. There are two additional β-strands relative to glutamine PRPP amidotransferase, one in each of the antiparallel β-sheets, at the C-terminal end of the domain. These are remote from the active site.

The structure provides a view of the product complex of a glutamine domain. Crystals were grown in very high glutamine concentration (100 mM), but the glutamine was hydrolyzed in the days-to-weeks of the crystallographic experiment, and the active site is occupied by the product glutamate. The residues implicated in glutamine specificity, in this case Arg-73 and Asp-123, are bound to the α-carboxyl and α-amino groups of glutamate, as in the active form of glutamine PRPP amidotransferase. Thus the means of substrate recognition and specificity is identical in these two Ntn amidotransferases.

In some significant details, the active site differs from that in glutamine PRPP amidotransferase. The Cys-1, Arg-26, and Asn-98 residues, all invariant in Ntn amidotransferases, are not oriented in a way that would support catalysis. The nucleophilic side chain of Cys-1 points away from the glutamine binding site. The side chain of Asn-98, an element of the oxyanion hole, is flipped to have its carbonyl group pointing into the glutamine pocket. The "anchor" residue Arg-26 does not interact with either Cys-1 or Asn-98. It was proposed that during catalysis invariant residues in the active site are arranged as in the structures of glutamine PRPP amidotransferase, and that a substrate-induced conformational change occurs in the active site upon glutamine binding. However, it is also possible that the unexpected orientations of Cys-1, Arg-26, and Asn-98 are due to the absence of the acceptor domain, which we presume occupies similar positions relative to the glutamine domain in glucosamine 6-phosphate synthase and in glutamine PRPP amidotransferase.

The active site of the isolated glutamine domain of glucosamine 6-phosphate synthase occurs in a cleft on the protein surface, unlike the internal-cavity active site in glutamine PRPP amidotransferase. This situation may differ in intact glucosamine 6-phosphate synthase because peptides from the acceptor domain of glutamine PRPP amidotransferase contribute to the glutaminase active site cavity.

Some details of glutamine recognition, beyond the invariant Ntn fingerprint, are shared by glucosamine 6-phosphate synthase and glutamine PRPP amidotransferase. The enzyme-bound product glutamate in glucosamine 6-phosphate synthase or substrate analogue DON in glutamine PRPP amidotransferase is covered by the conserved loop immediately following Arg-73, which makes a specific contact with glutamine. The Thr-76 residue forms a hydrogen bond with the α-amino group of glutamate/DON in both enzymes. The Arg-73 residue and the following loop are similarly configured in the active and inactive forms of *E. coli* glutamine PRPP amidotransferase, irrespective of substrate analogue binding in the active site, but the loop is differently folded in the inactive form of the *B. subtilis* enzyme. The Arg-73 residue is involved in signaling between catalytic domains in glutamine PRPP amidotransferase (Kim et al., 1996). If this residue has a similar function in glucosamine 6-phosphate synthease, then the conformation of the 74–77 peptide may be relevant to the status of the acceptor domain.

C. ASPARAGINE SYNTHETASE

Asparagine synthetase catalyzes the reaction given in Eq. 6. Two

$$\text{Aspartate} + \text{Glutamine} + \text{ATP} \rightarrow \text{Asparagine} + \text{Glutamate}$$
$$+ \text{AMP} + \text{PP}_i \tag{6}$$

asparagine synthetases have been described. One, a strictly NH_3-dependent enzyme was initially purified from *E. coli* (Cedar and Schwarz, 1969). Later, a second glutamine-dependent asparagine synthetase was detected in *E. coli* (Humbert and Simoni, 1980). The *E. coli* genes *asnA* (Nakamura et al., 1981) and *asnB* (Scofield et al., 1990) encode the strictly NH_3- and glutamine-dependent enzymes, respectively. There is no structural or evolutionary relationship between the two enzymes. Genes encoding glutamine-dependent asparagine synthetase have been isolated from a number of bacteria, yeast, plants, and animals. *Escherichea coli asnB* and a human cDNA have been overexpressed and the highly purified recombinant enzymes obtained (Boehlein et al., 1994a; Sheng et al., 1992). Recombinant AsnB has been crystallized and the structure determination is underway (Rayment, personal communication). A series of papers from the laboratories of Schuster and Richards, discussed in Section III.C.1, has focused on the mechanism for glutamine amide nitrogen transfer to β-aspartyl-AMP studied primarily using the *E. coli* enzyme.

1. Proposals for Direct Nitrogen Transfer

Two general mechanisms have been considered for asparagine synthetase: (a) attack of Cys-1 on the carboxamide of glutamine liberating NH_3 to react with aspartyl-AMP, and (b) schemes involving direct transfer of the glutamine amide. Three potential problems were noted by Boehlein et al. (1994b) for any mechanism in which NH_3 is used to mediate glutamine-dependent nitrogen transfer. First, NH_3 must remain bound by the enzyme to preclude formation of the nonnucleophilic NH_4^+ at cellular pH. Second, the enzyme must sequester NH_3 to prevent diffusion into solution prior to the reaction with aspartyl-AMP. Third, the enzyme must ensure that the two half-reactions that would generate and utilize NH_3 are synchronized for asparagine synthesis. Two lines of circumstantial evidence were cited to disfavor a mechanism of NH_3-mediated nitrogen transfer. First, the related Ntn amidotransferase, glucosamine 6-phosphate synthase has no detectable NH_3-dependent activity. Second, there is no evidence for a conserved histidine residue in asparagine synthetase and other Ntn hydrolases that could function in general acid–base catalysis. Moreover, there is no histidine in the structure of the inhibited form of *B. subtilis* glutamine PRPP amidotransferase that is positioned sufficiently close to Cys-1 to function as a general acid–base. A role for a histidine in glutamine PRPP amidotransferase had been proposed earlier by Mei and Zalkin (1989) based on sequence alignments of a limited set of enzymes and results of mutagenesis. As stated in Section III.A.4 on glutamine PRPP amidotransferase, the proposed role for histidine in catalysis is incorrect. It is now thought that the free NH_2-terminal α-amino group in the Ntn amidotransferases and hydrolases likely functions in acid–base catalysis. In terms of a glutamine PRPP amidotransferase tunnel mechanism, NH_3-dependent activity requires that NH_3 from solution have access to the tunnel. Possibly, NH_3 from solution cannot access the putative glucosamine 6-P synthase tunnel.

As a consequence of perceived limitations with a model for NH_3-mediated nitrogen transfer, Richards, Schuster, and colleagues have proposed alternative mechanisms based on direct attack of the glutamine amide on aspartyl-AMP (Richards and Schuster, 1992; Boehlein et al., 1994b, 1996; Stoker et al., 1996). Two of their alternative mechanisms are shown in Figure 5 (Boehlein et al., 1996). The common feature of these mechanisms is direct nucleophilic attack of either the amide of glutamine [Fig. 5(a)] or the -NH_2 of a cysteinyl-glutamine tetrahedral oxyanion [**4** in Fig. 5(b)] on aspartyl-AMP to yield imide intermediates **1** and **2**, which subsequently break down to asparagine and the Cys-1 glutamyl thioester **3**.

Three lines of evidence were reported to support direct nitrogen transfer. First, kinetic analyses of the inhibition by glutamine of the NH_3-dependent asparagine synthesis activity have been interpreted in terms of the formation of an abortive complex in which intermediate **1**, Figure 5(a), is a candidate for the abortive intermediate (Sheng et al., 1993; Boehlein et al., 1994a). These experiments were carried out with C1A and C1S mutants of human and *E. coli* asparagine synthetase. Since the Cys-1 nucleophile was replaced, there was no glutamine-dependent synthesis of asparagine and no hydrolysis of glutamine. However, the enzymes retained NH_3-dependent activity that was inhibited by glutamine. Direct evidence for an imide intermediate is awaited.

A second series of experiments investigated alternative nitrogen donors for *E. coli* asparagine synthetase and the results were interpreted as being inconsistent with NH_3-mediated nitrogen transfer from glutamine (Boehlein et al., 1996). In the presence of ATP, γ-

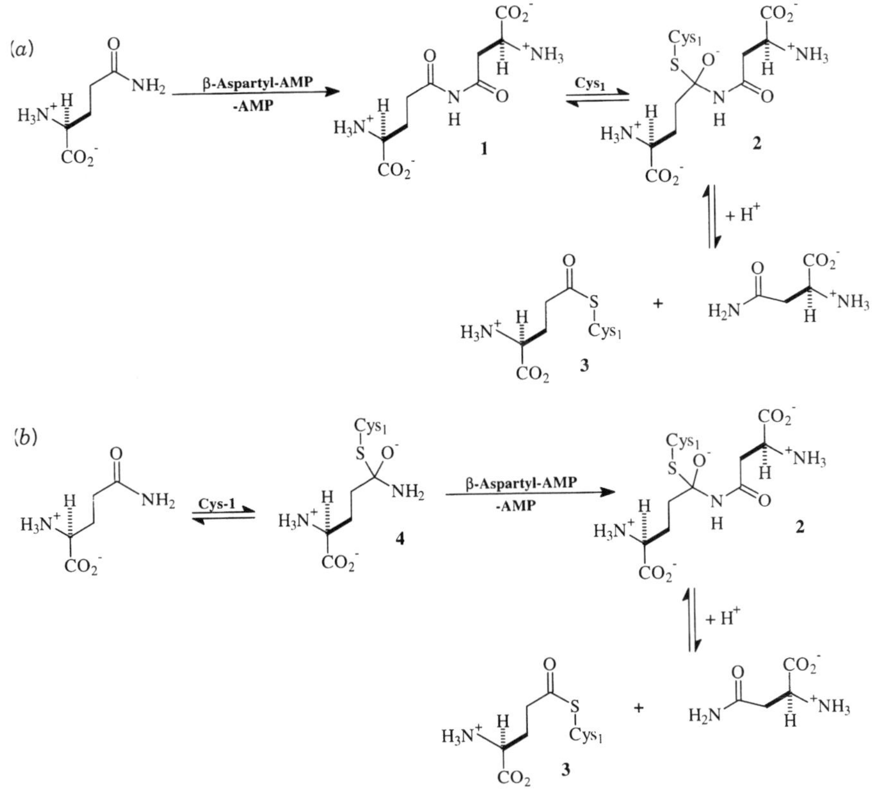

Figure 5. Alternative mechanisms for nitrogen transfer in asparagine synthetase involving a glutamine-aspartyl-AMP imide intermediate. (a) Direct attack of glutamine on aspartyl-AMP forms the imide intermediate 1. Attack of the Cys-1 thiolate yields the tetrahedral oxyanion 2. Collapse of the oxyanion yields asparagine and the γ-glutamyl thioester enzyme intermediate 3. (b) Formation of a tetrahedral glutamine oxyanion 4. precedes direct attack of the amide on aspartyl-AMP to yield the imide intermediate 2. The imide breaks down as in (a). [From Boehlein et al. 1996.]

glutamyl hydroxamate was hydrolyzed to glutamate and hydroxylamine with a k_{cat} slightly faster than that for hydrolysis of glutamine. The corresponding values for k_{cat}/K_m favored glutamyl hydroxamate by a factor of 2. In addition, k_{cat}/K_m values favored NH_2OH by 2.4-fold over NH_3 as a nitrogen source in the reaction with aspartyl-AMP. Thus, if the mechanism of nitrogen transfer involved sequential hydrolysis of glutamyl hydroxamate to yield NH_2OH, followed by reaction of NH_2OH with aspartyl-AMP, the catalytic efficiency for the reaction with the analogue would be expected to be at least equal to that with glutamine as nitrogen donor. However, the k_{cat} for glutamine was seven-fold higher than that for glutamyl hydroxamate and k_{cat}/K_m was 2.5 times higher with glutamine in the synthetase reaction. It was therefore concluded that NH_2OH is transferred directly from glutamate hydroxamate to aspartyl-AMP, giving aspartate β-monohydroxamate. There were, however, two issues that complicate the interpretation. First, in the reaction of glutamate hydroxamate with aspartyl-AMP there was a high amidohydrolase activity such that the stoichiometry between the products, glutamate and aspartate β-monohydroxamate, was not 1:1 as expected. Thus, it can be visualized that the hydrolysis of glutamate hydroxamate was at least partially uncoupled from production of aspartyl β-monohydroxamate. In the absence of the usual tight coupling, the mechanism of N-transfer may be difficult to evaluate. Second, glutamyl hydroxamate was tested as an inhibitor of the NH_3-dependent activity of the *E. coli* C1A mutant. Although glutamate hydroxamate was a strong inhibitor having a K_I of 1.9 μM, the pattern of lines in a double reciprocal plot of $1/v$ versus $1/NH_3$ was different than that obtained with glutamine as inhibitor and cannot be interpreted in terms of formation of an abortive complex. It is therefore uncertain whether the subtle differences in reactivity of glutamine and glutamate hydroxamate impact upon the analysis of the nitrogen-transfer mechanism.

Experiments based on heavy-atom isotope effects, carried out by Stoker et al. (1996), provide a third line of evidence offered in support of direct N-transfer from glutamine to β-aspartyl-AMP. The approach was to compare heavy atom kinetic isotope effects (KIE) determined from reactions with glutamine containing [^{15}N]amide or ^{13}C in the C5 position. Two key reactions, glutaminase and asparagine synthesis, were compared. Values of KIE for [5-^{13}C]glutamine were 1.0256 ± 0.0019 and 1.0231 ± 0.0020 for glutaminase and asparagine synthetase, respectively, consistent with a common mechanism for C—N cleavage in the two reactions. On the other hand, KIE values for [^{15}N]glutamine were 1.0065 ± 0.0032 for glutaminase and 1.0222 ± 0.0023 for synthetase. Given that KIE for [^{15}N]amide represents changes in chemical bonding about the nitrogen as the reactant enters into the transition state, the transition states giving rise to KIE must be different for the two reactions. This observation led to the conclusion that the two reactions, glutamine hydrolysis and N-transfer to β-aspartyl-AMP, proceed by different mechanisms. For this reason, Stoker et al. favor a variation of the mechanism shown in Figure 5(b), in which the glutaminyl tetrahedral intermediate **4** is linked to β-aspartyl-AMP. It was suggested that the KIE for [^{15}N]glutamine and [5-^{13}C]glutamine would arise from C—N bond breakage of the latter complex oxyanion.

2. Analyses of Glutamine Domain Residues

A number of invariant glutamine domain residues in asparagine synthetase have been replaced and the mutant enzymes characterized. Replacement of Cys-1 with Ala or Ser resulted in complete loss of glutamine-dependent asparagine synthetase activity as well as the glutaminase activity (Boehlein et al., 1994a). However, as noted in Section III.C.1, the Cys-1 mutant enzymes bound glutamine with high affinity. Replacements of nonconserved histidines and a nonconserved aspartate in the glutamine domain had negligible effects on activity with glutamine or NH_3 as expected.

Replacements of Arg-30 and Asn-74 (Boehlein et al., 1994b) are best interpreted in terms of the glutamine PRPP amidotransferase mechanism. The Arg-30 residue corresponds to glutamine PRPP amidotransferase Arg-26 and, based on the X-ray structure of the latter, should interact with the side chain of Asn-74, the backbone O of Cys-1 and the α-NH_2 group of Cys-1 (Kim et al., 1996). The main effects of R30A and R30K mutations were to increase the K_m for glutamine and abolish the capacity of ATP to stimulate glutamine hydrolysis. These results support the idea that Arg-30 is important for maintaining the structure of the glutamine site. Communication with the acceptor domain was impaired in Arg-30 mutants having pertubations in glutamine site structure. These results for the Arg-30 mutants are particularly important because replacement of the corresponding residue, Arg-26 in glutamine PRPP amidotransferase, results in an unstable enzyme that is not amenable to study (Mei and Zalkin, 1989; Kim et al., 1996).

The main effects of Asn-74 replacements were to decrease k_{cat} for glutamine, although there were also significant K_m changes for

N74A and N74Q mutations. In the case of the N74A enzyme, the glutamine K_m was decreased more than 10-fold such that the k_{cat}/K_m for the mutant was similar to the wild type enzyme. Nevertheless, two roles for Asn-74 in catalysis were proposed based on the direct transfer mechanisms favored by these authors. One role of Asn-74 in catalysis was suggested to involve stabilization of a glutamine oxyanion intermediate [4 in Fig. 5(b)] in order to increase the nucleophilicity of the amide nitrogen for attack on β-aspartyl-AMP. A variation of this idea involved H-bonding of the side chain of Asn-74 with the carboxamido of glutamine to polarize the amide bond or to stabilize a hydroxyimine tautomer that might be needed for direct attack on β-aspartyl-AMP to form imide intermediate 1 in Figure 5(a). This latter role for Asn-74 was preferred over one that involves stabilization of a tetrahedral oxyanion intermediate [4 in Fig. 5(b)] even in simple glutamine hydrolysis, because k_{cat} values for glutamine hydrolysis were identical for N74A and N74Q mutants.

Given that asparagine synthetase Asn-74 corresponds to glutamine PRPP amidotransferase Asn-101, we believe that the role of this residue is most clearly accounted for by the glutamine PRPP amidotransferase mechanism. The side chain of Asn-101 contributes together with the backbone N of the next residue, Gly-102, to formation of the oxyanion hole required for stabilization of the two tetrahedral oxyanions in the glutamine amide-transfer mechanism (Fig. 3).

D. GLUTAMATE SYNTHASE

Glutamate synthase coupled with glutamine synthetase provides a high-affinity route for ammonia assimilation and glutamate synthesis. A source of reducing equivalents, NAD(P)H or ferredoxin, is required. The NAD(P)H-dependent reaction is given in Eq. 7.

$$2\text{-Oxoglutarate} + \text{Glutamine} + \text{NAD(P)H} + \text{H}^+$$

$$\rightarrow 2 \text{ Glutamate} + \text{NAD(P)}^+ \tag{7}$$

There are three forms of the enzyme. Bacteria such as *E. coli* (Miller and Stadtman, 1972) and *Azospirillum brasilense* (Ratti et al., 1985) contain an NADPH-dependent $\alpha_4\beta_4$ enzyme. As shown schematically in Figure 6, there is a large α subunit and a smaller β subunit.

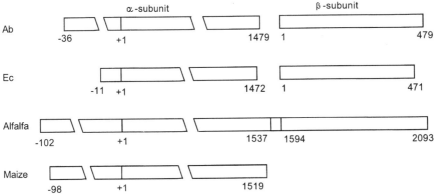

Figure 6.   Schematic representation of three forms of glutamate synthase. The three forms with representative examples are the $\alpha_4\beta_4$ bacterial enzymes (*A. brasilense*, Ab, and *E. coli*, Ec), the monomeric NADH-dependent enzymes from plants (alfalfa), *Chlamydomonas* and *Neurospora* corresponding to an $\alpha-\beta$ fusion, and the monomeric ferredoxin-dependent enzymes (Maize) corresponding to the α subunit. Amino acid residues at termini and junctions are numbered. The $+1$ is the processing site for cleavage of an NH_2-terminal propeptide or targeting sequence. A 58 amino acid connector is shown for the alfalfa enzyme. Slashes indicate that a portion of the protein is not shown.

Plants contain two different glutamate synthases. All higher plants studied thus far, as well as *Chlamydomonas reinhardtii* and *Neurospora crassa*, contain an NADH-dependent enzyme consisting of a single protein chain (see Gregerson et al., 1993). These are monomeric enzymes corresponding to putative fusions of the bacterial α and β subunits. Ferredoxin (Fd)-dependent glutamate synthase is a monomeric enzyme found in plant chloroplasts, cyanobacteria (Marqués et al., 1992), green algae (Galván et al., 1984), as well as nonphotosynthetic plant tissues (see Sakakibara et al., 1991). The Fd-dependent glutamate synthases are homologous to the α subunit of the bacterial enzymes but lack sequences corresponding to the β subunit.

Sequences derived from cloned genes have been determined for the enzymes shown in Figure 6. All of the enzymes are synthesized with NH_2-terminal targeting and/or propeptide sequences that must be processed to yield the mature, active enzymes having the characteristic NH_2-terminal active site cysteine. The *E. coli* and *A. brasi-*

lense enzymes have a relatively short propeptide, whereas the plant enzymes have longer targeting sequences for subcellular localization. The plant NH_2-terminal sequences are possibly removed in two steps, an initial clipping of the targeting segment of the prosequence after localization followed by cleavage of the residual propeptide to expose the NH_2-terminal cysteine and activate the enzyme. In the case of *E. coli* glutamate synthase proenzyme, there are two possible positions for translation initiation that could produce propeptides of 43 or 11 amino acids. An 11 amino acid propeptide is shown in Figure 6 because mutational analysis of the alternative translation start sites supports the downstream ATG codon as the preferred translational start site (Velázquez et al., 1991).

1. The [Fe–S] and Flavin Components

The three types of glutamate synthase outlined in Figure 6 are complex [Fe–S] flavoproteins. Binding motifs for [Fe–S] centers and flavin cofactors have been deduced from alignments with well-characterized enzymes (Pelanda et al., 1993). In some cases, flavins were identified by chemical and spectroscopic methods and [Fe–S] centers characterized by absorption, electron paramagnetic resonance (EPR), resonance Raman, and circular dichroism spectroscopies. By this methodology, the bacterial αβ protomer contains 1 flavin mononucleotide (FMN), 1 flavin adenine dinucleotide (FAD), and three [Fe–S] centers (a [3Fe–4S] center designated [Fe–S$_I$] and two [4Fe–4S] centers, [Fe–S$_{II}$] and [Fe–S$_{III}$]) (Vanoni et al., 1992). Studies with the native *Azospirillum* enzyme and recombinant α and β subunits made in *E. coli* led to the conclusion that the α subunit contains FMN and the [3Fe–4S] center [Fe–S$_I$], whereas the β subunit contains sites for FAD and NADPH (Vanoni et al., 1996, 1997). Since neither subunit made individually contains the two [4Fe–4S] centers that are components of the αβ promoter, these [Fe–S] centers may be assembled with cysteine ligands from both subunits. Evidence for a slow rate of posttranslational cluster assembly and a direct correlation between [4Fe–4S] content and activity have suggested to Vanoni et al. (1996) that assembly of the [4Fe–4S] centers into the recombinant enzyme is the rate-limiting step in enzyme maturation.

A two-site flavin model for the *Azospirillum* glutamate synthase is shown in Figure 7. The role of the β subunit is to provide electrons

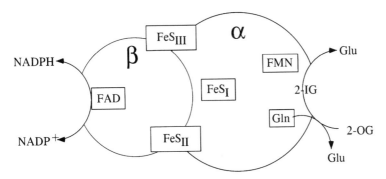

Figure 7. Schematic representation for *A. brasilense* αβ glutamate synthase pro-tomer showing sites for substrates and cofactors. The abbreviations are 2-OG -2-oxo-glutarate and 2-IG -2-imino glutarate. [Adapted from Vanoni, 1997.]

for reduction of 2-iminoglutarate to glutamate, a reaction catalyzed by the α subunit. The FAD bound to the β subunit is reduced by NADPH. Reduced FMN in the α subunit is the immediate electron donor for reduction of 2-iminoglutarate to glutamate. The [Fe–S$_I$] and [Fe–S$_{II}$] clusters are believed to mediate electron transfer be-tween the two flavins, whereas the role of FeS$_{III}$ is not known. A number of studies have provided evidence to support this model. The FAD bound to the β subunit is reduced by NADPH under pre-steady-state conditions at a rate faster than that for electron transfer to artificial acceptors or faster than the overall reaction with native enzyme (Vanoni et al., 1996). Reduction of enzyme with NADPH results in EPR signals characteristic of reduced [3Fe–4S] and [4Fe–4S] centers. Upon partial reduction of the enzyme with NADPH, [Fe–S$_I$] was reduced to a greater extent than [Fe–S$_{II}$]. The [Fe–S$_{III}$], with a still lower potential, could be reduced photo-chemically but not by NADPH. Thus, the role of [Fe–S$_{III}$] is not established. Electrons from reduced FMN are used in the final step of the reaction. Studies with an isolated α-subunit provide evidence for reduction of 2-iminoglutarate by reduced FMN. Thus, the α-subunit catalyzes a slow rate of glutamate oxidation, with the reverse reaction occurring when using artificial electron acceptors (Vanoni et al., 1997). This reaction, which is unaffected by NADP$^+$, an NADP$^+$ analogue, or glutamine, is competitively inhibited by 2-

oxoglutarate. Finally, FMN bound to the α-subunit but not the [Fe–S] center is reduced anaerobically by excess glutamate. This model is consistent with the two site ping–pong uni–uni bi–bi kinetic mechanism determined earlier (Vanoni et al., 1991).

Limited analyses of Fd-dependent glutamate synthase indicate the close similarity of [Fe–S] and flavin components with those in the corresponding α subunit of the bacterial enzyme. The only flavin in the Fd-dependent enzyme from *Synechococcus* is a molecule of FMN (Marques et al., 1992) and a [Fe–4S] cluster is the only [Fe–S] center in the Fd-dependent spinach enzyme (Knaff et al., 1991). These results support a model in which the β-subunit of bacterial αβ-type enzymes functions in NAD(P)H-dependent reduction of FAD followed by [Fe–S]-mediated electron transfer to [Fe–S$_I$] in the α-subunit. Lacking a β-subunit, the Fd-dependent enzymes do not require FAD and [4Fe–4S] centers. It appears that Fd can directly reduce the [3Fe–4S] cluster in the monomeric protein that corresponds to the bacterial α-subunit.

2. *Glutamine Domain*

Identification of the Ntn amidotransferase sequence fingerprint in glutamate synthase required an understanding of the glutaminase active site and the catalytic functions of elements of the fingerprint, which was possible only with knowledge of the structure of at least one Ntn glutamine domain. Specifically, the Asn-Gly oxyanion hole was expected to be conserved in all the Ntn amidotransferases because of its role in catalysis. The oxyanion hole was identified at Asn-231-Gly-232, as were the glutamine-recognition elements at Arg-210 and Asp-273 and the *cis*-proline structural element at Pro-224 (residue numbering for the *A. brasilense* α-subunit). This new alignment of the sequences of glutamate synthase to those of the other Ntn amidotransferases also identifies a 150-residue insertion encompassing residues 50–200, which may replace the third β-strand and following two helices of the Ntn glutamine domain [residues 40–60 of *E. coli* glutamine PRPP amidotransferase, upper right-hand corner of Fig. 1(*b*)]. Glutamate synthase appears to have an entire structural domain at this highly variable site in the Ntn glutamine domain. This region of the glutamine domain can accommodate considerable structural variability, from the addition of a structural domain in glutamate synthase, to deletion of the β-strand and both helices in asparagine synthetase. We infer that this part of the fold has been adapted to the function of each individual Ntn amidotransferase.

E. PERSPECTIVE

A mechanism for glutamine hydrolysis by the Ntn amidotransferases is consistent with the three-dimensional structures, sequence conservation, biochemical studies, and site-directed mutagenesis experiments. This information is summarized in Figure 3 and was described above in Section III.A on glutamine PRPP amidotransferase. The mechanism specifies a role for each invariant residue of the Ntn amidotransferase family. One group of invariant residues is catalytic. The nucleophile is Cys-1 Sγ. The proton donor is Cys-1 α-amino to the NH_3 leaving group. Asn-101 $N\delta_2$ and Gly-102 N form the oxyanion hole. A second group of residues recognizes glutamine. The Arg-73 residue forms a salt bridge with the glutamine α-carboxyl group and Asp-127 with the glutamine α-amino group. A third group of invariant residues is structural. The Arg-26 residue anchors Cys-1 and Asn-101 through three hydrogen bonds. Both Gly-27 and Gly-32 line the glutamine-binding cavity in tightly packed spaces where side chains cannot be accomodated. The *cis*-Pro-87 forms one wall of the glutamine binding pocket (residue numbering for *E. coli* glutamine PRPP amidotransferase). These specific functions for the invariant residues are consistent with biochemical and mutagenesis results for all the Ntn amidotransferases.

The study of Ntn amidotransferase structure and function has been enriched by the recent discovery of a larger homologous family, the Ntn hydrolases (Brannigan et al., 1995), of which the Ntn amidotransferases represent one corner. The Ntn hydrolases are thought to be homologous based on a common three-dimensional structure and the location of key elements of the active site within the core structure. All family members employ an N-terminal nucleophilic side chain, require a free α-amino terminus for catalysis, and possess a common oxyanion hole. However, they are too distantly related for any relic of common ancestry to be detectable in their primary sequences. Thus recognition of the family was possible only after the first three-dimensional structures were solved. These include glutamine PRPP amidotransferase (Smith et al., 1994), penicillin acylase (Duggleby et al., 1995), and the proteasome (Löwe et al., 1995). Essentially, the same structure-based catalytic mechanism has been proposed for each of the Ntn hydrolases (Duggleby et al., 1995; Smith, 1995; Löwe et al., 1995; Oinonen et al., 1995; Isupov et al., 1996). Interestingly, many of the Ntn hydrolases are subject to post-translational cleavage to generate the N-terminal catalytic nucleophile. In a few cases, this modification has been shown to be autocatalytic. Thus it may be that self-processing is a feature of the maturation of some Ntn amidotransferases, such as described above and proposed earlier (Mäntsälä and Zalkin, 1984) for glutamine PRPP amidotransferases with [Fe–S] clusters.

A major advance in understanding the coupling of catalysis in the two domains of Ntn amidotransferases was the observation of a tunnel between the glutamine and acceptor active sites of glutamine PRPP amidotransferase (Krahn et al., 1997). Key questions of amidotransferase catalysis can be answered with the new information. First, the overall transformation catalyzed by Ntn amidotransferases (Eq. 1) is indeed the sum of a glutaminase (Eq. 2) and an NH_3 transfer (Eq. 3). This settles the long-standing question of exactly what nitrogen species attacks the acceptor substrate. The answer is NH_3, channeled through the solvent-inaccessible tunnel. Taking the simplest possibility as a working hypothesis, we conclude that the same situation holds for the other Ntn amidotransferases. This conclusion is supported by the conserved structural features and biochemical properties of the glutaminase active sites of Ntn amidotransferases and by the structural and biochemical connection to other Ntn hydrolases.

A second fundamental question about amidotransferase catalysis is the means by which binding of the acceptor substrate stimulates the glutamine active site. We now understand that, in glutamine PRPP amidotransferase, binding of the acceptor substrate PRPP is followed by ordering of the flexible allosteric loop, which creates the tunnel between catalytic sites and signals the glutamine domain that the acceptor substrate is enzyme bound. The kinetic result is a reduced K_m for glutamine. The relevant K_m change is due in part to the orientation of Arg-73, whose side chain interacts with the glutamine analogue DON in the active (Krahn et al., 1997), but not in the inactive (Kim et al., 1996) conformation of the enzyme. The Arg-73 residue may signal acceptor substrate binding for the other Ntn amidotransferases as well because it is an Ntn-invariant residue and is located on a loop facing the acceptor domain.

New questions can also be posed by knowledge of the first Ntn amidotransferase structure in a catalytically competent conformation. First is the identity of a binding site for NH_3. Most of the amidotransferases exhibit NH_3-dependent activity under appropriate conditions. Glucosamine 6-phosphate synthase is the sole exception. Does this amidotransferase perhaps lack an NH_3 binding site, or is NH_3 from the solvent unable to enter the NH_3 tunnel?

Creation of an NH_3 tunnel is an elegantly simple means to carry out the complex reaction of nitrogen transfer from glutamine to an acceptor. Each of the amidotransferases contains a homologous glutamine domain and a unique acceptor domain designed for optimal coupling. Some of the acceptor domains also belong to homologous enzyme families. The Ntn amidotransferases are thus "complex" enzymes in which active sites are catalytically separate. Catalytic activities of the two active sites are coupled by signaling between them and not by merger into a single, complex active site. Fine tuning a catalytic domain for optimal activity, depending on availability of substrates or inhibitors, is a far simpler task in molecular design than is merger of two precise active sites into one. It is a simple matter for one domain to sense a regulatory signal and to either form or disrupt the channel needed for coupling of active sites. It remains for future work to determine if channels are indeed a common feature of Ntn amidotransferases, as predicted, and if so, to determine the signaling mechanism for each enzyme.

IV. Triad Amidotransferases

A. GMP SYNTHETASE

The guanosine monophosphate (GMP) synthetase catalyzes the second and final reaction in the two-step pathway from IMP to GMP (Eq. 8).

$$XMP + Glutamine + ATP + H_2O \rightarrow$$

$$GMP + Glutamate + AMP + PP_i \qquad (8)$$

A recently determined X-ray structure for *E. coli* GMP synthetase is the first for a Triad amidotransferase (Tesmer et al., 1996). The structure defines two modular catalytic domains, one required for glutaminase activity, and the other for an ATP-PPase activity. The protein fold and active site structure of the N-terminal glutaminase domain is a model for the related Triad amidotransferases, whereas the C-terminal ATP-PPase domain is expected to be a prototype for three related ATP-PPases, NAD synthetase, asparagine synthetase, and argininosuccinate synthetase.

1. Enzyme Structure and Mechanism

The structure of the N-terminal glutamine domain of GMP synthetase, the prototype for the homologous family of Triad amidotransferases, includes a central β-sheet of mixed polarity flanked by α-helices and a β-ribbon [Fig. 1(c)]. The catalytic residue Cys-86 occurs in a very tight connection between a central strand of the β-sheet and an α-helix, in a structural motif known as the "nucleophile elbow," which is common to a group of catalytic-triad enzymes known as α/β hydrolases (Ollis et al., 1992). The α/β hydrolases are homologous enzymes with a central parallel β-sheet. The GMP synthetase is not homologous with the α/β hydrolases because the topologies of the β-sheets differ and the catalytic residues occur in a different order in the primary sequences. If GMP synthetase belongs to an enzyme family larger than the triad amidotransferases, the family has not yet been identified.

The side chains of Cys-86, His-181, and Glu-183 are at the active site of the glutamine domain of GMP synthetase (Fig. 8). As expected, these side chains are arranged as a classic catalytic triad, very similar to the arrangement of the Cys-His-Gln catalytic triad of the cysteinyl protease papain. From the rich structural data base of hydrolytic enzymes with catalytic triads and from consideration of sequence identities, the oxyanion hole near the Cys-86 nucleophile consists of the backbone nitrogens of Gly-59 and Tyr-87. The functional groups that form the catalytic triad and oxyanion hole superimpose on the analogous groups of papain with a root-mean-square deviation of 0.6 Å. Thus, for the purposes of catalysis, the active site seems to be fully formed. However, in absence of the acceptor substrate, GMP synthetase is a very poor glutaminase. The missing element of the glutamine domain seems to be a specific site for recognition of glutamine. No specificity pocket is apparent in the structure of the glutamine domain. Even when Cys-86 is alkylated by the gluta-

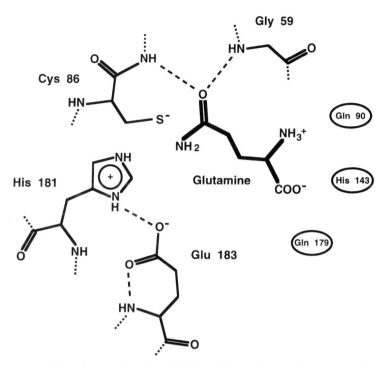

Figure 8.   Glutaminase active site of Triad amidotransferases. Interactions of the substrate glutamine are shown. The catalytic triad includes Cys-86, His-181, and Glu-183. The Cys-86 residue is the nucleophile and His-181 is the proton donor in the hydrolytic reaction. The Glu-183 residue maintains the orientation and tautomeric state of His-181. The Gln-90, His-143, and Gln-179 residues are proposed to recognize glutamine by making specific hydrogen bonds with the α-amino and α-carboxyl groups. Confirmation awaits a structure for the active conformation of a Triad amidotransferase. Residue numbering is for *E. coli* GMP synthetase.

mine analogue acivicin, only the atom covalently bonded to Cys-86 Sγ is ordered.

In addition to catalysis of NH_3-dependent GMP synthesis, the acceptor domain of GMP synthetase catalyzes 5′-adenylation of the substrate xanthosine monophosphate (XMP) to produce an unstable activated XMP adenylate, which is the actual acceptor substrate. The structure of the acceptor domain is a twisted parallel β-sheet surrounded by α-helices, a common fold for enzymes that bind nucleotides. The crystal structure is of the product complex with AMP,

PP_i, and Mg^{2+}, which serves to identify a ''PP_i loop'' between the first β-strand and first α-helix of the domain. The NAD synthetase, asparagine synthetase, and argininosuccinate synthetase were proposed to have catalytic domains homologous to the ATP-PPase domain of GMP synthetase based on conserved sequences for the functionally relevant PP_i loops and common catalytic reactions (Tesmer et al., 1996). All four enzymes catalyze the formation of activated adenylate intermediates, from which AMP is displaced by a nitrogen species. Asparagine synthetase and at least some NAD synthetases are glutamine amidotransferases. Homology with GMP synthetase was subsequently demonstrated in the structure NAD synthetase (Rizzi et al., 1996; see Section IV.D).

Our working hypothesis for overall catalysis by GMP synthetase is based on the crystal structure and biochemical data. In this model, binding of ATP, Mg^{2+}, and XMP to the ATP-PPase domain induces a structural change concomitant with formation of the XMP adenylate. The structural change (a) sequesters the XMP adenylate in the enzyme (b) ''signals'' the glutamine domain that the true acceptor substrate is bound by creating a specificity pocket for glutamine binding, and (c) creates a tunnel for NH_3 between the active sites.

A substrate-induced disorder–order transition in the acceptor domain of GMP synthetase is strongly supported by both biochemical and structural data. The GMP synthetase is very sensitive to trypsin cleavage in the absence of XMP (Zyk et al., 1969). In the crystal structure, a 22-residue peptide, including three potential trypsin cleavage sites and six invariant residues, is highly disordered although not cleaved. The substrate XMP, which is present in crystals, is similarly disordered beyond the 5'-phosphate. The products AMP and PP_i in the active site of the crystal structure were produced from ATP and XMP in a process requiring formation of the XMP adenlyate, which was presumably hydrolyzed during the days/weeks of crystal growth and diffraction experiments.

The catalytic domains of GMP synthetase are connected by a hinge, which is trapped in an open conformation by lattice contacts in the crystal. In this position, the catalytic sites are 30-Å apart, and the ATP-PPase active site is highly solvent exposed. The disordered peptide in the ATP-PPase domain is adjacent to the glutamine domain. When ordered, this peptide may impact the position of the hinge between domains and/or may form part of an NH_3 tunnel. Production of free NH_3 during normal turnover of the enzyme is strongly supported by observation of an intact catalytic triad in the glutamine domain. All the catalytic machinery for amide hydrolysis is present except a specific binding site for glutamine. Candidate residues for glutamine recognition are Gln-90 and His-143, which are part of the Triad amidotransferase fingerprint (Fig. 1). The side chain of Gln-90 points into the active site cleft, whereas that of His-143 is close to the catalytic center, but points away from the active site cleft. Rotation of the side chain brings it into the cleft. Other elements of the recognition site for glutamine are not apparent, but could involve the side chain of Gln-179 as well as backbone atoms in the protein.

2. Human GMP Synthetase

The human enzyme has been purified to homogeneity from T-lymphoblastoma cells (Nakamura and Lou, 1995) and a recombinant version from a Baculovirus system (Lou et al., 1995). The native and recombinant enzymes were indistinguishable based on molecular weight, oligomeric state, isoelectric points, amino acid sequence of tryptic peptides, and kinetic properties. Two enzyme forms, differing by about 0.1 pH unit, were resolved by isoelectric focusing. These two forms were indistinguishable by all the criteria used for characterization and the basis for their separation is not known. The enzyme is a monomer of molecular mass 77360 ± 20 Da for the reduced and alkylated species determined by electrospray ionization mass spectrometry. Positive cooperativity for saturation by XMP (Hill coefficient of 1.5) suggests interacting XMP sites but the enzyme was monomeric in the presence of different combinations of substrates. Thus the explanation for cooperativity is not known. The enzyme was inhibited by decoyinine and psicofuranine, two adenosine antibiotics that selectively inhibit GMP synthetase (Fukuyama, 1966). The Mg^{2+} ion is required for activity beyond the concentration required for ATP chelation and thus free Mg^{2+} appears to have a role.

The reactions catalyzed by the tightly coupled glutaminase and acceptor domains of the human enzyme can be uncoupled similar to E. coli GMP synthetase (Zalkin and Truitt, 1977) and to other amidotransferases (Zalkin, 1993). The maximal glutaminase activity, which is twofold greater than the synthetic rate, requires binding of XMP, PP_i, and Mg^{2+} to the acceptor domain (Nakamura et al.,

1995). Omission of any of the three ligands results in 2% or less of the maximal activity. It is likely that binding of these ligands to the acceptor domain is required to form the glutamine binding site in the glutaminase domain. Alkylation of Cys-104 by the glutamine affinity analogue acivicin also requires XMP, PP_1, and selectively blocks glutamine to serve as a N-donor. This reaction is similar to, but more selective than, that observed previously using DON (Patel et al., 1977; Zalkin and Truitt, 1977) and other glutamine analogues. Although all of the glutamine analogues alkylated the catalytic cysteine of GMP synthetase, the reaction was second order. Saturation kinetics were not observed, meaning that there is no evidence for analogue binding to the glutamine site to form an enzyme–inhibitor complex prior to alkylation of the active site cysteine.

Overall, the properties of the human enzyme can be accommodated by the structural model for *E. coli* GMP synthetase with one exception. A 13-residue peptide, which encompasses residues 454–466 and maps to a loop in the C-terminal dimerization domain of *E. coli* GMP synthetase, includes 6 invariant and 3 conservatively substituted residues in the 11 known sequences of GMP synthetase. The apparent sequence conservation strongly suggests an important functional role for this loop, as does its position in the XMP binding site of the apposing subunit of dimeric *E. coli* GMP synthetase. However, this conclusion is not consistent with a monomeric human enzyme. Three possibilities could resolve this enigma. (a) The sequence alignment in this region may not be meaningful. (b) The human enzyme may, in fact, be monomeric. (c) The domain may be folded differently in human GMP synthetase such that the loop plays the same role in a single subunit of the monomeric enzyme as it does in the partner subunit of the dimeric enzymes. The latter possibility is plausible because three insertions totaling 110 residues in the higher eukaryotic sequences of GMP synthetase map to sites in the dimerization domain of *E. coli* GMP synthetase.

B. CARBAMYL PHOSPHATE SYNTHETASE

Carbamyl phosphate is an intermediate in the synthesis of arginine, pyrimidine nucleotides, and urea. Three different types of carbamyl phosphate synthetase (CPS) have been distinguished based on the N-donor (glutamine or NH_3) and allosteric activator. Carbamyl phosphate synthetase I is a strictly NH_3-dependent mitochondrial enzyme in the livers of ureotelic vertebrates that is required for the urea cycle (Eq. 9). *N*-Acetylglutamate is an essential allosteric activator. Carbamyl phosphate synthetase I from bullfrog has been crystallized (Marina et al., 1995).

$$NH_3 + HCO_3^- + 2\ ATP \rightarrow Carbamyl\ phosphate + 2\ ADP + P_i$$

$$(9)$$

Type II enzymes use glutamine as the physiological N-donor for synthesis of arginine and pyrimidine nucleotides (Eq. 10).

$$Glutamine + HCO_3^- + 2\ ATP \rightarrow$$

$$Carbamyl\ phosphate + 2\ ADP + 2\ P_i \qquad (10)$$

The properties of CPS III are very similar to those of the Type I enzyme except glutamine is the N-donor. The role of this enzyme in invertebrates and ureoosmotic elasmobranch fishes (sharks, skates, and rays) is in the urea cycle.

1. Type II Enzymes

This section focuses on recent developments in the glutamine amide transfer mechanism, particularly communication between the glutaminase and synthetase sites. There are two prototype enzymes, one from *E. coli* and a mammalian CAD protein. These enzymes differ primarily in oligomeric composition and allosteric control. *Escherichia coli* CPS contains a 382 amino acid glutaminase subunit and a 1073 residue synthetase subunit. In CAD (trifunctional carbamyl phosphate synthetase-aspartate transcarbamylase-dihydrorotase), the glutaminase and synthetase domains are fused. The enzyme mechanism is independent of this structural difference.

The synthesis of carbamylphosphate takes place in four steps:

1. Glutamine + H_2O → Glutamate + NH_3
2. ATP + HCO_3^- → Carboxyphosphate + ADP
3. Carboxyphosphate + NH_3 → Carbamate + P_i
4. Carbamate + ATP → Carbamylphosphate + ADP

In addition to the overall reaction, Eq. 10, the enzyme catalyzes three partial reactions.

$$ATP + H_2O \xrightarrow{HCO_3^-} ADP + P_i \tag{11}$$

$$MgADP + Carbamylphosphate \rightarrow MgATP + CO_2 + NH_3 \tag{12}$$

$$Glutamine + H_2O \rightarrow Glutamate + NH_3 \tag{13}$$

The partial reactions reflect steps in the mechanism, and each step has been correlated with a specific enzyme domain, as is shown schematically in Figure 9. Sequence identities between NH_2- and CO_2H-terminal halves of the synthetase subunit support the idea that the two halves are homologous (Nyunoya and Lusty, 1983), and each half contains sequences ascribed to an ATP binding site (Lusty et al., 1983). Each site is linked to one of the phosphorylation steps in the mechanism. The HCO_3-dependent ATPase (Eq. 11) reflects Step 2) in the mechanism. Two conserved glycines, Gly-176 and Gly-180 (*E. coli* synthetase subunit), are required for this activity, indicating that the proximal ATP site in the synthetase subunit is utilized for phosphorylation of HCO_3^- to carboxy phosphate (Post et al., 1990). Gly-722, a conserved residue in the distal ATP side of the synthetase subunit is involved in Step 4) of the mechanism, phosphorylation of carbamate to yield carbamyl phosphate. The glutaminase partial reaction, Step 1) in the mechanism, generates NH_3 for reaction with carboxyphosphate. The Cys-269 and His-353 residues in the glutaminase subunit function as nucleophile and the acid–base group, respectively, to generate NH_3 for reaction with carboxyphosphate (Rubino et al., 1986; Miran et al., 1991). The Glu-355 residue completes the Cys-His-Glu catalytic Triad. The His-243 residue in the synthetase subunit is critical for the reaction of NH_3 with carboxyphosphate Step 3, although its role is not known (Miles et al., 1993).

The work just cited by Post et al. (1990), provides strong evidence that each of the homologous halves of the synthetase subunit carries out one of the two phosphorylation steps in the overall reaction. However, Guy and Evans (1996) recently reported that each of the synthetase subdomains can catalyze both partial reactions resulting in the synthesis of carbamyl phosphate from HCO_3^-, 2ATP, and

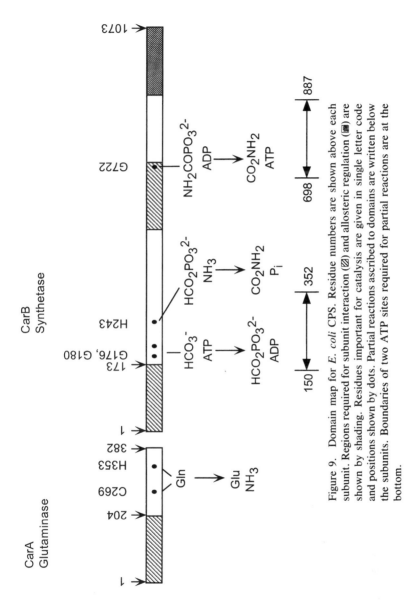

Figure 9. Domain map for *E. coli* CPS. Residue numbers are shown above each subunit. Regions required for subunit interaction (▨) and allosteric regulation (▨) are shown by shading. Residues important for catalysis are given in single letter code and positions shown by dots. Partial reactions ascribed to domains are written below the subunits. Boundaries of two ATP sites required for partial reactions are at the bottom.

NH_3. For this work, deletions of hamster CAD were constructed in which either the proximal (designated CPS.A) or distal (CPS.B) half of the synthetase was excised. These enzymes, having the structure GLN-CPS.A and GLN-CPS.B, in which the intact glutaminase domain is fused to either CPS.A or CPS.B, were produced in *E. coli* and purified. The gene encoding GLN-CPS.A complemented an *E. coli carA carB* mutant, indicating *in vivo* function. Following purification, both of the truncated enzymes catalyzed the two ATP-dependent partial reactions and the overall synthesis of carbamyl phosphate using glutamine or NH_3 as N-donor. Furthermore, synthetase ligands stimulated glutaminase activity of both enzymes, indicative of coupling between the two domains. Thus, either synthetase domain could contribute to formation of the glutamine site. Turnover numbers for GLN-CPS.A and GLN-CPS.B of 0.17 and 0.34 s^{-1}, respectively, were less than fivefold lower than 0.79 s^{-1}, the value for the intact CPS. One functional distinction between the two synthetase domains is that the carboxyl end of the CPS.B domain is essential for allosteric regulation. Indeed, the allosteric effectors UTP and PRPP modulated the activity of GLN-CPS.B but not GLN-CPS.A. Most important, the enzymes were dimeric, $(GLN-CPS.A)_2$ and $(GLN-CPS.B)_2$. Thus, each enzyme had two identical synthetase subdomains, one functioning in each of the ATP-dependent partial reactions. To reconcile these results with the mutagenesis studies, which indicate that the function of CPS.A and CPS.B are fixed in the native enzyme and do not alternate between bicarbonate activation and carbamate phosphorylation, Guy and Evans suggested that either subunit of the mutant dimer may catalyze activation of HCO_3^-, but once carboxy phosphate is formed by one subunit and reacts with the NH_3 to make carbamate, the other subunit phosphorylates carbamate by default.

Even though the separated glutaminase and synthetase subunits of the *E. coli* enzyme have catalytic activity, catalysis is dependent on subunit interactions (see discussion in Zalkin, 1993; Lusty and Liao, 1993; Mareya and Raushel, 1994). A common theme for the amidotransferase glutamine and acceptor domains is repeated. Namely, binding of ligands to the synthetase subunit/domain is required to form the glutaminase site. There are at least two important questions: How is the availability of carboxyphosphate communicated to the glutamine site? How are the activities of the two domains coupled?

Two residues have been identified in the *E. coli* enzyme that are required for coupling, Glu-841 in the synthetase subunit (Lusty and Liao, 1993) and Cys-248 in the glutaminase subunit. The Glu-841 residue is in the distal ATP site of the synthetase. An E841K mutation inactivates the overall synthesis of carbamyl phosphate with either glutamine or NH_3 as N-donor (Guillou et al., 1992). Defects in the three partial reactions appear to account for the essentially complete loss of enzyme activity. Two of the defects were in partial reactions catalyzed by the synthetase subunit (Eqs. 11 and 12). In addition, there were several pertubations in glutamine subunit function resulting from loss of communication with the synthetase subunit (Lusty and Liao, 1993). These include a 40-fold decrease in glutaminase k_{cat}/K_m, complete loss in capacity of MgATP and HCO_3^- to stimulate glutaminase, decreased capacity to generate the γ-glutamyl thioester enzyme intermediate, greater than 100-fold reduction in reaction of the cysteine nucleophile with a glutamine affinity analogue, and a glutaminase pH profile of mutant changed from that of native enzyme to that of glutaminase subunit. It is therefore apparent that the E841K mutation uncouples the communication between synthetase sites and the glutamine site. Whether the uncoupling results from failure of the synthetase to generate a signal (perhaps carbamate or carbamyl phosphate) or to communicate the signal is not known.

A cysteine residue was identified in the glutaminase subunit, Cys-248, that also was proposed to have a role in communication (Mareya and Raushel, 1994). A C248D replacement resulted in 70-fold increased glutaminase but the NH_3 released was unavailable for carbamyl phosphate synthesis. Glutamine-dependent carbamyl phosphate synthesis was below the limit of detection although the capacity to use external NH_3, at a slower rate, was retained. Mareya and Raushel (1994) suggested that this replacement of Cys-248 mimics a conformation change produced by MgATP and HCO_3^- that enhances the reactivity of Cys-248 with *N*-ethylmaleimide. The derivatization of Cys-248 with *N*-ethylmaleimide or its replacement by mutagenesis, was suggested to lock the enzyme into a conformation that can hydrolyze glutamine but cannot relax to one needed for utilization of the released NH_3. Whatever the explanation, the C248D replacement effectively uncoupled reactions catalyzed by the two subunits. However, since Cys-248 is not conserved in all CPS II sequences a direct role in coupling is unlikely. If a channel connecting the glutamine and acceptor sites were required for N transfer, mutation of Cys-248 might damage the channel and disrupt normal passage of NH$_3$.

Guy and Evans (1995) reported another aspect of the coupling mechanism. The glutamine domain of CAD, which is homologous to the corresponding subunit of *E. coli* CPS, consists of two halves, an interaction subdomain, residues 1–174, and a catalytic subdomain, residues 175–365. This dissection parallels the earlier separation of the *E. coli* glutaminase subunit into an interaction subdomain, residues 1–123, and a catalytic subdomain, residues 204–382 (Guillou et al., 1989). Using slightly different boundaries than deduced for the *E. coli* subunit, Guy and Evans showed that the isolated catalytic subdomain has an extraordinarily active glutaminase, with a k_{cat} 350-fold higher and K_m 40-fold lower than the intact CAD glutaminase domain. The CAD catalytic subdomain interacted with and coupled to the synthetase from *E. coli*, although coupling was inefficient. A 12:1 ratio of CAD catalytic subdomain to *E. coli* synthetase was required for maximal glutamine-dependent carbamyl phosphate synthetase activity. This is in contrast to the 1:1 stoichiometry obtained with the intact CAD glutaminase domain and *E. coli* synthetase. In addition, the isolated CAD interaction subdomain interacted with the *E. coli* synthetase. The important conclusion is that the interaction subdomain has two functions: (1) interaction with the synthetase to promote coupling of glutaminase and synthetase active sites, and (2) suppression of the high glutaminase activity in the absence of catalysis at synthetase sites. This interaction enables the synthetase domain to modulate glutaminase activity. The glutaminase interaction subdomain can thus be considered one component of the coupling mechanism. It will be interesting to see how the interaction subdomain and the synthetase fine-tune the glutamine site. Finally, the functional coupling between hamster CAD glutamine domain and *E. coli* synthetase subunit demonstrates the remarkable structural conservation of the enzyme in these species. This work provides an excellent example of how domain swapping can be used to probe enzyme function. A structural foundation for the CPS II systems should soon result from ongoing crystallographic analyses of the *E. coli* enzyme (Thoden et al., 1995; Lusty, personal communication).[2]

[2] After completion of this article, the x-ray structure of *E. coli* CPS was published (Thoden et al., 1997). An update is included in Section 4F.

2. *Carbamyl Phosphate Synthetase III, an Evolutionary Link between the NH_3-Dependent Enzyme of the Urea Cycle and the Glutamine-Dependent Biosynthetic Enzyme*

Carbamyl phosphate synthetase I (CPS I) is an NH_3-dependent enzyme, requiring activation by N-acetylglutamate (reviewed in Zalkin, 1993). It is found primarily in the liver and small intestine of ureotelic vertebrates. The enzyme, a single protein chain of about 1500 residues, is localized in mitochondria and functions in the urea cycle. There is about 50% amino acid sequence identity between rat CPS I and hamster CPS II. Invertebrates and some fish lack CPS I but have a glutamine-dependent, acetylglutamine-activated CPS III for urea cycle function. The amino acid sequence of dog fish shark CPS III is over 70% identical with human, rat and tadpole CPS I and exhibits between 53–57% identity with CPS II from yeast, *Dictostelium discoideum* and hamster (Hong et al., 1994). Thus CPS III is more closely related to CPS I than to CPS II. The evolutionary relationship diagrammed in Figure 10 has been proposed to account for the different forms of CPS (see Hong et al., 1994). Replacement of the cysteine nucleophile (*E. coli* glutaminase subunit Cys-269) with serine in rat CPS I could explain loss of glutamine function in the rat enzyme. However, the cysteine nucleophile is retained in the tadpole CPS I (Helbing and Atkinson, 1994), implying that other amino acid replacements must also contribute to loss of glutamine function in the CPS I from tadpole and perhaps from the other sources. However, Helbing and Atkinson emphasized that evidence is lacking about whether the tadpole enzyme can utilize glutamine as well as NH_3 as a substrate. If the tadpole CPS can use glutamine, the original classification of the tadpole CPS is called into question.

C. IMIDAZOLE GLYCEROL PHOSPHATE SYNTHASE

Imidazole glycerol phosphate (IGP) synthase catalyzes the fifth step in the histidine biosynthetic pathway (Eq. 14).

Escherichia coli IGP synthase has been characterized by Klem and Davisson (1993). This enzyme is a 1:1 complex of HisH and HisF subunits. The HisH subunit provides the glutamine function and HisF an NH_3-dependent IGP synthase. The NH_3-dependent IGP synthase activity is independent of the HisH subunit. The k_{cat} and K_m values for NH_4Cl and N'-(5'-phosphoribulosyl)-formimino-5-aminoimidazole-4-carboxamide ribonucleotide (PRFAR) are vir-

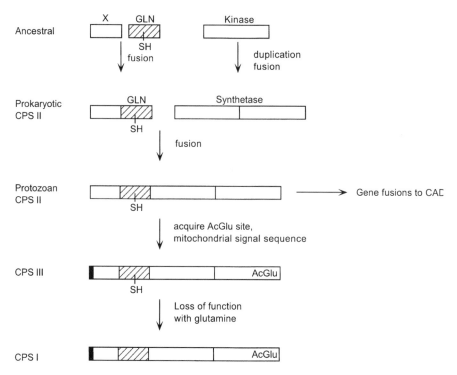

Figure 10.   Hypothetical relationship between different forms of CPS. Bacterial CPS II is assumed to be derived from fusions of genes for a glutaminase (GLN) (hatched region with SH), subunit interaction (X) and a kinase. Fusion of glutaminase and synthetase functions resulted in eukaryotic CPS II and CAD. CPS III acquired an allosteric site for acetylglutamate (AcGlu) and a mitochondrial targeting sequence (black region). Loss of the cysteine nucleophile and other residues required for glutamine-dependent activity led to CPSI.

tually identical for HisF and the complex with HisH. The HisH, on the other hand, does not have detectable glutaminase activity unless combined with HisF. The IGP and N'-[(5'-phosphoribosyl)-formimino]-5-aminoimidazole- 4-carboxamide ribonucleotide (5'-ProFAR), a nucleotide biosynthetic precursor to the normal substrate, stimulated turnover nearly 40-fold. This ligand-mediated signal from the HisF to HisH subunits, typical for the amidotransferases, serves to coordinate the reactions at the glutamine and acceptor sites and may result in a channel for delivery of NH_3 from the glutamine to the acceptor site. Yeast *HIS7* encodes an IGP synthase corresponding to a HisH-HisF fusion (Kuenzler et al., 1993).

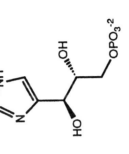

Ribose-5-PO$_4^{-2}$

5'-(5-Aminoimidazole-4-carboxamide) ribonucleotide

glutamine   H$_2$O

glutamate

Imidazole glycerol-3-phosphate   (14)

OPO$_3^{-2}$

N^1-(5'-Phosphoribulosyl)-formimino--5-aminoimidazole-4-carboxamide ribonucleotide

Ribose-5-PO$_4^{-2}$

$^{-2}$O$_3$PO

133

D. NAD SYNTHETASE

The NAD synthetase catalyzes the conversion of nicotinate adenine dinucleotide (NaAD) to NAD (Eq. 15)

$$NaAD + Glutamine + ATP \rightarrow NAD + Glutamate + AMP + PP_i$$

$$(15)$$

This purified yeast enzyme has comparable activity with glutamine or NH_3 as N-donor (Yu and Dietrich, 1972). The main developments since the previous review (Zalkin, 1993) are the cloning of genes encoding NAD synthetase and a structure determination for the *B. subtilis* enzyme. The NAD synthetase genes were cloned from *E. coli* (Allibert et al., 1987), *Rhodobacter capsulatus* (EMBL accession number X59399) and *B. subtilis* (Nessi et al., 1994). Upon purification, enzymes encoded by these cloned genes catalyzed NH_3-dependent synthesis of NAD (Willison and Tissot, 1994; Nessi et al., 1994). Glutamine was not utilized as an N donor although glutamine-dependent activity was detected in extracts of *E. coli*. The expectation is therefore that *E. coli* and other organisms contain an NAD synthetase having subunits for glutamine and acceptor (NH_3-dependent synthase) functions, and that in the case of the bacterial enzymes the glutamine subunit has thus far eluded isolation. We provisionally classify NAD synthetase as a Triad amidotransferase based on the fact that there are presently no examples of Ntn amidotransferases with glutamine and acceptor domains on separate subunits.

According to this reasoning, the yeast genome should contain genes for the glutamine and synthetase subunits. A search of the recently sequenced yeast genome (Goffeau et al., 1996) has identified candidates for these genes. The synthetase subunit (accession number P38795) is a protein chain of molecular mass of 80.7 kDa with 714 amino acids, a portion of which is 22% identical to the 271-residue *B. subtilis* NH_3-dependent enzyme. By using a BLAST search, we identified a single open reading frame (ORF) of unknown function in the yeast genome having a Triad amidotransferase sequence fingerprint (accession U53877). This candidate for the glutamine subunit has 251 amino acids and molecular mass 28.8 kDa. Subunit masses of 65 and 80 kDa were previously identified for yeast NAD synthetase (Yu and Dietrich, 1972), but not one of about 28 kDa. The possible role of this ORF for glutamine-dependent NAD synthetase requires experimental verification. Additional genes for NH_3-dependent NAD synthetase have been identified from *Methanococcus jannaschii* (NCBI gi/1591995), *Synechocystis* sp. (NCBI gi/1653472), *Mycoplasma genitalium* (NCBI gi/1361857), and *Mycoplasma capricolum* (NCBI gi/62920).

1. Structure and Mechanism

The NAD synthetase catalyzes activation of its substrate NaAD by 5′-adenylation of the carboxyl group, from which AMP is displaced by NH_3. The fold of NAD synthetase was predicted to be the same as that of the acceptor domain of GMP synthetase based on this biochemical similarity and similar sequences for the PP_i loops (Tesmer et al., 1996). The crystal structure of *B. subtilis* NAD synthetase (Rizzi et al., 1996) reveals the expected similarity in the structural core, especially in the PP_i loop. Away from the core, the structures differ substantially. More details of substrate binding are available in the NAD synthetase structure than is the case for GMP synthetase. The structure is of the complex with AMP, PP_i, Mg^{2+}, and ATP. The AMP, PP_i, and Mg^{2+} bind at the PP_i loop as in GMP synthetase. A striking difference at the PP_i loop is in the mode of adenine base recognition, which involves quite different residues and functional groups in the two enzymes. Order–disorder transitions are an element of substrate interaction with NAD synthetase, as in GMP synthetase. Two loops that enclose the AMP-PP_i-Mg^{2+} binding site are disordered in the absence of nucleotide. A plausible model of NaAD binding was based on ATP binding at the adenine nucleotide end of the NaAD binding site. While no binding site for NH_3 was identified in the structure of NAD synthetase, the site of a mutation that strongly affects the K_m for NH_3 maps to a position adjacent to the AMP-PP_i-Mg^{2+} binding site. Interestingly, this site, Gly-156, occurs at the position of the peptide in GMP synthetase whose conformation is thought to become ordered upon substrate binding.

E. AMINODEOXYCHORISMATE SYNTHASE AND ANTHRANILATE SYNTHASE

Three proteins are required for the synthesis of *p*-aminobenzoate (PABA) from chorismate (Huang and Gibson, 1970; Ye et al., 1990;

Green and Nichols, 1991). Aminodeoxychorionate synthase, a 1:1 complex of PabA and PabB subunits, catalyzes the glutamine- or NH_3-dependent synthesis of 4-amino 4-deoxychorismate (ADC), and ADC lyase is required for

	Chorismate	4-Amino-4-deoxychorismate	PABA

(16)

elimination of the pyruvyl side chain and aromatization to yield PABA (Eq. 16). The PabB subunit catalyzes NH_3-dependent synthesis of ADC and PabA confers the capacity to use glutamine as N donor. The PabA subunit contains all of the conserved residues that define the Triad glutamine site. Both PabB and PabA subunits of ADC synthase are homologous with anthranilate synthase I (AS I) and II (AS II), respectively, that catalyze the synthesis of anthranilate (o-aminobenzoate) from chorismate and glutamine (Goncharoff and Nichols, 1984; Kaplan et al., 1985). During the past few years, the oligomeric state of ADC synthase has been documented and the importance of subunit communication for catalysis demonstrated. The results support the general theme developed from structural analyses of other amidotransferases. Namely, interaction with PabB is required to form a functional PabA glutamine site.

Three lines of evidence indicate that the glutamine site in PabA is not functional prior to interaction with PabB (Roux and Walsh, 1992). First, using standard assay conditions, PabA had no detectable glutaminase activity in the absence of PabB. Titration of PabA with PabB yielded a 1:1 complex of the two subunits having a glutaminase k_{cat} of 17 min^{-1}, an estimated 7000-fold activation by PabB. Binding of chorismate to PabB further increased the glutaminase activity by twofold. Second, a PabA γ-glutamyl acylenzyme intermediate was formed from [^{14}C]glutamine and isolated using an acid quench technique. The intermediate was formed only when PabA was complexed with PabB. Although not shown directly, the acylenzyme intermediate is most likely a Cys-79 thioester. The Cys-79 residue is the conserved nucleophile. Finally, the rate of inactivation of PabA by affinity labeling of Cys-79 with DON was increased at least 100-fold by the presence of PabB. Competition by glutamine decreased the rate of inactivation by DON and supports the interpretation that DON binds to the glutamine site.

These results parallel those obtained earlier for anthranilate synthase (Goto et al., 1976). The glutaminase activity of isolated AS II was stimulated 30-fold by interaction with AS I plus chorismate, although not by AS I alone. Affinity labeling of AS II by DON was likewise dependent on formation of the complex with AS I and binding of chorismate, supporting the earlier work of Queener et al. (1973). The results indicated that formation of the functional AS II glutamine site required not only AS I but also binding of chorismate. It was not determined directly whether binding of chorismate to AS I was required for a functional subunit interaction with AS II or rather for site–site communication. Acceptor site–glutamine site communication is favored because chorismate is known to be required for activation of the AS II glutamine site in oligomeric anthranilate synthase enzymes having avidly associated subunits (Zalkin, 1993).

Although a number of experiments have provided evidence for a 1:1 complex of PabA and PabB (Roux and Walsh, 1992; Viswanathan et al., 1995), direct isolation by gel filtration was achieved only recently (Rayl et al., 1996). Subunit interaction is increased by glutamine and by incubation at 37°C. Incubation at low temperatures reduced complex formation. Thus, PabA and PabB interact to form a weak complex in solution, which generates a functional PabA glutamine site. Binding of glutamine to the nascent site strengthens the subunit interactions. Glutamine had been shown earlier to be required for interaction of the *Bacillus* and *Pseudomonas* AS I and AS II subunits (Holmes and Kane, 1975; Queener et al., 1973).

F. PERSPECTIVE

The recent x-ray structure determination of *E. coli* CPS (Thoden et al., 1997) provides direct evidence for NH_3-mediated N transfer via a channel that connects the glutamine site in the small CarA α subunit to the first phosphorylation site in the large CarB β subunit, where synthesis of carboxyphosphate, and presumably carbamate, takes place. Remarkably, this channel continues on for delivery of carbamate to the second phosphorylation site and synthesis of carbamylphosphate. In the active $(\alpha\beta)_4$ enzyme there is an extensive interface of the α subunit with the first half of the β subunit, in keeping with the requirement for N transfer between subunits. The α subunit is bilobal, containing an NH_2-terminal interaction domain and a COOH-terminal glutamine domain. Essential features of the glutamine domain and its active site are identical in CPS and GMP synthetase. The glutamine domains have the same topology, and superimpose with a root-mean-square deviation (rmsd) of 1.7 Å between 129 structurally equivalent α-carbon atoms. The Cys-His-Glu triad characteristic of Triad amidotransferases is formed similarly in the two enzymes. The specificity pocket for glutamine appears to be unformed in CPS, as in GMP synthetase. Conformational changes associated with catalysis are proposed for both enzymes in order to sequester NH_3 following glutamine hydrolysis. The β-subunit of CPS is folded into homologous halves, corresponding to residues 1–553 and 554–1073, which include the phosphorylation sites for HCO_3^- and carbamate, respectively. The two β-subunit halves superimpose with an rmsd of 1.1 Å for 255 equivalent α-carbon atoms and are related by pseudo-twofold rotation axis. Each half β-subunit consists of four well defined domains. In the reported structure, the carboxyphosphate site contained ADP, P_i, and two Mn ions bound in a pocket between the B- and C-domains, and the carbamyl phosphate site contained ADP and a pair of metal ions bound to similar structural motifs in the second half of the β-subunit. The channel connecting the glutamine and carboxyphosphate active sites in CPS is 35 Å long and is lined by a number of polar side chains. In contrast, the 20-Å NH_3 channel of glutamine PRPP amidotransferase is lined by hydrophobic amino acids, which are thought to exclude water from the channel as it forms during the catalytic cycle (Krahn et al., 1997). The channel connecting the carboxyphosphate and carbamyl phosphate active sites of CPS crosses the pseudo-twofold axis and is lined primarily with backbone atoms and a few polar and hydrophobic side chains. The channel through the interior of CPS must sequester and stabilize NH_3, carbamate, and carboxyphosphate. The CPS structure thus provides convincing support for NH_3-mediated N transfer and provides the first physical explanation for the coupling of the four CPS partial reactions given in Section IV.B.1.

Given the finding of NH_3 channels in one Ntn and one Triad amidotransferase it appears likely that the other homologous amidotransferases in these families have similar mechanisms for N transfer. There are, however, subtle distinctions between the Ntn and Triad amidotransferases that suggest important differences in N transfer. First, some Triad amidotransferases have interacting glutamine and acceptor subunits in contrast to the invariant fused domains of Ntn enzymes. Second, some Triad acceptor subunits have full catalytic activity with NH_3 in the absence of the glutamine subunit. NH_3-dependent activity thus far has not been detected for an isolated Ntn amidotransferase acceptor domain. This suggests that the channel connecting the two active sites is needed for NH_3-dependent activity in the Ntn enzymes but perhaps not in the Triad amidotransferases.

Curnow et al. (1997) recently reported the cloning of three *B. subtilis* genes required for glutamyl-tRNA^{Gln} amidotransferase, providing the first connection to primary sequence for the enzyme. Properties of this glutamine amidotransferase have been summarized (Zalkin, 1993). Despite several functional similarities to other amidotransferases there is no obvious sequence relationship to Triad or Ntn amidotransferases. On the other hand, there is significant amino acid sequence similarity between one subunit of the *B. subtilis* enzyme and a family of amidases (Kobayashi et al., 1997), particularly in a signature region of nearly 40 amino acids. This relationship between glutamyl-tRNA^{Gln} amidotransferase and a family of amidases defines a third family of amidotransferases, in which an amidase activity is used to hydrolyze glutamine to glutamate plus NH_3. NH_3 is then used to amidate Glu-tRNA^{Gln} to Gln-tRNA^{Gln}. There is no evidence to indicate the amidotransferase family to which NAD synthetase may belong.

Acknowledgments

We thank the following colleagues for providing preprints prior to publication and unpublished information: C. J. Lusty, I. Rayment, A. Teplyakov, and M. A. Vanoni. JLS thanks co-workers J. M.

Krahn, C. R. A. Muchmore, J. J. G. Tesmer, and D. R. Tomchick for many helpful discussions during the development of ideas about structure and function of amidotransferases. Work from our laboratories was supported by United States Public Health Service Grants DK42303 (to JLS) and GM24658 (to HZ).

References

Allibert, P., Willison, J. C., and Vignais, P. M., *J. Bacteriol.*, **169**, 260–271 (1987).

Amuro, N., Paluh, J. L., and Zalkin, H., *J. Biol. Chem.*, **260**, 14844–14849 (1985).

Boehlein, S. K., Richards, N. G. J., and Schuster, S. M., *J. Biol. Chem.*, **269**, 7450–7457 (1994a).

Boehlein, S. K., Richards, N. G. J., Walworth, E. S., and Schuster, S. M., *J. Biol. Chem.*, **269**, 26789–26795 (1994b).

Boehlein, S. K., Schuster, S. M., and Richards, N. G. J., *Biochemistry*, **35**, 3031–3037 (1996).

Brannigan, J. A., Dodson, G., Duggleby, H. J., Moody, P. C. E., Smith, J. L., Tomchick, D. R., and Murzin, A. G., *Nature (London)*, **378**, 416–419 (1995).

Buchanan, J. M., *Adv. Enzymol. Related Areas Mol. Biol.*, **39**, 91–183 (1973).

Cedar, H., and Schwarz, J. H., *J. Biol. Chem.*, **244**, 4112–4121 (1969).

Chen, S., Tomchick, D. R., Wolle, D., Hu, P., Smith, J. L., Switzer, R. L., and Zalkin, H., *Biochemistry*, **36**, 10718–10726 (1997).

Curnow, A. W., Hong, K-W., Yuan, R., Kim, S-I., Martins, O., Winkler, W., Henkin, T. M., and Söll, D., *Proc. Nat. Acad. Sci. USA*, **94**, 11819–11826 (1997).

Denisot, M. A., Le Goffic, F., Badet, B., *Arch. Biochem. Biophys.*, **288**, 225–230 (1991).

Duggleby, H. J., Tolley, S. P., Hill, C. P., Dodson, E. J., Dodson, G., and Moddy, C. E., *Nature (London)*, **373**, 264–268 (1995).

Eads, J. C., Scapin, G., Xu, Y., Grubmeyer, C., and Sacchettini, J. C., *Cell*, **78**, 325–334 (1994).

Fernández-Herrero, L. A., Badet-Denisot, M. A., Badet, B., and Berenguer, J., *Mol. Microbiol.*, **17**, 1–12 (1995).

Fukuyama, T. T., *J. Biol. Chem.*, **241**, 4745–4749 (1966).

Galván, F., Márquez, A. J., and Vega, J. M., *Planta*, **162**, 180–187 (1984).

Gergerson, R. G., Miller, S. S., Twary, S. N., Gantt, J. S., and Vance, C. P., *Plant Cell*, **5**, 215–226 (1993).

Goffeau, A., Barrell, B. G., Bussey, H., Davis, R. W., Dujon, B., Feldmann, H., Galibert, F., Hoheisel, J. D., Jacq, L., Johnston, M., Louis, E. J., Mewes, H. W., Murakami, Y., Phillipsen, P., Tettelin, H., and Oliver, S. G., *Science*, **274**, 546–567 (1996).

Goncharoff, P. and Nichols, B. P., *J. Bacteriol.*, **159**, 57–62 (1984).

Goto, Y., Zalkin, H., Keim, P., and Heinrickson, R. L., *J. Biol. Chem.*, **251**, 941–949 (1976).

Green, J. M. and Nichols, B. P., *J. Biol. Chem.*, **266**, 12971–12975 (1991).

Guillou, F., Liao, M., Garcia-Espana, A., and Lusty, C. J., *Biochemistry*, **31**, 1656–1664 (1992).

Guillou, F., Rubino, S. D., Markovitz, R. S., Kinney, D. M., and Lusty, C. J., *Proc. Nat. Acad. Sci. USA*, **86**, 8304–8308 (1989).

Guy, H. I., and Evans, D. R., *J. Biol. Chem.*, **270**, 2190–2197 (1995).

Guy, H. I. and Evans, D. R., *J. Biol. Chem.*, **271**, 13762–13769 (1996).

Hartman, S. C., *J. Biol. Chem.*, **238**, 3024–3035 (1963).

Helbing, C. C. and Atkinson, B. G., *J. Biol. Chem.*, **269**, 11743–11750 (1994).

Holmes, W. H. and Kane, J. F., *J. Biol. Chem.*, **250**, 4462–4469 (1975).

Hong, J., Salo, W. L., Lusty, C. J., and Anderson, P. M., *J. Mol. Biol.*, **243**, 131–140 (1994).

Hove-Jensen, B., Harlow, K. W., King, C. J., and Switzer, R. L., *J. Biol. Chem.*, **261**, 6765–6771 (1986).

Huang, M. and Gibson, F., *J. Bacteriol.*, **102**, 767–773 (1970).

Humbert, R. and Simoni, R. D., *J. Bacteriol.* **142**, 212–220 (1980).

Isupov, M. N., Obmolova, G., Butterworth, S., Badet-Denisot, M. A., Badet, B., Polikarpov, I., Littlechild, J. A., and Teplyakov, A., *Structure*, **4**, 801–810 (1996).

Kaplan, J. B., Merkel, W. K., and Nichols, B. P., *J. Mol. Biol.*, **183**, 327–340 (1985).

Kim, J. H., Krahn, J. M., Tomchick, D. R., Smith, J. L., and Zalkin, H., *J. Biol. Chem.*, **271**, 15549–15557 (1996).

Klem, J. J. and Davisson, V. J., *Biochemistry*, **32**, 5177–5186 (1993).

Knaff, D. B., Hirasawa, M., Ameyibor, E., Fu, W., and Johnson, M. K., *J. Biol. Chem.*, **266**, 15080–15084 (1991).

Kobayashi, M., Fujiwara, Y., Goda, M., Komeda, H., and Shimizu, S., *Proc. Nat. Acad. Sci. USA*, **94**, 11986–11991 (1997).

Krahn, J. M., Kim, J. H., Burns, M. R., Parry, R. J., Zalkin, H., and Smith, J. L., *Biochemistry*, **36**, 11061–11068 (1997).

Kuenzler, M., Balmelli, T., Egli, C. M., Paravicini, G., and Braus, G. H., *J. Bacteriol.*, **175**, 5548–5558 (1993).

Leff, R. L., Itakura, M., Udom, A., and Holmes, E. W., *Adv. Enz. Regul.*, **22**, 403–411 (1984).

Li, H.-C. and Buchanan, J. M., *J. Biol. Chem.*, **246**, 4720–4726 (1971).

Lou, L., Nakamura, J., Tsing, S., Nguyen, B., Chow, J., Straub, K., Chan, H., and Barnet, J., *Protein Express Purif*, **6**, 487–495 (1995).

Löwe, J., Stock, D., Jap, B., Zwickl, P., Baumeister, W., and Huber, R., *Science*, **268**, 533–539 (1995).

Lusty, C. J. and Liao, M., *Biochemistry*, **32**, 1278–1284 (1993).

Lusty, C. J., Widgren, E. E., Broglie, K. E., and Nyunoya, H., *J. Biol. Chem.*, **258**, 14466–14472 (1983).

Mäntsälä, P. and Zalkin, H., *J. Biol. Chem.*, **259**, 14230–14236 (1984).

Mareya, S. M. and Raushel, F. M., *Biochemistry*, **33**, 2945–2950 (1994).

Marina, A., Bravo, J., Fita, I., and Rubio, V., *Proteins: Structure, Function, Genet.*, **22**, 193–196 (1995).

Marqués, S., Florencio, F. J., and Candau, P., *Eur. J. Biochem.*, **206**, 69–77 (1992).

Marshall, S., Garvey, W. T., and Traxinger, R. R., *FASEB J*, **5**, 3031–3036 (1991).

Mattevi, A., Rizzi, M., and Bolognesi, M., *Curr. Op. Structure Biol.*, **6**, 824–829 (1996).

Mei, B. and Zalkin, H., *J. Biol. Chem.*, **264**, 16613–16619 (1989).

Messenger, L. J. and Zalkin, H., *J. Biol. Chem.*, **254**, 3382–3392 (1979).

Miles, B. W., Mareya, S. M., Post, L. E., Post, D. J., Chang, S. H., and Raushel, F. M., *Biochemistry*, **32**, 232–240 (1993).

Miller, R. E. and Stadtman, E. R., *J. Biol. Chem.*, **247**, 7407–7419 (1972).

Miran, S. G., Chang, S. H., and Raushel, F. M., *Biochemistry*, **30**, 7901–7907 (1991).

Muchmore, C. R. A., Krahn, J. M., Kim, J. H., Zalkin, H., and Smith, J. L., *Protein Science* **7**, 39–51 (1998).

Nagano, H., Zalkin, H. and Henderson, E. J., *J. Biol. Chem.*, **245**, 3810–3820 (1970).

Nakamura, J. and Lou, L., *J. Biol. Chem.*, **270**, 7347–7353 (1995).

Nakamura, J., Straub, K., Wu, J., and Lou, L., *J. Biol. Chem.*, **270**, 23450–23455 (1995).

Nakamura, M., Yamada, M., Hirota, Y., Sugimoto, K., Oka, A., Takanami, X., *Nucleic Acids Res.*, **9**, 4669–4676 (1981).

Nessi, C., Albertini, A. M., Speranza, M. L., and Galizza, A., *J. Biol. Chem.*, **270**, 6181–6185 (1994).

Nyunoya, H., and Lusty, C. J., *Proc. Nat. Acad. Sci., USA*, **80;** 4629–4633 (1983).

Obmolova, G., Badet-Denisot, M. A., Badet, B., and Teplyakov, A., *J. Mol. Biol.*, **242**, 703–705 (1994).

Oinonen, C., Tikkanen, R., Rouvinen, J., and Peltonen, L., *Nature Structural Biol.*, **2**, 1102–1108 (1995).

Ollis, D. L., Cheah, E., Cygler, M., Dijkstra, B., Frolow, F., Franken, S. M., Harel, M., Remington, S. J., Silman, I., Schrag, J., Sussman, J. L., Verschueren, K. H. G., and Goldman, A., *Protein Eng.*, **5**, 197–211 (1992).

Patel, N., Moyed, H. S., and Kane, J. F., *Arch. Biochem. Biophys.*, **178**, 652–661 (1977).

Pelanda, R., Vanoni, M. A., Perego, M., Piubelli, L., Galizzi, A., Curti, B., and Zanetti, G., *J. Biol. Chem.*, **268**, 3099–3106 (1993).

Perutz, M. F., *Q. Rev. Biophys.*, **22**, 139–236 (1989).

Post, L. E., Post, D. J., and Raushel, F. M., *J. Biol. Chem.*, **265**, 7742–7747 (1990).

Queener, S. W., Queener, S. F., Meeks, J. R., and Gunsalus, I. C., *J. Biol. Chem.*, **248**, 151–161 (1973).

Ratti, S., Curti, B., Zanetti, G., and Galli, E., *J. Bacteriol*, **163**, 724–729 (1985).

Rayl, E. A., Green, J. M., and Nichols, B. P., *Biochim. Biophys. Acta.*, **1295**, 81–88 (1996).

Richards, N. G. J., and Schuster, S. M., *FEBS Lett.*, **313**, 98–102 (1992).

Rizzi, M., Nessi, C., Mattevi, A., Coda, A., Bolognesi, M., and Galizzi, A., *EMBO J*, **15**, 5125–5134 (1996).

Roux, B., and Walsh, C. T., *Biochemistry*, **31**, 6904–6910 (1992).

Rowe, P. B., and Wyngaarden, J. B., *J. Biol. Chem.*, **243**, 6373–6383 (1968).

Rubino, S. D., Nyunoya, H., and Lusty, C. J., *J. Biol. Chem.*, **261**, 11320–11327 (1986).

Sakakibara, H., Watanabe, M., Hase, T., and Sugiyama, T., *J. Biol. Chem.*, **266**, 2028–2035 (1991).

Sayeski, P. P. and Kudlow, J. E., *J. Biol. Chem.*, **271**, 15237–15243 (1996).

Scapin, G., Grubmeyer, C., and Sacchettini, J. C., *Biochemistry*, **33**, 1287–1294 (1994).

Scapin, G., Ozturk, D. H., Grubmeyer, C., and Sacchettini, J. C., *Biochemistry*, **34**, 10744–10754 (1995).

Schumacher, M. A., Carter, D., Roos, D. S., Ullman, B., and Brennan, R. G., *Nature Structural Biol.*, **3**, 881–887 (1996).

Scofield, M. A., Lewis, W. S. and Schuster, S. M., *J. Biol. Chem.*, **265**, 12895–12902 (1990).

Sheng, S., Moraga-Amagor, D. A., van Heeke, G., Allison, R. D., Richards, N. G. J., and Schuster, S. M., *J. Biol. Chem.*, **268**, 16771–16780 (1993).

Sheng, S., Morago, D. A., van Heeke, G., and Schuster, S. M., *Protein Expression Purification*, **3**, 337–346 (1992).

Smith, J. L., *Biochem. Soc. Trans.*, **23**, 894–898 (1995).

Smith, J. L., Zaluzec, E. J., Wery, J. P., Niu, L., Switzer, R. L., Zalkin, H., and Satow, Y., *Science*, **264**, 1427–1433 (1994).

Stoker, P. W., O'Leary, M. H., Boehlein, S. K., Shuster, S. M., and Richards, N. G. J., *Biochemistry*, **35**, 3024–3030 (1996).

Switzer, R. L., *BioFactors*, **2**, 77–86 (1989).

Tesmer, J. J. G., Klem, T. J., Deras, M. L., and Davisson, V. J., *Nature Structural Biol.*, **3**, 74–86 (1996).

Thoden, J. B., Raushel, F. M., Mareya, S., Tomchick, D., and Rayment, I., *Acta Crystallogi.*, **D51**, 827–829 (1995).

Thoden, J. B., Holden, H. M., Wesenberg, G., Raushel, F. M., and Rayment, I., *Biochemistry*, **36**, 6305–6316.

Tomchick, D. R., Smith, J. L., Wolle, D., and Zalkin, H., *Specificity of inhibition: The* B. subtilis *glutamine PRPP amidotransferase example*, ACA Annual Meeting, Series 2, Vol. 23, (1995), p. 170.

Vanoni, M. A., Edmondson, D. E., Zanetti, G., and Curti, B., *Biochemistry*, **31**, 4613–4623 (1992).

Vanoni, M. A., Nuzzi, L., Rescigno, M., and Zanetti, G., *Eur. J. Biochem.*, **202**, 181–189 (1991).

Vanoni, M. A., Verzotti, E., Fischer, F., Coppola, M., Ferretti, S., Zanetti, G., and Curti, B., In *Flavins and Flavoproteins 1996* Stevenson, K., Williams, C. H., and Massey, V., Eds., City: Publisher, (1997). pp. 879–888.

Vanoni, M. A., Verzotti, E., Zanetti, G., and Curti, B., *Eur. J. Biochem.*, **263**, 937–946 (1996).

Velázquez, L., Camarena, L., Reyes, J. L., and Bastarrachea, F., *J. Bacteriol.*, **173**, 3261–3264 (1991).

Viswanathan, V. K., Green, J. M., and Nichols, B. P., *J. Bacteriol.*, **177**, 5918–5923 (1995).

Vollmer, S. J., Switzer, R. L., and Debrunner, P. G., *J. Biol. Chem.*, **258**, 14284–14293 (1983).

Willison, J. L. and Tissot, G., *J. Bacteriol.*, **176**, 3400–3402 (1994).

Ye, Q.-Z., Liu, J., and Walsh, C. T., *Proc. Natl. Acad. Sci., USA*, **87**, 9391–9395 (1990).

Yu, C. K. and Dietrich, L. S., *J. Biol. Chem.*, **247**, 4794–4802 (1972).

Zalkin, H., *Adv. Enzymol. Relat. Areas Mol. Biol.*, **66**, 203–309 (1993).

Zalkin, H., and Truitt, C. D., *J. Biol. Chem.*, **252**, 5431–5436 (1977).

Zhou, G., Smith, J. L., and Zalkin, H., *J. Biol. Chem.*, **269**, 6784–6789 (1994).

Zyk, N., Citri, N., and Moyed, H. S., *Biochemistry*, **8**, 2787–2794 (1969).

MECHANISTIC ISSUES IN ASPARAGINE SYNTHETASE CATALYSIS*

By NIGEL G. J. RICHARDS, *Department of Chemistry, University of Florida, Gainesville, FL 32611* and SHELDON M. SCHUSTER, *Department of Biochemistry and Molecular Biology, University of Florida, Gainesville, FL 32610, and Interdisciplinary Center for Biotechnology Research, University of Florida, Gainesville, FL 32611*

CONTENTS

Advances in Enzymology and Related Areas of Molecular Biology, Volume 72: Amino Acid Metabolism, Part A, Edited by Daniel L. Purich
ISBN 0-471-24643-3 ©1998 John Wiley & Sons, Inc.

* This paper is dedicated to the memory of Alton Meister (1922–1995), a pioneer in the study of asparagine synthetase and other glutamine-dependent amidotransferases.

I. Introduction

Asparagine is biosynthesized in a deceptively simple fashion from aspartic acid by the enzyme asparagine synthetase (AS), using an adenosine triphosphate (ATP)-dependent reaction and either glutamine or ammonia as a source of nitrogen (Buchanan, 1973; Zalkin, 1993). As formation of the side-chain amide is thermodynamically unfavorable, ATP breakdown not only activates the carboxylate but also provides energy for the overall transformation. Despite the biological importance of asparagine in plant metabolism (Lam et al., 1996), and the role of this amino acid as a glycosylation site in eukaryotic glycoproteins (Kornfeld and Kornfeld, 1985), relatively few studies have focused upon the structure and mechanism of asparagine synthetases from plants and animals. In part, this has been due to difficulties in obtaining highly purified enzyme from various sources, and recent studies have only been possible with the advent of suitable expression systems for the production of recombinant AS (Van Heeke and Schuster, 1990; Sheng et al., 1993). Interest in asparagine synthetases is rooted in two separate, but related, metabolic issues. First, asparagine has a relatively high N/C ratio and it is therefore likely that the cellular metabolism of this amino acid is linked to nitrogen homeostasis and protein biosynthesis in all organisms. Second, removal of this amino acid by the enzyme L-asparaginase is widely used in chemotherapeutic protocols for treating acute lymphoblastic leukemia (ALL) in children (Ertel et al., 1979), although the underlying mechanism of this therapy remains to be fully elucidated.

To date, two families of AS enzymes have been described in prokaryotes and eukaryotes that do not appear to possess any sequence similarity. Members of the first enzyme family, which have been isolated solely from prokaryotes, employ ammonia as the sole nitrogen source (Scheme 1) and appear to be evolutionarily related to

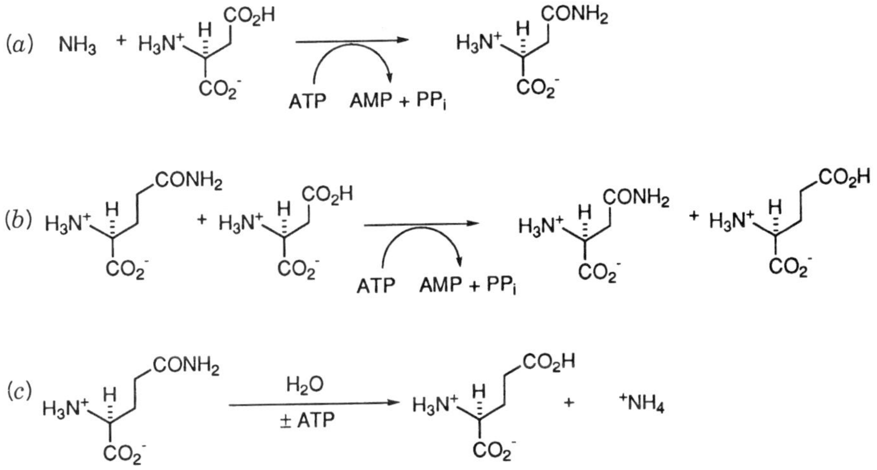

Scheme 1. Reactions catalyzed by glutamine-dependent asparagine synthetases. (a) Ammonia-dependent synthesis, catalyzed by both ammonia- and glutamine-dependent AS; (b) Glutamine-dependent synthesis; (c) Glutaminase activity observed in the absence of aspartic acid. Hydrolysis is stimulated by ATP although no AS catalyzed breakdown of the nucleotide itself is observed.

amino-acyl tRNA synthetases (Hinchman et al., 1992). The asparagine synthetase encoded by the *asnA* gene of *Escherichia coli* is a representative member (AS-A) of this group of enzymes (Hinchman and Schuster, 1992). For the second class of asparagine synthetases, glutamine is the preferred nitrogen source, although these enzymes can also employ ammonia as an alternate substrate (Scheme 1). Glutamine-dependent asparagine synthetases have been cloned and/or isolated from yeast (Ramos and Waime, 1979), plants (Tsai and Coruzzi, 1990; Lam et al., 1994), bacteria (Burchall et al., 1964; Ravel et al., 1962; Reitzer and Magasanik, 1982), and mammalian sources (Hongo and Sato, 1983; Huang and Knox, 1975; Luehr and Schuster, 1985; Andrulis et al., 1987). The human glutamine-dependent AS is encoded by a single gene located in region q21.3 on chromosome 7 (Heng et al., 1994). Sequence analyses have indicated that these enzymes are all members of the Class II (formerly *purF*) glutamine amidotransferase (GAT) superfamily (Smith, 1995; Zalkin, 1993), which also includes glutamine 5′-phosphoribosyl-1-pyrophosphate amidotransferase (GPA) (Tso et al., 1982), glutamine fructose-6-phosphate amidotransferase (GFAT) (Badet-Denisot et al., 1993) and glutamate synthase (Vanoni et al., 1991a). In contrast to bacteria, eukaryotic organisms appear to lack an ammonia-dependent AS, presumably because of the need to maintain cellular concentrations of ammonia at very low levels.

Whereas the kinetic and chemical mechanisms employed by ammonia-dependent asparagine synthetases, such as *E. coli* AS-A, appear to be well understood (Cedar and Schwartz, 1969a,b), there are several key issues with respect to the structure and mechanistic enzymology of the glutamine-dependent enzymes that remain to be defined. These can be roughly divided into four main categories. First, what is the structural organization of the active site(s) involved in glutamine utilization and aspartate activation, and the corresponding molecular basis for coordination of the two half-reactions that occur during asparagine synthesis? Second, how are glutamine-dependent asparagine synthetases related to other amidotransferases, proteases, and synthetases in terms of their evolution, mechanism, and active site design? The third category includes questions concerning the mechanism by which ATP is used to activate aspartate, and the roles of specific residues in stabilizing key intermediates and transition state structures. Finally, what is the detailed molecular mechanism by which nitrogen is transferred from the side-chain amide of glutamine to the activated aspartic acid, and what is the role of the kinetic mechanism in exerting control over reactions leading to the futile hydrolysis of ATP and/or glutamine? We note that answers to the mechanistic issues raised in this fourth category will probably have more general implications for work on other glutamine-dependent amidotransferases. In this chapter, we will outline recent progress on several of these mechanistic problems, emphasizing recent results from experiments involving asparagine synthetase B (AS-B), the enzyme encoded by the *asnB* gene of *E. coli* (Scofield et al., 1990).

II. Background

A. GENETIC CONTROL OF ASPARAGINE SYNTHETASE

Asparagine synthetase is a key enzyme in plant, as well as in mammalian, metabolism, and its genetic regulation has therefore been the focus of much study (Gong and Basilico, 1990; Guerrini et al., 1993). In plants, asparagine is thought to play a major role in nitrogen transport, and it appears that peas (Tsai and Coruzzi, 1990), maize (Chevalier et al., 1996), and lotus (Waterhouse et al., 1996), have two, differentially expressed genes encoding glutamine-dependent asparagine synthetases. In contrast, arabidopsis only appears to possess a single gene coding for AS (Lam et al., 1994). Research in this area is, however, complicated by the fact that plants can also synthesize asparagine by enzyme-catalyzed hydration of β-cyanoalanine (Castric et al., 1972). Yeasts also possess two genes for glutamine-dependent asparagine synthetases, but the metabolic importance of this observation, or their cellular roles, remains to be determined (Ramos and Waime, 1979). In mammalian cells, it has been observed that a temperature sensitive AS mutant blocks progression through the G1 phase of the cell cycle (Gong and Basilico, 1990), and further work has shown that AS expression is controlled at both the transcriptional and translational levels (Gong et al., 1991; Guerrini et al., 1993; Chakrabati et al., 1993; Hutson et al., 1997).

B. METABOLIC IMPORTANCE OF ASPARAGINE BIOSYNTHESIS IN PLANTS AND ANIMALS

There appear to be two major metabolic fates for asparagine. The first is the use of this amino acid in protein biosynthesis, particularly as an attachment point for complex carbohydrate structures (Kaplan et al., 1987). The other, perhaps quantitatively more major, function involves conversion into citric acid cycle intermediates, and hence glucose, with concomitant use of nitrogen in the synthesis of other amino acids (Scheme 2). While hydrolysis of asparagine by cellular asparaginase(s) likely takes place, the flux through this pathway is probably less than through other catabolic routes. For example, in rat liver, the predominant conversion of asparagine carbon into oxaloacetate is via an initial transamination to give α-ketosuccinamide followed by hydrolysis (Scheme 2) (Maul and Schuster, 1986a,b; Schuster, 1982), although the detailed regulation of these reactions remains poorly understood. Indeed, this pathway may be the only means of asparagine breakdown that is utilized by mitochondria (Schuster, 1982). There appear to be two different aminotransferases in rat liver cells, mitochondrial and cytosolic, that mediate the synthesis of α-ketosuccinamide from asparagine (Maul and Schuster, 1986a,b) and that appear to be differentially regulated by calcium, although it is still not known how this control is exerted. One possi-

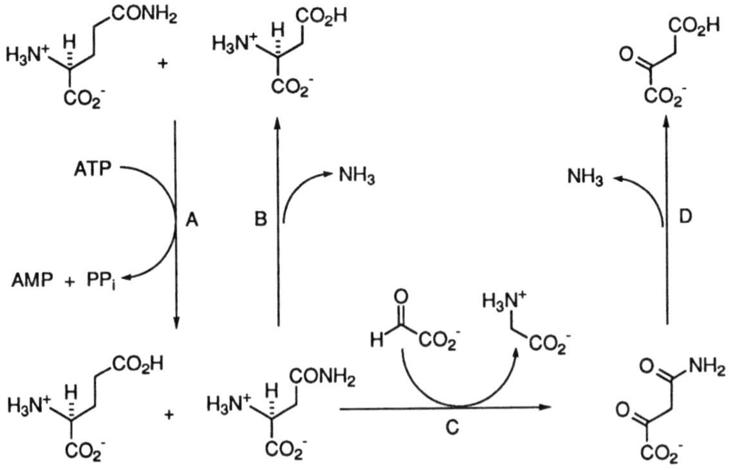

Scheme 2. Metabolic pathways for asparagine formation and degradation. Enzymes: A = asparagine synthetase; B = L-asparaginase; C = asparagine-glyoxylate aminotransferase; D = ω-amidase.

bility is that the controlling factor in flux regulation is the availability of glyoxylate or pyruvate in either the mitochondrial or cytosolic compartments. This pathway, however, is the main means of converting asparagine to glucose, rather than via direct hydrolysis to aspartic acid. As the nitrogen acceptor in the transmination of asparagine to α-ketosuccinamide is glyoxylate, yielding glycine, one possible mechanism by which asparagine depletion might cause cellular damage is through increases in the levels of glyoxylate.

Such a depletion in asparagine can occur when cells are treated with L-asparaginase, often with toxic consequences due to disruption of the complex interactions between amino acids involved in mediating asparagine homeostasis. L-Asparaginase treatment decreases asparagine levels in susceptible, and resistant, tumor cells as well as those in the liver, spleen, thymus, and kidney of treated animals (Broome, 1968). It has also been observed that this gives rise to increased amounts of aspartate, glutamate, and glutamine in L-asparaginase sensitive tumor cells after treatment. These changes do not, however, appear to be the basis for the toxicity of L-asparaginase to susceptible tumors. One observation that might explain the effectiveness of L-asparaginase as a therapeutic agent in childhood ALL is that glycine concentrations are decreased in tumor cells that are sensitive to L-asparaginase treatment relative to those in resistant tumors (Ryan and Dworak, 1970; Ryan and Sorenson, 1970). Glycine depletion might have severe consequences for purine biosynthesis (Ryan and Sorenson, 1970) and its importance has been shown by animal model experiments in which intraperitoneal injection of glycine, or asparagine, overcomes the antitumor effect of L-asparaginase. An alternative view is that decreased glycine levels reflect a reduction in the transaminase-catalyzed conversion of glyoxylate, which has been observed in sensitive cells after L-asparaginase treatment. Increased glyoxylate could then cause toxic effects by indiscriminate reaction with a variety of proteins to give stable, covalent adducts (Keefer et al., 1985a,b). This hypothesis remains untested.

Clinical observations also suggest an interesting potential relationship between C1 and C2 metabolism that involves asparagine. In the treatment of ALL, the combination of methotrexate, an inhibitor of dihydrofolate reductase (DHFR), and L-asparaginase can be either synergistic or antagonistic, depending on administration and drug dosage (Sur et al., 1987; Vadlamudi et al., 1973). During DNA synthesis, tetrahydrofolate is involved in deoxythymidine monophosphate (dTMP) synthesis, the C1 unit being obtained by conversion of serine to glycine. Methotrexate treatment might therefore result in treated cells having an increased need for asparagine-mediated glycine production from glyoxylate. Indeed, increased asparagine synthetase activity is observed in the liver of rats that have been treated with methotrexate (Maul and Schuster, 1982), although detailed experiments that directly show a causal relationship have yet to be reported.

C. CHEMOTHERAPEUTIC IMPORTANCE OF AS INHIBITORS

Interest in the clinical importance of asparagine synthetases was initially stimulated by the observation that certain types of leukemias can be treated by lowering the levels of asparagine in circulating blood. For example, L-asparaginase is routinely employed as a single agent in chemotherapy, and when used with prednisone, vincristine, and doxorubicin, the remission induction rate in childhood ALL is improved to nearly 95% (Sanz et al., 1986). On the other hand, this therapeutic approach is far less beneficial in adult ALL due to the occurrence of tumors that are resistant to L-asparaginase therapy. Given that L-asparaginase treatment is probably effective as a result of decreased levels of circulating asparagine, killing tumor cells that cannot obtain sufficient amounts of this amino acid, it is possible that AS inhibitors might prove useful as antileukemic agents. Such compounds would be especially valuable for cells displaying resistance to L-asparaginase treatment, which probably express endogenous AS. We also note that T-cell based immunosuppression is often observed in patients undergoing initial treatment with L-asparaginase (Hersh, 1971), raising the possible utility of AS inhibitors as immunosuppressive agents. The observation that L-asparaginase may be effective in treating a variety of solid tumors also provides additional stimulus to developing AS inhibitors. Such compounds might act by reducing the potential of hepatic tissue as an asparagine reservoir, enhancing the sensitivity of tumors in other organs to L-asparaginase chemotherapy (Uren et al., 1977).

Several hundred analogues of glutamine, ATP, and aspartic acid have been screened as AS inhibitors in a variety of assay systems (Mokotoff et al., 1975; Jayaram et al., 1975, 1976; Handschumacher et al., 1968; Cooney et al., 1976; Cooney et al., 1980). To date, only a few compounds have been identified that have any inhibitory activity in either *in vivo* or *in vitro* assays, and all of these are either nonspecific, or are inhibitors in the millimolar range. For example, albizziin, **1** (Fig. 1), has been shown to inhibit AS *in vitro*, albeit with a K_I of 1–5 mM, causing amplification of the AS gene (Andrulis et al., 1983, 1990). This compound, however, has little potential as an antitumor agent as it inhibits many other enzymes for which glutamine is a substrate. Another AS inhibitor, β-aspartyl methylamide,

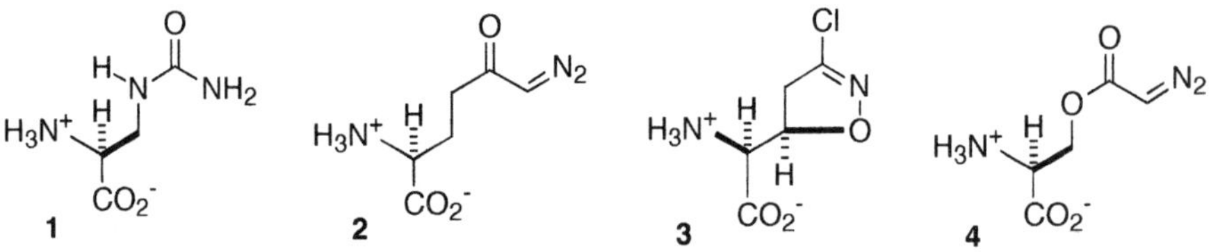

Figure 1. Asparagine synthetase inhibitors. These glutamine analogues irreversibly inhibit the enzyme, presumably by formation of a covalent attachment to the thiolate of Cys-1, but do not exhibit selectivity for AS.

did increase the lifespan of mice carrying L-asparaginase resistant tumors by 30–70%, but was found to be a better inhibitor of L-asparaginase. Attention has therefore turned to developing mechanism-based AS inhibitors as these are likely to exhibit tight binding and high specificity, in a similar manner to those developed for human immunodeficiency virus (HIV) protease (Baldwin et al., 1995; Silva et al., 1996). A prerequisite of such an approach, however, is detailed information on both the structure and mechanism of asparagine synthetase.

III. Structure of Glutamine-Dependent Asparagine Synthetases

Genes encoding glutamine-dependent asparagine synthetases have been cloned from a wide variety of prokaryotic and eukaryotic organisms, and a number of these enzymes have been isolated and characterized from mammalian sources such as rat, cow, and chick embryos (Hongo et al., 1992; Luehr and Schuster, 1985; Arfin, 1967). In addition, AS has been purified from Novikoff hepatomas (Patterson and Orr, 1968), as has a similar enzyme from RADA1 murine leukemia cells resistant to L-asparaginase (Horowitz and Meister, 1972). All mature forms of asparagine synthetase are characterized by a conserved N-terminal cysteine residue (Cys-1) that is produced by posttranslational removal of methionine or in some cases, such as *Bacillus subtilis* AS, a leader peptide. The presence of Cys-1 and the primary structure of the AS N-terminal domain places AS in the family of the Class II amidotransferases (Smith, 1995; Zalkin, 1993). Expression systems have been reported for the production of human AS in yeast (Van Heeke and Schuster, 1990) and AS-B in *E. coli* (Boehlein et al., 1994a). Both of these systems yield active enzyme and sequencing has confirmed that the recombinant material is processed correctly so as to give cysteine as the N-terminal residue. The participation of Cys-1 in catalyzing glutamine hydrolysis was first suggested by studies in which mammalian asparagine synthetases were covalently modified by 5-diazo-oxo-norleucine (DON) **2** (Fig. 1), and subsequently confirmed by experiments employing AS mutants in which Cys-1 was replaced by either alanine (C1A) or serine (C1S) residues (Van Heeke and Schuster, 1989; Sheng et al., 1993; Boehlein et al., 1994a).

Multiple sequence alignment of members of the Class II amido-transferase family suggests that the N-terminal regions of GPA, GFAT, and AS are homologous (Fig. 2). The results of this sequence analysis, in combination with the domain structure seen in the crystal structures of GPA (Smith et al., 1994; Kim et al., 1996), support the hypothesis that AS consists of an N-terminal, glutamine-utilizing (GAT) domain linked to a larger C-terminal region containing the active site responsible for aspartate activation. Experiments employ-ing monoclonal antibodies that specifically and selectively inhibit the glutaminase and synthetase activities of bovine AS (Pfeiffer et al., 1986), and which were carried out prior to the availability of sequence information, provide direct support for location of the two active sites in different domains of the enzyme (Pfeiffer et al., 1987). Although the AS GAT-domain appears to be considerably smaller than those observed in GPA or GFAT, the core regions of the N-terminal domains in all three proteins likely possess an identical fold. On the other hand, the AS C-terminal domain does not appear to possess any significant homology with the synthetase domains of other Class II amidotransferases, and its detailed three-dimensional structure remains undefined.

Understanding the mechanism of glutamine binding and utiliza-tion has been greatly aided by the availability of the crystallographic coordinates for the covalent adduct of DON and *E. coli* GPA (Kim et al., 1996) [Fig. 3(*a*)] and the complex between L-glutamic acid γ-monohydroxamate (LGH) **5** and an N-terminal proteolytic fragment of *E. coli* GFAT (Isupov et al., 1996). Although there is evidence that organization of the GAT-domain active site is dependent on

---

Figure 2.   Sequence alignment of the GAT domains of a partial set of Class II amido-transferases. Residue numbering corresponds to human AS. This alignment was ob-tained using a progressive alignment method (Feng and Doolittle, 1987) implemented in PILEUP, a module in the GCG sequence analysis software suite (Devereux et al., 1984). A gap weight of 3.0 was used in obtaining these results. LtHmAS = *Cricetulus longicaudatus* AS; GnHmAS = *Mesocricetus auratus* AS; RatAS = *Rattus norvegi-cus* AS; MurAS = *Mus musculus* AS; HumAS = *Homo sapiens* AS; ScerGA = *Saccharomyces cerevisiae* GPA; SkluGA = *Saccharomyces kluyveri* GPA; SpomGA = *Schizosaccharomyces pombe* GPA; MtbGS = *Mycobacterium tuberculosis* GFAT; HumGS = *Homo sapiens* GFAT; ScerGS = *Saccharomyces cerevisiae* GFAT; CalbGS = *Candida albicans* GFAT.

```
                                                                                        1        10         20         30                 40          50
LtHmAS   cgiwalf......gsddclsvqclsamkiahrgpdafrfe.................nvngytnccfgfhrlavvdp
GnHmAS   cgiwalf......gsddclsvqclsamkiahrgpdafrfe.................nvngytnccfgfhrlavvdp
RatAS    cgiwalf......gsddclsvqclsamkiahrgpdafrfe.................nvngytnccfgfhrlavvdp
MurAS    cgiwalf......gsddclsvqclsamkiahrgpdafrfe.................nvngytnccfgfhrlavvdp
HumAS    cgiwalf......gsddclsvqclsamkiahrgpdafrfe.................nvngytnccfgfhrlavvdp
ScerGA   cgilgiv...lanqttpvapelcdgciflqhrgqdaagiatcgsrgriy.........qckgngmardvftqqrvs...g
SkluGA   cgilgia...ladqssvvapelfdgslflqhrgqdaagmatcgergrly.........qckgngmardvftqhrms...g
SpomGA   cgilalm...ladphqqacpeiyeglyslqhrgqdaagivtagnkgrly..........qckgsqmvadvfsqhqlr...q
MtbGS    cgivgyv.....grrpayvvvmdalrrmeyrgydssgialvdg........gtltvrragrlanleeavaem...psta
HumGS    cgifaylnyhvprtrreiletlikglqrleyrgydsagvfdggndkdw...eanacktqlikkkgkvkaldeevhkqqdmdldie
ScerGS   cgifgycnylversrgeiidtlvdglqrleyrgydstgiaidg.........deadstfiykqigkvsalkeei.tkqnprdvt
CalbGS   cgifgyvnflvdksrgeiidnliegqlrleyrgydsagiavdgkkltkdpsngdeeymdsiivkttgkvkvlkqki.iddqidrsai


                                                                                        55   60          70         80         90        100
LtHmAS   lfgmqpirvkypy................lwlcyngeiynhkalqqrfe....feyqtnvdgeiilhlyd.............
GnHmAS   lfgmqpirvkkypy...............lwlcyngeiynhkalqqrfe....feyqtnvdgeiilhlyd.............
RatAS    lfgmqpirvrkypy...............lwlcyngeiynhkalqqrfe....feyqtnvdgeiilhlyd.............
MurAS    lfgmqpirvrkypy...............lwlcyngeiynhkalqqrfe....feyqtnvdgeiilhlyd.............
HumAS    lfgmqpirvkkypy...............lwlcyngeiynhkkmqqhfe...feyqtkvdgeiilhlyd.............
ScerGA   lagsmgiahlryptagssanseaqpfyvnspyginlahngnlvntaslkrymdedvhrhintdsdselllnifaaelekhnkyrvnn
SkluGA   lvgsmgiahlryptagscanseaqpfyvnspygiclshngtlvntslrsyldevvhrhintdsdselllnvfaaelerhnkyrvnn
SpomGA   lvgsmgighlryptagscahseaqpfyvnspyglvlghngnlingpelrrfldteahrvntgsdselllnifayelqrldkfrine
MtbGS    lsgttglghtrwathgrptdrnahphr.daagkiavvhngiienfavlrreletagv.efasdtdtevaahlvarayr..hgetad.
HumGS    fdvhlgiahtrwathgepspvnshpqrsdknnefivihngiitnykdlkkfleskgy.dfesedtetiaklvkymyd..n..resq
ScerGS   fvshcgiahtrwathgrpeqvnchpqrsdpedqfvvvhngiitnfrelktllinkgy.kfesdtdteciaklylhlyn..tnlqngh
CalbGS   fdnhvgiahtrwathgqpktenchphksdpkgefivvhngiitnyaalrkyllskgh.vfesetdteciaklfkhfyd..lnvkagv
```

155

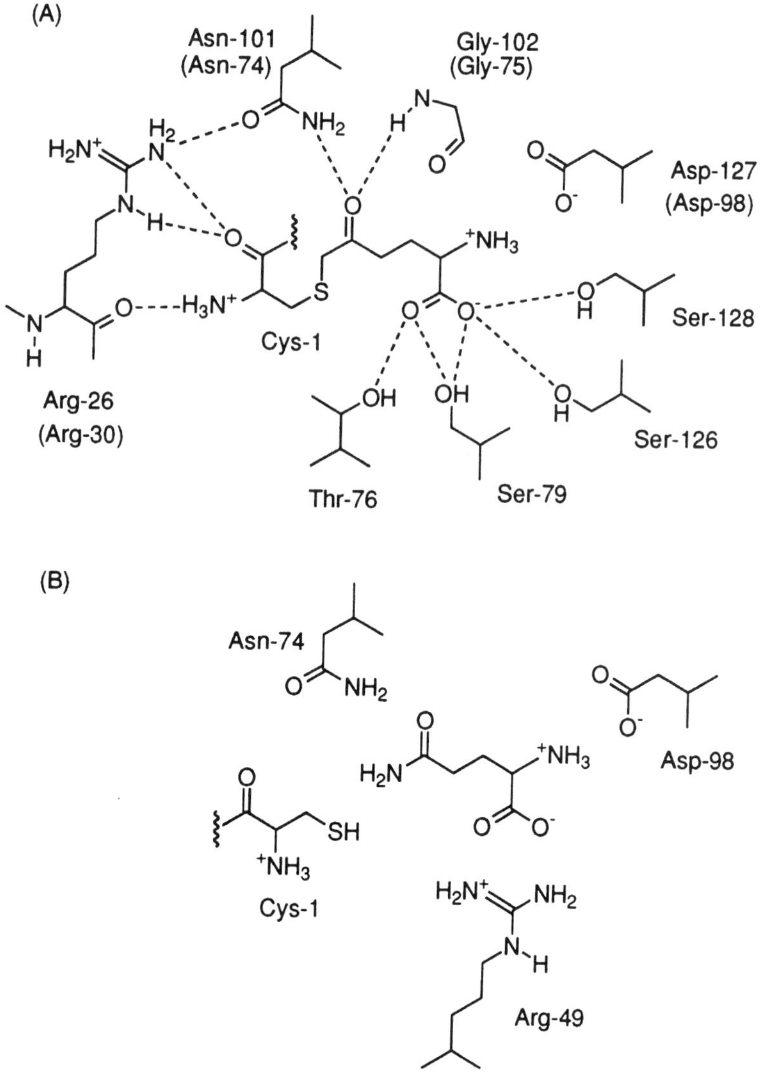

(A)

(B)

Figure 3. Schematic representation of the glutamine-binding sites of two *E. coli* Class II amidotransferases. (*A*) Intermolecular interactions observed in the crystal structure of the covalent adduct between DON 2 and GPA (Kim et al., 1996). Residue numbers in brackets refer to the cognate residues in AS-B; (*B*) Current model of putative interactions between L-glutamine and residues conserved in the AS-B GAT domain.

substrate binding in these two amidotransferases, similar structural changes may not be required in the AS GAT-domain given its significantly higher glutaminase activity. In the absence of high-resolution structural information, therefore, homology modeling methods (Greer, 1990), in combination with the stochastic search approaches for ligand docking (Guida et al., 1992; Montgomery et al., 1993) implemented in the MacroModel/BATCHMIN software package (Mohamadi et al., 1990), have been employed to construct a hypothetical model of the AS-B glutamine-binding site [Fig. 3(b)] (Richards, NGJ, unpublished results). The results are consistent with the glutamine/protein interactions observed for GPA and GFAT, which are primarily mediated by GAT-domain residues that are conserved throughout all Class II amidotransferases. Only the interaction between the α-carboxylate group of glutamine and the GAT-domain remains somewhat ill defined. Our AS-B model suggests that this substrate moiety interacts with the Arg-49 side chain in AS-B, the cognate interaction involved also being present in the LGH/GFAT complex. On the other hand, the α-carboxylate of DON appears to interact with a series of serine and threonine residues in the active site of GPA (Kim et al., 1996). Although this might reflect the minor structural differences between DON and bound glutamine, site-directed mutagenesis studies that delineate the exact role of this conserved arginine residue in GPA have not been reported (see below). The qualitative model of the AS-B GAT-domain also suggests that glutamine must be bound in a fully extended conformation and therefore rationalizes the observation that asparagine cannot be employed by AS-B as a substrate, even though asparagine inhibits the enzyme by competing for the glutamine binding site (Habibzadegah-Tari, unpublished results). Product inhibition also removes the requirement for specific regulatory sites in the enzyme, such as those that are observed in GPA (Kim et al., 1995).

An important mechanistic aspect of glutamine-dependent asparagine synthesis involves coordination of the two half-reactions that must take place in the active site of each AS domain. Structural reorganization of the protein upon substrate binding currently appears to be the most likely method by which this problem can be solved. Such a conformational change also has the advantage that water could be excluded from the active site, minimizing futile ATP hydrolysis.

IV. Mechanism of Aspartic Acid Activation

A. β-ASPARTYL-AMP IS A REACTION INTERMEDIATE

Activation of the β-carboxylate of aspartic acid toward nucleophilic attack is required for asparagine biosynthesis. The participation of ATP in the synthetase reaction was demonstrated in early studies on enzymes from murine and bovine sources, using ^{18}O-labeled aspartic acid as a substrate (Cedar and Schwartz, 1969a; Luehr and Schuster, 1985). The observation that the label was transferred from aspartic acid to the product AMP is consistent with the formation of a β-aspartyl-AMP intermediate on the enzyme that can then undergo reaction with a nucleophilic nitrogen species. Use of an AMP derivative rather than an acylphosphate for carboxylate activation, as observed, for example, in glutamine synthetase (Abell and Villafranca, 1991; Yamashita et al., 1989), is consistent with proposals that amidotransferase synthetase domains and aminoacyl tRNA synthetases might be evolutionary related (Hinchman and Schuster, 1992; Di Giulio, 1993). While the formation of β-aspartyl-AMP is theoretically a reversible reaction, no ATP/inorganic pyrophosphate (PP$_i$) exchange has been demonstrated for any glutamine-dependent AS. This result implies that pyrophosphate is tightly bound within the synthetase active site, presumably by residues in the P-like loop motif (Bork and Koonin, 1994; Tesmer et al., 1996). Recent experiments in which [3H] labeled ATP is incubated with AS-B suggest that futile hydrolysis of this metabolite does not take place in the absence of aspartate and/or glutamine (Boehlein et al., unpublished results). Either attack of aspartate on the α-phosphorus, or pyrophosphate release, might therefore require the presence of glutamine (or glutamate) in the GAT-domain. Although detailed stereochemical investigations have not yet been reported, the mechanism of aspartate activation probably proceeds via a pentacoordinate phosphorus intermediate in a reaction that requires Mg^{2+} ions for electrophilic catalysis (Villafranca, 1991) (Scheme 3). Recombinant AS-B and human AS exhibit surprising differences in their nucleotide selectivity (Boehlein et al., 1994a; Sheng et al., 1993). For example, when CTP is employed as an alternate substrate, human AS retains 90% of its activity with ATP, while AS-B activity is decreased to very low, albeit detectable, levels. Similarly, AS-B cannot employ ITP as a substrate in contrast to human AS. On the other hand, both

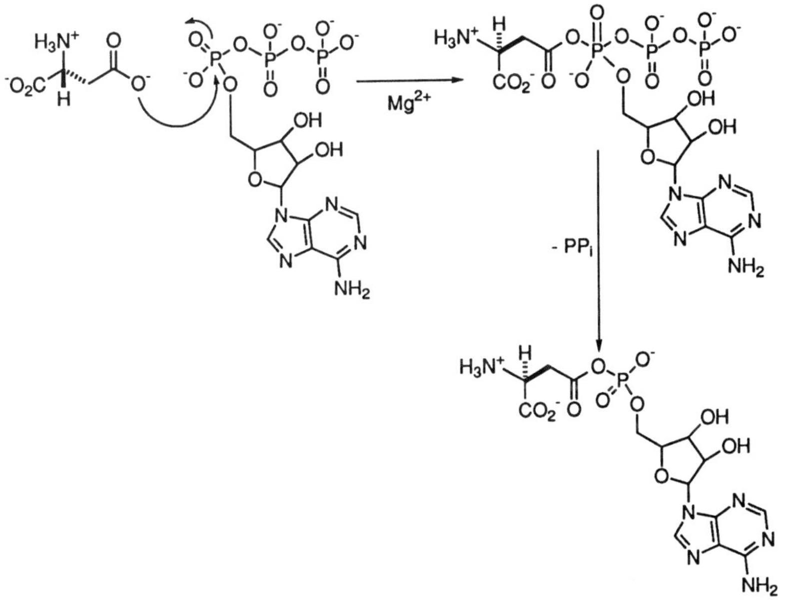

Scheme 3. Mechanism for formation of β-aspartyl-AMP. This reaction is thought to proceed in two steps, with initial attack of the carboxylate anion on the α-phosphate of ATP to form a pentacovalent intermediate. Release of PP_i then yields β-aspartyl-AMP. Both steps require Mg^{2+} for activation of the electrophic center and stabilization of the leaving group.

enzymes can discriminate between ATP and GTP, and can use dATP as an alternative to ATP. The ability of ATP analogues to stimulate the glutaminase activity of AS-B (see below) also mirrors the ability of the enzyme to employ these compounds in place of ATP.

B. THE BOUND CONFORMATION OF ASPARTIC ACID IN AS-B

The ability of AS-B to employ aspartic acid analogues (Fig. 4) in asparagine synthesis has been investigated, albeit to a limited extent. These nonnatural amino acids, prepared by stereoselective alkylation of an L-aspartate diester (Parr et al., 1996; Baldwin et al., 1989), do not function as substrates in AS-B synthetase activity. Both diastereoisomers of β-methylaspartate, 6 and 7, and of pyrrolidine-2,3-dicarboxylic acid, 8 and 9, act as weak, competitive inhibitors of

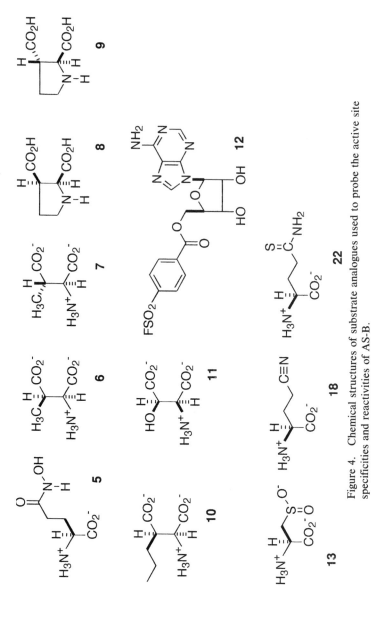

Figure 4. Chemical structures of substrate analogues used to probe the active site specificities and reactivities of AS-B.

160

aspartate with WT AS-B, although the propyl derivative **10** does not inhibit the enzyme. The steric bulk of the substituent at C3 therefore appears limited by the shape of the aspartate-binding pocket in the synthetase domain to that of a methyl group. This result is consistent with the observation that *threo*-β-hydroxyaspartate **11** is a weak, competitive inhibitor of mammalian AS (Mokotoff et al., 1975). Further, **8** inhibits AS-B synthetase activity about 30-fold better than **9**, suggesting that aspartic acid initially binds to AS in a conformation in which all of the polar functional groups are located on one face of the molecule. The C—H bonds then define a hydrophobic surface that contacts the enzyme (Fig. 5). On the other hand, none of these aspartate analogues participate in nitrogen transfer or cause pyrophosphate production when incubated with AS-B in the presence of glutamine and ATP. Evidence that they bind in the aspartic acid binding site has been provided (Parr et al., 1996; Boehlein et al., 1997c), however, by measurements of their ability to protect AS-B against chemical inactivation by 5'-*O*-[*p*-(fluorosulfonyl)benzoyl]-adenosine, **12**, (FSBA) (Fig. 4), an ATP analogue that can covalently label proteins (Colman, 1990). For example, in the presence of ATP, β-methylaspartate, **6**, protects WT AS-B against inactivation to a similar extent as aspartic acid, consistent with the hypothesis that these two compounds interact with the same aspartate-binding site (Parr et al., 1996). The inability of analogues **6** and **8** to participate in formation of the corresponding AMP derivatives also suggests

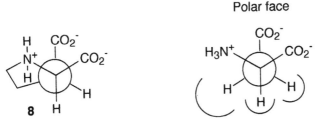

Figure 5.   Schematic representation of AS-B aspartic acid binding pocket. The putative bound conformation of aspartate is based on the ability of the conformationally constrained analogue **8** to inhibit AS-B. The amino acid displays a nonpolar surface to the enzyme.

that repositioning of the β-carboxylate anion of aspartate must occur after initial binding for its reaction with ATP. Such a conformational rearrangement could be triggered by the presence of glutamine in the AS-B GAT domain, and would provide a mechanism for preventing futile ATP hydrolysis due to inappropriate formation of the β-aspartyl-AMP intermediate. Finally, cysteine sulfinic acid (CSA) **13** (Foote et al., 1985) has also been shown to be a competitive inhibitor of aspartate for WT AS-B, and enhances the ability of the enzyme to catalyze the futile hydrolysis of ATP to AMP and PP_i (Parr et al., 1996).

C. IDENTIFICATION OF THE ASPARTIC ACID BINDING SITE

Several regions in the C-terminal domain of asparagine synthetases from plants, mammals, yeast, and bacteria are defined by extensive sequences of residues that are highly conserved (Fig. 6). Within the contiguous region comprising AS-B residues Glu-317 to Met-330, alanine scanning mutagenesis studies suggest that Thr-322, Thr-323, and Arg-325 play significant roles in aspartate binding and activation (Boehlein et al., 1997a). Indeed, the R325A and R325K AS-B mutants have no detectable synthetase activity while retaining their ability to hydrolyze glutamine with a catalytic efficiency (k_{cat}/K_M) similar to that of WT AS-B. The importance of Arg-325 in aspartate utilization has also been demonstrated by the observation that guanidinium hydrochloride (GdnHCl) rescues the synthetase activity

---→

Figure 6.   Sequence alignment of known asparagine synthetases showing only highly conserved motifs in the GAT- and C-terminal synthetase domains. Conserved residues are in boldface type. Residue numbering corresponds to AS-B. This alignment was obtained using a progressive alignment method (Feng and Doolittle, 1987) implemented in PILEUP, a module in the GCG sequence analysis software suite (Devereux et al., 1984). A gap weight of 3.0 was used in obtaining these results. LtHmAS = *Cricetulus longicaudatus* AS; GnHmAS = *Mesocricetus auratus* AS; RatAS = *Rattus norvegicus* AS; MurAS = *Mus musculus* AS; HumAS = *Homo sapiens* AS; SoyAS = *Glycine max* AS; LotAS = *Lotus japonicus* AS; FavaAS = *Vicia faba* AS; PeaNAS = *Pisum sativa* AS (nodule); AlflAS = *Medicago sativa* AS; PeaRAS = *Pisum sativa* AS (root); AspaAS = *Asparagus officinalis* AS; ArabAS = *Arabidopsis thaliana* AS; BrssAS = *Brassica oleracea* AS; RiceAS = *Oryza sativa* AS; MaizAS = *Zea mays* AS; ScerAS = *Saccharomyces cerevisiae* AS; CeleAS = elegans AS; EcoAS = *Escherichia coli* AS.

(*a*) GAT-domain

```
LtHmAS   cgiwal<10>lsamkiahrgpd<14>fgfhr lav<7>qp<12>yngeiynhkal<10>nvdgei
GnHmAS   cgiwal<10>lsamkiahrgpd<14>fgfhr lav<7>qp<12>yngeiynhkal<10>nvdgei
RatAS    cgiwal<10>lsamkiahrgpd<14>fgfhr lav<7>qp<12>yngeiynhkal<10>nvdgei
MurAS    cgiwal<10>lsamkiahrgpd<14>fgfhr lav<7>qp<12>yngeiynhkal<10>nvdgei
HumAS    cgiwal<10>lsamkiahrgpd<14>fgfhr lav<7>qp<12>yngeiynhkkm<10>kvdgei
SoyAS    cgilav<16>.lsrrlkhrgpd<10>lahqr lai<7>qp<11>vngeiynheel<11>gsdcdv
LotAS    cgilav<16>.lsrrlkhrgpd<10>lahqr lai<7>qp<11>vngeifnheel<11>gcdcdv
FavaAS   cgilav<16>.lsrrlkhrgpd<10>lahqr lai<7>qp<11>vngeiynheel<11>qcdcdv
PeaNAS   cgilav<16>.lsrrlkhrgpd<10>lahqr lai<7>qp<11>vngeiynheel<11>qcdcdv
AlflAS   cgilav<16>.lsrrlkhrgpd<10>lahqr lai<7>qp<11>vngeiynhedl<11>qcdcdv
PeaRAS   cgilav<16>.lsrrlkhrgpe<10>laqqr lai<7>qp<11>vngeiynhedl<11>gsdcdv
AspaAS   cgilav<16>.lsrrlkhrgpd<10>lshqr lai<7>qp<11>vngeiynheel<11>gsdcev
ArabAS   cgilav<16>.lsrrlrhrgpd<10>lahqr lav<7>qp<11>vngeiynheel<11>gsdcev
BrssAS   cgilav<16>.lsrrlrhrgpd<10>lahqr lai<7>qp<11>vngeiynheel<11>gsdcdv
RiceAS   cgilav<16>.lsrrlrhrgpd<10>lahqr lai<7>qp<11>vngeiynheel<11>gsdcev
MaizAS   cgilav<16>.lsrrlrhrgpd<10>lahqr lai<7>qp<11>vngeiynheel<11>asdcev
ScerAS   cgifaa<15>.lskrirhrgpd<11>fvher lai<7>qp<11>vngeiynhiql<11>lsdcep
CeleAS   cgvfsi<18>.lsrrqhhrgpd<13>lvher lai<6>qp<11>hngeiynhqel<11>hcdsev
EcoAS    csifgv<16>.lsrlmrhrgpd<11>laher lsi<7>qp<11>vngeiynhqal<11>gsdcev
         1           30            49             74  79           98
```

(*b*) Synthetase-domain

```
LtHmAS   lmtdrrigcllsggldsslva<37>yigsehhev<15>sletydittvr asvgmy
GnHmAS   lmtdrrivcllsggldsslva<37>yigsehhev<15>sletydittvr asvgmy
RatAS    lmtdrrigcllsggldsslva<37>yigsehhev<15>pletydittvr asvgmy
MurAS    lmtdrrigcllsggldsslva<37>yigsehhev<15>sletydittvr asvgmy
HumAS    lmtdrrigcllsggldsslva<37>higsehyev<15>sletydittvr asvgmy
SoyAS    lmtdvpfgvllsggldsslva<41>yigtvhhef<15>hietydvttir asipmf
LotAS    lmtdvpfgvllsggldsslva<41>yigtvhhef<15>hvetydvttir agtpmf
FavaAS   lmtdvpfgvllsggldsslva<41>flgtvhhef<15>htetydvttir aatpmf
PeaNAS   lmtdvpfgvllsggldsslva<41>flgtvhhef<15>htetydvttir aatpmf
AlflAS   lmtdvpfgvllsggldsslva<41>flgtvhhef<15>htetydvttir aatpmf
PeaRAS   lmtdvpfgvllsggldsslva<41>ylgtvhhef<15>hvetydvtsir astpmf
AspaAS   lmtdvpfgvllsggldsslva<41>ylgtvhhef<15>hietydvttir astpmf
ArabAS   lmtdvpfgvllsggldsslva<41>ylgtvhhef<15>hvetydvttir astpmf
BrssAS   lmtdvpfgvllsggldsslva<41>ylgtvhhef<15>hvetydvttir astpmf
RiceAS   lmtdvpfgvllsggldsslva<41>ylstvhhef<15>hietydvttir astpmf
MaizAS   lmtdvpfgvllsggldsslva<41>ylgtvhhel<15>hvetydvttir astpmf
ScerAS   lmaevpygvllsggldsslia<64>figsihheh<15>hletydvttir astpmf
CeleAS   lmsdapigvllsggldsslvs<34>figtthhef<15>hlesydvtsir astpmy
EcoAS    lmsdvpygvllsggldssiis<45>hlgtvhhei<17>hietydvttir astpmy
         223         234  240               325
```

of the R325A AS-B mutant to about 15% of that of the WT enzyme (Boehlein et al., 1997a). Chemical rescue is specific, and is consistent with the hypothesis that guanidinium ion can bind within a pocket formed by the absence of the arginine side chain in the R325A AS-B mutant. This mechanistic proposal is supported by similar studies on carboxypeptidase (Phillips et al., 1992; Perona et al., 1994). In AS-B, Arg-325 may be involved either in binding aspartic acid in a catalytically competent manner, or in stabilizing the transition state leading to formation of the pentacovalent phosphorus intermediate during the synthesis of β-aspartyl-AMP (Scheme 3). Stabilization of similar intermediates by arginine residues has been observed in a number of other enzymes (Dahnke et al., 1992; Ghosh et al., 1991a, 1991b; Lu and Hill, 1994). The Thr-322 residue also appears to partic- ipate in catalysis since mutants in which this residue is replaced by alanine, serine, or valine exhibit decreased turnover numbers rela- tive to WT enzyme without any significant change in the apparent K_m for aspartate (Boehlein et al., 1997a). The exact functional role of this residue has been obscured, however, by its apparent absence in the primary structures of glutamine-dependent asparagine synthe- tases deduced from the cloned genes in B. subtilis (Yoshida et al., 1995), Saccharomyces pombe (Yoshioka et al., unpublished results), and Methanococcus jasmonii (Bult et al., 1996). As replacement of Thr-322 decreases the dissociation constant, K_d, for the complex between AS-B and ATP (Boehlein et al., 1997a), ATP binding may cause a structural change in the aspartate binding site through its interaction with Thr-322. As discussed, such a conformational rear- rangement might be involved in positioning the β-carboxylate of as- partic acid for subsequent reaction with the α-phosphate of ATP. Finally, Thr-323 seems to mediate aspartate binding given that the steady-state kinetic properties of AS-B mutants appear to reflect the size and shape of the residue that replaces threonine at this position (Boehlein et al., 1997a).

The kinetic effects of mutating Thr-322, Thr-323, and Arg-325 in AS-B are strikingly similar to those observed in cognate experiments on methionyl tRNA aminoacyl synthetase (MetRS) (Carter, 1993) involving active site arginine (Arg-322) and tyrosine (Tyr-258) resi- dues (Ghosh et al., 1991a, 1b). Hence, replacement of Arg-322 by glutamine yields a mutant MetRS that exhibits a 60,000-fold decrease in k_{cat}/K_m for ATP-PP$_i$ exchange (Ghosh et al., 1991a), consistent with the idea that Arg-322 stabilizes the pentacoordinate intermediate. Crystallographic evidence also indicates that Tyr-258 binds to the α-phosphate of ATP in the transition state leading to α-methionyl-AMP (Ghosh et al., 1991b). Given the probable mechanistic similarities in the reactions leading to β-aspartyl-AMP and α-methionyl-AMP, the detailed structural comparison of MetRS with the AS-B synthetase domain is eagerly awaited.

The functional importance of other residues in the putative aspartic acid binding segment have also been evaluated by alanine scanning mutagenesis (Boehlein et al., 1997a). For example, Val-318 may form the enzyme surface that contacts the hydrophobic face of the bound substrate. While mutation of Asp-320 does not affect AS-B synthetase activity, this residue may be implicated in molecular mechanisms underlying the ability of AS-B to distinguish structural differences between CSA and aspartic acid. In addition, the diminished glutaminase activity of the E317A AS-B mutant suggests that Glu-317 participates in maintaining the overall, multidomain fold of the enzyme. Finally, there is evidence from mutagenesis and chemical modification experiments that suggests that Cys-523, which is not conserved throughout the AS family, plays a role in binding aspartate in a configuration suitable for interaction with enzyme-bound ATP (Boehlein et al., 1997c).

D. IDENTIFICATION OF A PYROPHOSPHATE BINDING MOTIF IN AS-B

Many proposals for the molecular evolution of soluble enzymes, catalyzing metabolically important reactions, postulate the existence of modular segments that confer a particular function on the protein in which they are located (Bork and Koonin, 1996). Examples of such motifs are those for binding nicotinamide dinucleotides (Schulz, 1992) and ATP (Walker et al., 1982). Recent algorithms for detecting homologous elements in protein sequences suggest that the contiguous sequence defined by AS-B residues Ser-234 to Ser-239 (Fig. 6) represents a pyrophosphate binding element (Bork and Koonin, 1994). This segment exhibits high conservation throughout all known asparagine synthetases, and is present in other enzymes in which ATP is hydrolyzed to AMP and PP_i, including GMP synthetase (GMPS) (Mäntsälä and Zalkin, 1992), argininosuccinate synthetase (Ratner, 1973; Surh et al., 1988), and ATP sulfurylase (Leyh et al.,

1992). The X-ray crystal structure of GMP synthetase complexed with 2-iodo-AMP and inorganic pyrophosphate (Tesmer et al., 1996) has validated this sequence-based hypothesis, revealing a series of interactions between residues in the N-type P-loop motif and the pyrophosphate anion (Fig. 7). In GMPS, the pyrophosphate-binding loop connects a β-strand with an α-helix, the main chain (ϕ, ψ) angles of the two conserved glycines being distorted to accommodate the secondary structural packing in this region of the enzyme. The principal interactions between GMPS and PP_i appear to involve hydrogen bonds from the backbone amide of Val-238 (GMPS numbering), and the side chains of Ser-235, Ser-240, and Lys-381, the latter resi-

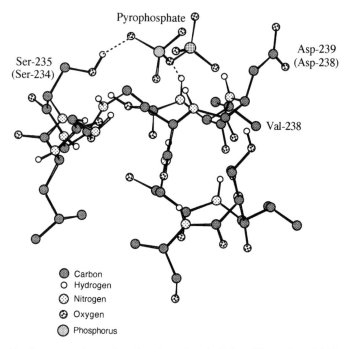

Figure 7.   Structure of pyrophosphate bound to the P-loop-like region of GMP synthetase. This loop structure lies between a β-strand and an α-helix. Residue numbers in brackets correspond to amino acids in the AS-B P-loop-like domain. Putative hydrogen bonds between the protein residues and inorganic pyrophosphate are represented by dashed lines. Atomic coordinates are taken from the crystal structure of GMP synthetase complexed with AMP and PP_i (1GPM) (Tesmer et al., 1996).

due being located in an adjacent structural domain. Alanine scanning mutagenesis experiments (Boehlein et al., unpublished observations) are consistent with the hypothesis that similar interactions are present in the putative pyrophosphate binding motif of AS-B (Table 1). For example, large effects on the kinetic parameters for aspartic acid and ATP are observed for the G236A and D238A AS-B mutants. Replacement of Gly-236 by alanine probably disrupts the loop structure preventing formation of suitable hydrogen bonds to the β- and/ or γ-phosphate groups of ATP. The functional role of Asp-238 in AS-B, however, appears more complicated than initially apparent from the location of the cognate residue (Asp-239) in the crystal structure of the GMPS/I-AMP/PP_i complex (Tesmer et al., 1996). The side chain of Asp-239 in GMPS makes an electrostatic interaction with Lys-381 that positions the latter residue for hydrogen bonding to PP_i. A slight adjustment of the Asp-239 carboxylate, however, places it close to a Mg^{2+} ion that can be modeled between the two phosphate groups of bound PP_i. Therefore, this residue might also be involved in metal binding, and catalysis of the breakdown of the pentacovalent phosphorus intermediate (Scheme 3). Resolution of this mechanistic question awaits detailed studies of the metal dependence of AS-B synthetase activity and structural information to locate residues in other structural domains that might participate in pyrophosphate binding.

V. Molecular Mechanism of Glutamine-Dependent Nitrogen Transfer

A. MECHANISTIC PROPOSALS FOR GLUTAMINE-DEPENDENT NITROGEN TRANSFER IN ASPARAGINE SYNTHETASE

The major mechanistic question that remains to be answered in asparagine biosynthesis concerns the molecular details underlying nitrogen transfer from glutamine to β-aspartyl-AMP (Richards and Schuster, 1992). The structural basis by which the half-reactions that take place in the separate active sites in the GAT- and C-terminal synthetase domains are coordinated is also an intriguing chemical problem. A number of plausible mechanisms for nitrogen transfer have been proposed (Richards and Schuster, 1992; Stoker et al., 1996; Mei and Zalkin, 1989), the simplest of which involves generation, and subsequent reaction, of enzyme-bound ammonia. This

TABLE 1

(A) Steady-State Kinetic Constants for the Glutamine-Dependent Synthetase Activity of WT AS-B and Selected AS-B Mutants That Suggest a Role for Conserved Residues Ser-234 and Asp-238 in Pyrophosphate Binding and Activation[a]

	Glutamine			Aspartic Acid			ATP		
	K_M (mM)	k_{cat} (s^{-1})	k_{cat}/K_M (M^{-1}s^{-1})	K_M (mM)	k_{cat} (s^{-1})	k_{cat}/K_M (M^{-1}s^{-1})	K_M (mM)	k_{cat} (s^{-1})	k_{cat}/K_M (M^{-1}s^{-1})
wt AS-B	0.69 ± 0.07	1.01 ± 0.05	1463	0.68 ± 0.07	1.05 ± 0.04	1544	0.18 ± 0.01	1.10 ± 0.03	6111
S234A	0.49 ± 0.77	0.35 ± 0.02	714	18.5 ± 2.5	0.54 ± 0.03	29.2	0.53 ± 0.04	0.43 ± 0.11	811
G236A	0.35 ± 0.024	0.23 ± 0.005	657	43.7 ± 3.6	0.361 ± 0.013	8.21	6.95 ± 0.53	0.40 ± 0.015	57.5
D238E	0.59 ± 0.04	0.28 ± 0.01	475	54.8 ± 5.0	0.31 ± 0.01	5.6	4.19 ± 0.76	0.35 ± 0.02	84

(B) Steady-State Kinetic Constants for the Glutaminase Activity of WT AS-B and the S234A, G236A, D238A, and D238E AS-B Mutants

	Glutamine (ATP Absent)			Glutamine (5 mM ATP Present)		
	K_M (mM)	k_{cat} (s^{-1})	k_{cat}/K_M (M^{-1}s^{-1})	K_M (mM)	k_{cat} (s^{-1})	k_{cat}/K_M (M^{-1}s^{-1})
wt AS-B	1.90 ± 0.20	0.80 ± 0.03	412	1.39 ± 0.07	1.38 ± 0.02	993
S234A	1.58 ± 0.10	0.52 ± 0.01	329	1.12 ± 0.08	0.60 ± 0.01	535
G236A	1.29 ± 0.16	1.23 ± 0.05	953	1.55 ± 0.18	1.44 ± 0.05	929
D238A	1.67 ± 0.13	1.41 ± 0.03	844	1.13 ± 0.81	1.39 ± 0.03	1230
D238E	1.37 ± 0.04	0.50 ± 0.004	365	1.08 ± 0.07	0.68 ± 0.01	629

[a] The D238A AS-B mutant has no detectable activity at glutamine concentrations up to 100 mM.

mechanism is consistent with the observations that (a) all glutamine-dependent asparagine synthetases can employ ammonia as an alternate nitrogen source in asparagine synthesis, (b) the Cys-1 residue is absolutely required for both glutamine-dependent AS activities (Van Heeke and Schuster, 1989; Sheng et al., 1993), and (c) AS glutaminase activity is often substantial relative to synthetase activity, being stimulated in the presence of ATP (Boehlein et al., 1994b). The essential role of the Cys-1 in catalyzing glutamine-dependent nitrogen transfer also suggests that the AS GAT-domain is an amidohydrolase catalyzing amide bond cleavage in a similar manner to thiol proteinases (Brocklehurst et al., 1987), such as the cathepsins (Musil et al., 1991), papain (Storer and Ménard, 1994), and interleukin converting enzyme (ICE) (Thornberry and Molineaux, 1995). In such a mechanism, the amide initially reacts with the highly nucleophilic thiolate anion to give the tetrahedral intermediate **14** in which C—N bond cleavage can occur after N-protonation to yield ammonia and an acylthioenzyme intermediate **15** (Scheme 4). Subsequent hydrolysis of thioester **15** regenerates the enzyme and produces glutamate. Several of these steps require general acid–base catalysis, which is mediated by an active site histidine residue in the case of the thiol proteinases (Brocklehurst et al., 1987).

Early studies on the kinetic mechanism of mammalian asparagine synthetases provided support for ammonia-mediated nitrogen transfer (Milman et al., 1980; Markin et al., 1981). Although differing in important details, these experiments indicated that glutamine binds initially to the enzyme, with glutamate release occurring prior to binding of either aspartate or ATP. In this kinetic model, glutamine is hydrolyzed to yield an $E.NH_3$ complex [Fig. 8(a and b)], which can also be formed during asparagine synthesis when exogenous ammonia acts as the nitrogen source. Such a kinetic scheme for the nitrogen-transfer reaction does not mandate coupling of the two enzyme activities, and is therefore consistent with reports that the glutaminase activity of mammalian asparagine synthetases is independent of aspartate and ATP concentration (Horowitz and Meister, 1972; Milman and Cooney, 1979). On the other hand, ammonia-mediated nitrogen transfer raises a series of chemical problems that must be solved by the enzyme in order to prevent the futile consumption of glutamine and/or ATP (Richards and Schuster, 1992). First, ammonia has to be sequestered in the enzyme during glutamate re-

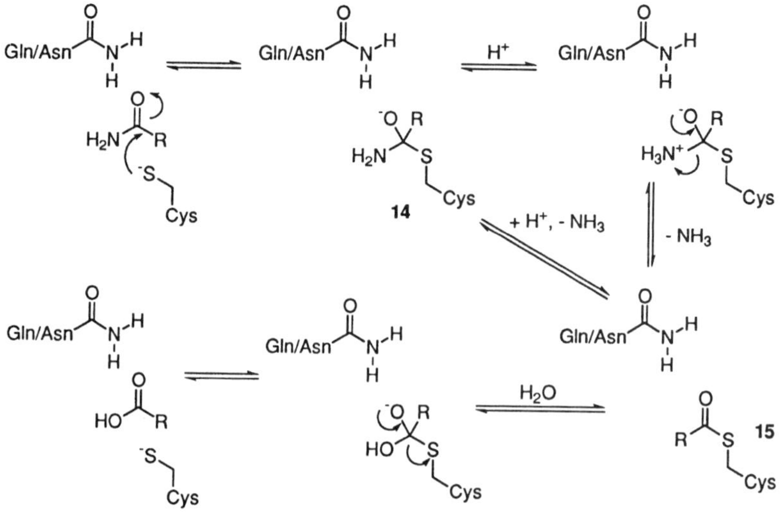

Scheme 4.   Mechanism of thiolate-catalyzed amide hydrolysis. Initial attack of the thiolate anion on the primary amide yields the tetrahedral intermediate 14. The protonation state of 14 is unknown and it is represented here as an oxyanion for convenience. Loss of ammonia then occurs after N-protonation to yield a thioacylenzyme intermediate 15. Protonation may occur at the same time as C-N bond cleavage. In papain, His-159 plays the role of a general acid catalyst for this step. In Class II amidotransferases, the N-terminal amino group of Cys-1 may have a similar catalytic function.

lease and subsequent substrate binding, so as to prevent its diffusion into solution. Second, ammonia must participate in the nucleophilic attack on the β-aspartyl-AMP intermediate in its neutral form. Even between pH 6.5–8.0, the range over which most asparagine synthetases exhibit optimum activity, protonation of the ammonia released by the GAT-domain can still occur to yield the unreactive ammonium ion. Therefore, the structure of the binding site must be such that the ammonia pK_a is significantly decreased.

These chemical problems have raised the question of whether alternate nitrogen-transfer mechanisms can occur which (a) do not require the presence of free ammonia, and (b) provide a simple molecular approach to coordinating the two half-reactions that must take place in asparagine synthesis. Any alternative mechanisms, however, must address the problem of activating the nonnucleophilic

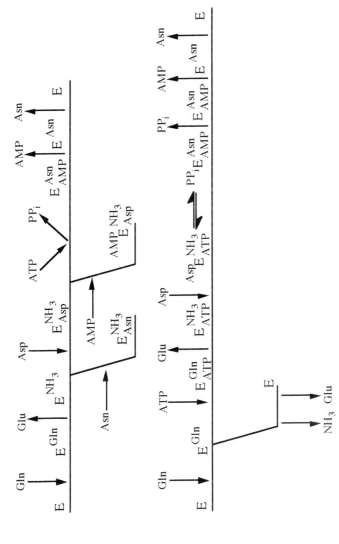

Figure 8. Proposed kinetic mechanisms for glutamine-dependent asparagine synthesis. (a) Murine AS (Milman et al., 1980). (b) Bovine AS (Markin et al., 1981); (c) *Escherichia coli* AS-B (Boehlein et al., unpublished). The exact product release order is tentative and is under experimental investigation.

171

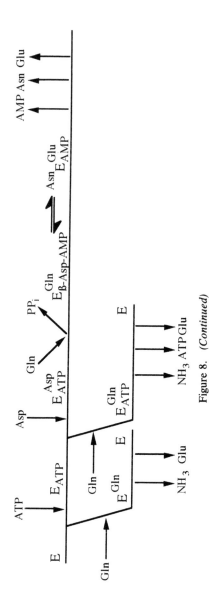

Figure 8. *(Continued)*

172 (c)

amide nitrogen for reaction with β-aspartyl-AMP. For example, due to resonance stabilization arising from delocalization of the nitrogen lone pair into the π*-orbital of the adjacent carbonyl group, electrophiles react on oxygen to give iminoether derivatives (Challis and Challis, 1970) and so O-acylation of glutamine by β-aspartyl-AMP would be expected to yield the iminoether **16** instead of imide **17** [Scheme 5(a)]. Although **16** could undergo rearrangement to **17**, this reaction is acid catalyzed with nitriles being produced by elimination under neutral, or basic, conditions (Davidson and Skovronek, 1958). Given that formation of nitrile **18** as a side product in asparagine synthesis has not been reported, such a mechanism is almost certainly not operative in nitrogen transfer. On the other hand, two mechanistic proposals that address this issue of amide nitrogen activation have been outlined (Stoker et al., 1996; Richards and Schuster, 1992). In the first of these, tetrahedral intermediate **14**, formed by addition of the Cys-1 thiolate to glutamine, is proposed to react with β-aspartyl-AMP to give the covalent intermediate **20** after release of AMP [Scheme 5(b)]. Collapse of **20** then gives thioacylenzyme **15** and asparagine. Regeneration of the active site thiolate can then take place as in the glutaminase reaction. This proposal has a number of chemically interesting features, such as the involvement of **14**, an intermediate that is formed in glutaminase activity, and the fact that C—N bond cleavage might not necessarily require N-protonation by a general acid. Further, glutamine hydrolysis and asparagine formation are closely coupled in this mechanism. On the other hand, there is little chemical precedent for the existence of **19**, in which there are unfavorable steric, and/or electrostatic, interactions between the proximal quaternary centers. The second mechanistic proposal for nitrogen transfer via a covalent intermediate postulates the existence of the imide intermediate **17** [Scheme 5(c)]. In this mechanism, the enzyme must promote direct reaction of the substrate amide with the activated carbonyl group of β-aspartyl-AMP and subsequent release of AMP. Imide **17** can then be broken down into asparagine and the thioacylenzyme **15** by reaction with the thiolate of Cys-1 as in the previous mechanism [Scheme 5(b)]. Although N-acylation of the glutamine side chain is kinetically unfavorable, there is chemical precedent for this proposal. For example, the rearrangement of asparagine residues to isoasparagine in vancomycin (Harris and Harris, 1982), and other peptides (Wright, 1991),

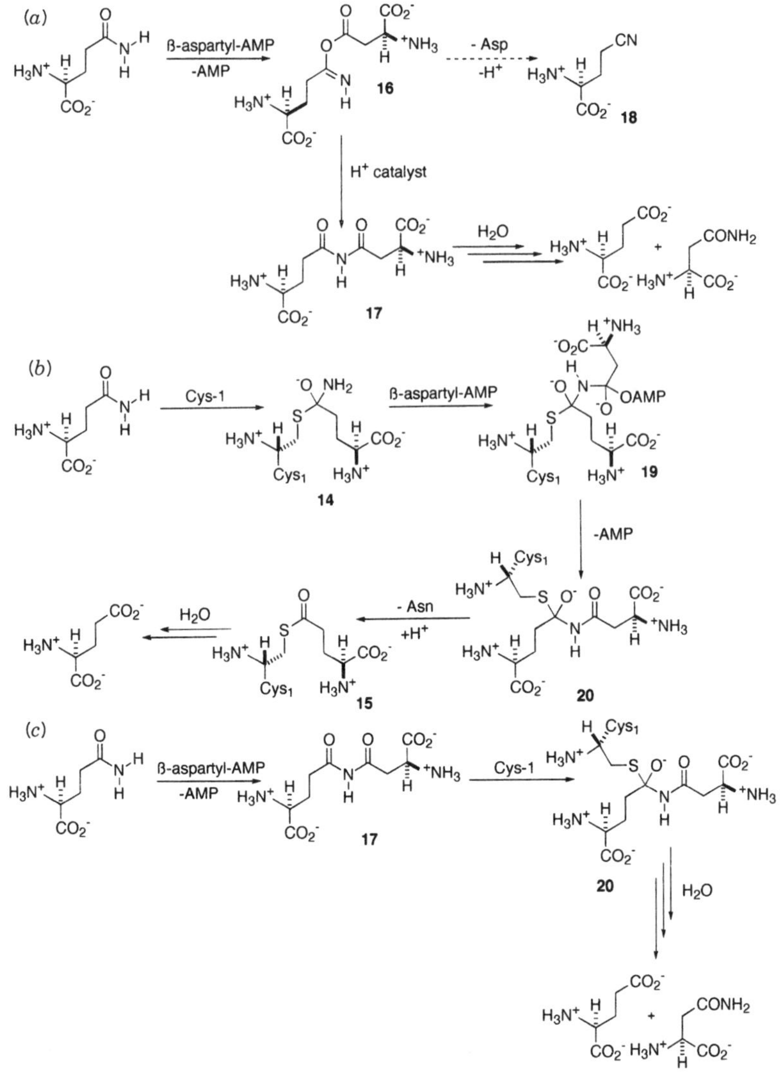

Scheme 5.   Alternate mechanisms for nitrogen transfer via formation of covalent intermediates. (*a*) *O*-Acylation of glutamine by β-aspartyl-AMP, followed by acid-catalyzed rearrangement to an imide **17** that can be hydrolyzed to glutamine and asparagine in a thiolate-catalyzed reaction; (*b*) nitrogen transfer by acylation of the tetrahedral intermediate **14** formed in the first step of glutamine hydrolysis. Breakdown of **20** could be acid catalyzed but cannot proceed via N-protonation; (*c*) imide-mediated nitrogen transfer.

has been observed, a process that almost certainly proceeds via cyclic imide intermediates. In addition, murine catalytic antibodies have been obtained by immunization with a ground-state analogue of the transition state leading to a cyclic imide intermediate. These antibodies catalyze the rearrangement of asparaginyl to isoaspartyl residues (Gibbs et al., 1992). It is also possible that the side-chain amide of substrate glutamine might be polarized by formation of low-barrier hydrogen bonds with Asn-74 in the AS-B active site (Cleland and Kreevoy, 1994; Gerlt and Gassman, 1993). On the other hand, generation of an imide intermediate in the nitrogen-transfer reaction requires that the Cys-1 thiolate does not initially react with the gluta-mine side chain as in the initial step of glutamine hydrolysis.

B. MECHANISTIC COMPARISON OF AS-B AND PAPAIN

The ability of AS-B, and other asparagine synthetases, to catalyze the conversion of glutamine into glutamate provides an opportunity to elucidate the functional roles of conserved residues in the AS GAT-domain using a single-substrate reaction. In thiol proteases, protonation of the leaving group nitrogen is a key step in facilitating subsequent C—N bond cleavage in the catalytic mechanism and these enzymes therefore possess an active site Cys-His diad (Storer and Ménard, 1994; Hol et al., 1978). In papain, for example, the His-159 side chain is specifically oriented so as to promote breakdown of the initial tetrahedral intermediate, and to stabilize the thiolate of Cys-25 (Kamphuis et al., 1984; Drenth et al., 1971). Experiments employing site-specific GPA mutants initially supported the idea that His-101 in the GAT-domain of GPA did play a role in catalyzing glutamine hydrolysis (Mei and Zalkin, 1989), and chemical modifica-tion studies using diethyl pyrocarbonate suggested that a histidine might be involved in catalyzing the synthesis of glucosamine 6-phos-phate by GFAT (Badet-Denisot and Badet, 1992). Recent crystal structures of GPA (Smith et al., 1994; Kim et al., 1996), and an N-terminal proteolytic fragment of GFAT (Isupov et al., 1996), how-ever, provide no evidence for the existence of a Cys-His diad in Class II amidotransferases. Extensive mutagenesis studies using re-combinant AS-B have also failed to identify catalytically important histidine residues (Boehlein et al., 1994a), although the H47N AS-B mutant does exhibit a 10-fold reduction in catalytic efficiency (k_{cat}/K_m) (Schnizer, unpublished results). Given that replacement of critical histidines in enzymes often decreases activity by several orders of magnitude, it is likely that the relatively small effect of mutating His-47 is due to perturbation of AS-B structure. On the other hand, if product release is rate limiting the overall effect of this mutation on the chemical reaction steps might be masked. Further examination of this question is therefore necessary in order to eliminate His-47 as a general acid–base catalyst in glutamine hydrolysis.

The apparent absence of a general acid–base catalyst in Class II amidotransferases has been resolved by recognizing these enzymes as members of the Ntn protease family (Brannigan et al., 1995), which also includes penicillin acylase (Duggleby et al., 1995), the proteasome (Groll et al., 1997; Lowe et al., 1995), and aspartylgluco-saminidase (Oinonen et al., 1995). In all of these enzymes, the nu-cleophilic catalyst is provided by an N-terminal residue (Ser, Thr, or Cys) that is obtained by posttranslational processing. General acid–base catalysis is then mediated by the N-terminal amino group, which may therefore exhibit an increased acidity due to the electro-static properties of the active site. Recent computational studies using the crystal structure of aspartylglucosaminidase support this hypothesis (Peräkylä and Kollman, 1997) for this enzyme in which catalysis is effected by an N-terminal threonine residue. On the other hand, experimental efforts to investigate this question, using mutants of GPA (Kim et al., 1996) and GFAT (Isupov et al., 1996) in which an additional N-terminal residue is present before the catalytic cysteine, have yielded contradictory results. In an effort to resolve the func-tional role of the N-terminal amine in AS-B, we have carried out a series of calculations employing the AM1 semiempirical model (Dewar et al., 1985). The effects of an aqueous environment were introduced using the COSMO solvation algorithm (Klamt and Schü-ürmann, 1993). Our results show that the N-terminal amine can rea-sonably act to stabilize the Cys-1 thiolate anion, although proton transfer to the nitrogen leaving group in the tetrahedral intermediate requires approximately 9 kcal/mol (Boone, unpublished results). A more intriguing role for this moiety is to facilitate nucleophilic attack of the thiolate on the side chain amide of the substrate (Scheme 6). Several computational studies (Howard and Kollman, 1988; Arad et al., 1990) have indicated that the oxyanion intermediate **14** does not represent a stable structure on the potential energy surface for this reaction, presumably as the anionic charge is more disperse on sulfur

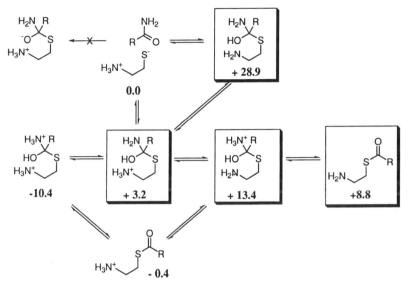

Scheme 6. Possible mechanisms for the participation of the N-terminal amino group of Cys-1 in general acid–base catalysis of amide bond hydrolysis. This analysis assumes that the tetrahedral intermediate must be protonated as all oxyanion structures are not stable minima on the AM1 semiempirical potential energy surface. Structures in the pathway for thioester formation in which the N-terminal amino group participates in acid–base catalysis are shown in boxes. Boldface numbers represent the heats of formation of each structure relative to free glutamine and the zwitterion representing cysteine, calculated using the AM1 semiempirical model in combination with the COSMO model for calculating the solvation energy in water.

relative to oxygen. The neutral intermediate **21** is a potential energy minimum, however, suggesting that thiolate attack can only occur with concomitant protonation of the amide. The catalytic role of the N-terminal amine may therefore be to provide the proton necessary to yield **21** as a stable intermediate. Breakdown of this intermediate by C—N bond cleavage would then require an additional proton from solvent, either directly or in a stepwise manner mediated by the N-terminal amine. Such a hypothesis is consistent with recent ^{15}N kinetic isotope effect (KIE) determinations for AS-B glutaminase activity (Stoker et al., 1996).

The observation of a dithioester chromophore in the UV–visible spectrum when papain is incubated with thiopeptides provides strong evidence for the existence of a thioacylenzyme in the reaction mechanism of thiol proteases (Lowe and Williams, 1964). Thioglutamine, 22 (Fig. 4), does not, however, yield thioglutamate when incubated with AS-B at pH 6.5. Similar observations have been reported for *E. coli* GFAT (Badet-Denisot et al., 1995), although the structural basis for this substrate selectivity remains unclear upon examination of the complex between glutamate and the N-terminal proteolytic fragment of this enzyme (Isupov et al., 1996). In contrast to glutamine, thioglutamine does not inhibit ammonia-dependent AS-B synthetase activity, suggesting that this analogue does not bind in the GAT-domain active site or is bound in an alternate, unreactive mode. Semiempirical calculations on acetamide and thioacetamide (Richards, unpublished results) indicate that while these functional groups differ somewhat in ground-state geometry, there appear to be substantial differences in the electrostatic potentials of the amide and thioamide moieties, which may allow discrimination of glutamine and thioglutamine during formation of the ES complex.

Direct evidence for the existence of a thioacylenzyme intermediate has been obtained for CAD (Chaparian and Evans, 1991) and anthranilate synthetase (Roux and Walsh, 1992), which are both Class I amidotransferases (Smith, 1995; Tesmer et al., 1996). Using similar methods, it has been found that incubation of AS-B with [^{14}C]glutamine at 4°C for short time periods yields radioactive protein (Schnizer, unpublished results). Up to 34% of the initial enzyme is labeled by substrate under these conditions and the effect is abolished when AS-B that has been inactivated by preincubation with DON is incubated with [^{14}C]glutamine under identical reaction conditions (Table 2). The thiolate sidechain of Cys-1 is therefore essential for covalent labeling of the protein, and the amount of labeled material saturates with increasing glutamine concentrations. These experiments provide the first direct evidence for the existence of a thioacylenzyme intermediate in the mechanism of AS-B catalyzed glutamine hydrolysis.

Relatively few residues are conserved throughout the GAT-domains of the family of Class II amidotransferases. As discussed above, in AS-B, excluding Cys-1, and conserved glycines that probably have a structural function, these correspond to Arg-49, Asn-74, Asn-79, and Asp-98. The Glu-100 residue is also highly conserved within asparagine synthetases, being replaced only by aspartic acid. A number of site-specific AS-B mutants have been expressed, puri-

TABLE 2

Isolation of a Radiolabeled WT AS-B Derivative by Filter Binding After
Incubation of the Enzyme (2 nmol) with ^{14}C-Gln[a−c]

Entry	Reaction Sample	Filter Radioactivity (dpm)
1	WT AS-B	6500 ± 180
2	WT AS-B + DON 2	190 ± 55
3	C1A AS-B mutant	300 ± 115
4	WT AS-B absent	400 ± 187

[a] Specific Activity 22,000 dpm/nmol at 25°C, pH 8 (Schnizer, unpublished observations).

[b] 1-mM ^{14}C-Gln sample was incubated with the appropriate enzyme (2 nmol) in 100-mM Tris-HCl, pH 8, for 30 s at 25°C (total volume: 100 μL). Reaction was quenched by the addition of 1 mL of a solution containing 8% TCA in 0.5 M NaOAc (pH 4), and 1-mg BSA. Samples were filtered through nitrocellulose (0.45-μm porosity) and the filter washed with 50 mL of 1 M HCl.

[c] Assays were performed in triplicate. (1) A ^{14}C-Gln sample was incubated with WT AS-B; (2) ^{14}C-Gln was incubated with WT AS-B after treatment of enzyme with DON 2 (Fig. 1); (3) ^{14}C-Gln was incubated with C1A AS-B mutant; (4) ^{14}C-Gln incubation in the absence of enzyme.

fied, and kinetically characterized to test the functional roles of these residues in glutamine-dependent activity (Table 3), as suggested by current models of the AS-B GAT-domain [Fig. 3(b)] (Boehlein et al., 1994a,b). Replacement of either Arg-49 or Asn-79 by alanine yields mutant enzymes with steady-state kinetic properties that are almost unchanged with respect to WT AS-B, although the turnover number for AS-B glutaminase activity is increased threefold in the R49K AS-B mutant. The reasons underlying the conservation of these GAT-domain residues throughout known Class II amidotransferases remain to be established. The hypothesis that Arg-49 makes critical interactions with the carboxylate group of substrate glutamine, however, remains uncertain on the basis of these steady-state experiments. On the other hand, mutation of Asp-98 to glutamic acid significantly increases in the apparent K_m for glutamine in AS-B glutaminase activity, and this residue likely forms a salt bridge to the α-amino group of the substrate. Replacement of Asp-98 by either asparagine or alanine residues gives AS-B mutants that exhibit kinetic behavior during glutamine hydrolysis such that double-reciprocal plots are nonlinear. There have not been any systematic studies of the role of Glu-100 in modulating the glutamine-dependent activi-

TABLE 1

(A) Steady-State Kinetic Constants for the Glutamine-Dependent Synthetase Activity of WT AS-B and Selected AS-B Mutants That Suggest a Role for Conserved Residues Ser-234 and Asp-238 in Pyrophosphate Binding and Activation[a]

	Glutamine			Aspartic Acid			ATP		
	K_M (mM)	k_{cat} (s^{-1})	k_{cat}/K_M (M^{-1}s^{-1})	K_M (mM)	k_{cat} (s^{-1})	k_{cat}/K_M (M^{-1}s^{-1})	K_M (mM)	k_{cat} (s^{-1})	k_{cat}/K_M (M^{-1}s^{-1})
wt AS-B	0.69 ± 0.07	1.01 ± 0.05	1463	0.68 ± 0.07	1.05 ± 0.04	1544	0.18 ± 0.01	1.10 ± 0.03	6111
S234A	0.49 ± 0.77	0.35 ± 0.02	714	18.5 ± 2.5	0.54 ± 0.03	29.2	0.53 ± 0.04	0.43 ± 0.11	811
G236A	0.35 ± 0.024	0.23 ± 0.005	657	43.7 ± 3.6	0.361 ± 0.013	8.21	6.95 ± 0.53	0.40 ± 0.015	57.5
D238E	0.59 ± 0.04	0.28 ± 0.01	475	54.8 ± 5.0	0.31 ± 0.01	5.6	4.19 ± 0.76	0.35 ± 0.02	84

(B) Steady-State Kinetic Constants for the Glutaminase Activity of WT AS-B and the S234A, G236A, D238A, and D238E AS-B Mutants

	Glutamine (ATP Absent)			Glutamine (5 mM ATP Present)		
	K_M (mM)	k_{cat} (s^{-1})	k_{cat}/K_M (M^{-1}s^{-1})	K_M (mM)	k_{cat} (s^{-1})	k_{cat}/K_M (M^{-1}s^{-1})
wt AS-B	1.90 ± 0.20	0.80 ± 0.03	412	1.39 ± 0.07	1.38 ± 0.02	993
S234A	1.58 ± 0.10	0.52 ± 0.01	329	1.12 ± 0.08	0.60 ± 0.01	535
G236A	1.29 ± 0.16	1.23 ± 0.05	953	1.55 ± 0.18	1.44 ± 0.05	929
D238A	1.67 ± 0.13	1.41 ± 0.03	844	1.13 ± 0.81	1.39 ± 0.03	1230
D238E	1.37 ± 0.04	0.50 ± 0.004	365	1.08 ± 0.07	0.68 ± 0.01	629

[a] The D238A AS-B mutant has no detectable activity at glutamine concentrations up to 100 mM.

ties of AS-B, although preliminary experiments using the E100Q AS-B mutant show that changes in the kinetic parameters for its glutaminase activity are similar to those observed for the D98E AS-B mutant. The mechanistic interpretation of this result awaits detailed structural information on the enzyme.

Replacement of Asn-74 by a variety of amino acids gives AS-B mutants with an impaired ability to catalyze the conversion of glutamine to glutamate (Boehlein et al., 1994b, 1996) (Table 3). In particular, the N74D AS-B mutant exhibits no detectable glutaminase activity in high-performance liquid chromatography (HPLC) based assays that can detect picomoles of product (Boehlein et al., 1997b). Experiments employing glutamic acid γ-monohydroxamate (LGH) 5 and 4-cyano-2-aminobutyric acid 18 (Ressler and Ratzkin, 1961; Brysk and Ressler, 1970) (Fig. 4) as substrate analogues, and the N74A AS-B mutant, suggest (a) that the interaction between the glutamine side chain and Asn-74 appears to be destabilizing in the enzyme–substrate complex (Boehlein et al., 1996), and (b) that Asn-74 may be involved in stabilizing the acylthioenzyme intermediate 15 (Boehlein et al., 1997b) (Scheme 4). The functional role of Asn-74 in AS-B glutaminase activity therefore appears strikingly similar to that of Gln-19 in the active site of papain (Ménard et al, 1995; Dufour et al, 1995b). In the thiol proteases, it is likely that the side chain of this conserved glutamine residue defines an oxyanion hole, stabilizing either tetrahedral intermediate 14 and/or the thioester 15 in the mechanism of thiolate-catalyzed peptide bond hydrolysis (Scheme 4) based on experiments employing aldehyde (Lewis and Wolfenden, 1977; Mackenzie et al, 1986), and nitrile (Hanzlik et al., 1990; Moon et al, 1986; Brisson, et al., 1986; Liang and Abeles, 1987) inhibitors of papain. Further evidence for the catalytic equivalence of Gln-19 in papain and Asn-74 in AS-B has been obtained using nitrile 18, which is a weak, noncompetitive inhibitor of AS-B glutaminase activity at pH 6.5 (Boehlein et al., 1997b). Despite exhibiting no detectable glutaminase activity, the N74D AS-B mutant can employ 18 as a substrate, catalyzing the addition of water across the C—N triple bond to give glutamine (Scheme 7). The pH dependence of the kinetics of the hydration reaction indicates that the pK_a of the functional group mediating this activity is 7.0 ± 0.4, assuming that catalysis is modulated by a single residue (Boehlein et al., 1997b). Although it is possible that this reflects a significant perturbation in the acidity

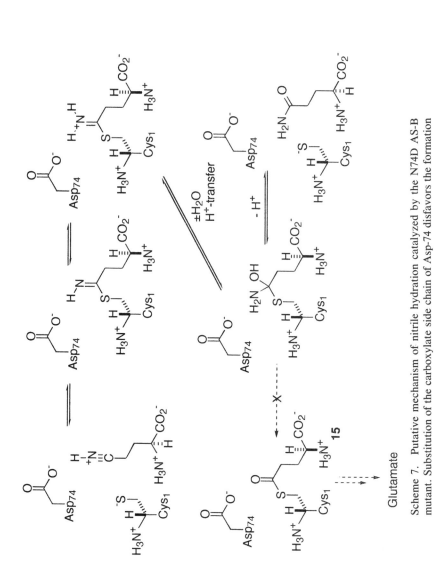

Scheme 7. Putative mechanism of nitrile hydration catalyzed by the N74D AS-B mutant. Substitution of the carboxylate side chain of Asp-74 disfavors the formation of the thioester intermediate **15** and so glutamine is released from the active site.

182

of the side chain of Asp-74 or the N-terminal amino group of Cys-1, this observation seems more consistent with the involvement of the Cys-1 thiolate in catalyzing nitrile hydration. Thus, protonation of **18** can occur after binding, due to the proximity of the Asp-74 carboxylate anion, and attack of the Cys-1 thiolate yields a thioimidate that can subsequently react with water to yield the tetrahedral intermediate **14** that is common to both the hydratase and glutaminase reaction mechanisms. In principle, **14** can partition to products via one of two pathways. In the first, which occurs exclusively in WT AS-B, C—N bond cleavage yields thioacylenzyme **15** that reacts with water to yield glutamate. In the N74D AS-B mutant, however, ejection of ammonia may be kinetically unfavorable due to the absence of the Asn-74 side chain, disfavoring formation of the thioester. Reaction of **14** therefore proceeds via C—S bond cleavage to give the resonance-stabilized amide moiety of glutamine.

Significantly, the Q19E papain mutant also employs peptide nitriles that are potent inhibitors of WT papain, as substrates, and exhibits a decreased ability to catalyze amide bond hydrolysis. For example, the Q19E papain mutant has a six-fold reduction in amidohydrolase activity (k_{cat}/K_m) relative to WT enzyme when CBZ-Phe-Ala-NH$_2$ is the substrate (Dufour et al., 1995a). These observations therefore provide the first direct chemical evidence for a close mechanistic relationship between the amidohydrolase mechanisms of papain and Class II Ntn amidotransferases. Sequence analysis does not appear to support the idea that these enzyme families are related by common ancestry and the three-dimensional folds of papain (Kamphuis et al., 1984) and the N-terminal glutamine-utilizing domain of glutamine PRPP amidotransferase (Smith et al., 1994; Kim et al., 1996) are clearly different. The active sites of these two families of enzymes appear to have arisen through mechanistic convergence (Doolittle 1994; Kraut, 1977), the functional equivalence of Gln-19 in papain and Asn-74 in AS-B being dictated by the chemical nature of the transition states and intermediates formed during thiol-catalyzed amide hydrolysis.

C. PROBING THE MECHANISM OF NITROGEN TRANSFER IN AS-B USING HEAVY-ATOM KINETIC ISOTOPE EFFECTS

Despite the clear mechanistic similarities between AS-B and papain, there are some interesting discrepancies in the catalytic activities exhibited throughout the Class II amidotransferase family (Zalkin, 1993). For example, asparagine synthetases and glutamine PRPP amidotransferases can both employ ammonia as an alternate nitrogen source and possess a glutaminase activity that is stimulated by the presence of other substrates (Boehlein et al., 1994b, Kim et al. 1995). In contrast, GFAT cannot employ exogenous ammonia in the synthesis of glucosamine 6-phosphate (Golinelli-Pimpaneau et al., 1989), and glutamate synthase from *Azospirillum brasilense* has no detectable glutaminase activity (Vanoni et al., 1991a,b).

Heavy atom KIEs have been widely used in the elucidation of enzyme mechanisms as they provide information about the first irreversible step as well as prior, moderately slow steps (O'Leary, 1980). The AS-B glutaminase and glutamine-dependent synthetase activities were therefore probed using multiply labelled glutamine analogs (Stoker et al., 1996), the small KIEs being determined by isotope ratio mass spectroscopy (O'Leary and Marlier, 1979). These experimental values were interpreted by reference to previous ^{15}N KIE determinations for papain-catalyzed peptide hydrolysis in which $^{15}(V/K)$ (Northrop, 1977) was observed to be 1.022 (O'Leary et al., 1974). This heavy-atom KIE was shown to be consistent with a mechanistic model in which C—N bond cleavage, giving the thioester intermediate **15** (Scheme 4), is the first irreversible step in the papain-catalyzed reaction. For AS-B, although $^{15}(V/K)_5$ was observed to be 1.022 ± 0.002 in the synthetase reaction, $^{15}(V/K)_5$ was significantly decreased in glutamine hydrolysis, being only 1.01 ± 0.004. These KIEs were expected to be identical for ammonia-mediated nitrogen transfer given that C—N bond cleavage should involve the same intermediates and transition states in the reaction catalyzed by the GAT-domain active site. In order to rule out the possibility that these differences in $^{15}(V/K)_5$ arose from modified commitment factors for amide hydrolysis in the AS-B glutaminase and synthetase activities, the effect of placing a ^{13}C label in the carbonyl group of the glutamine substrate was also examined. In this case, $^{13}(V/K)$ was found to be identical, within experimental error, for the two glutamine-dependent reactions. While the cognate KIE for papain-catalyzed amide hydrolysis does not appear to have been determined, experiments employing suitably labeled substrates and serine proteases indicate that $^{13}(V/K)$ has a value of 1.024 for rehybridization of the carbon atom during formation and breakdown of the tetrahedral intermediates in the mechanism of this enzyme-catalyzed reaction (O'Leary and Kluetz, 1970, 1972). The changes in $^{15}(V/K)_5$ appear unlikely to be due to significant changes in the commitment factors for breakdown of the tetrahedral intermediate. These data therefore appear to represent evidence against ammonia-mediated nitrogen transfer during asparagine biosynthesis, although it is possible that other effects such as conformational changes in the protein might give rise to alterations in the vibrational modes of the transition state structures for which these KIEs are observed. Given that $^{15}(V/K)_5$ is similar in magnitude to the ^{15}N KIE observed during papain-catalyzed amide hydrolysis, these data also support direct attack of the tetrahedral intermediate on β-aspartyl-AMP as being the most likely mechanism of nitrogen transfer that involves the formation of a covalent intermediate [Scheme $5(b)$], although the existence of an imide intermediate could not be eliminated solely on the basis of these measurements.

The small magnitude of $^{15}(V/K)_5$ observed for the AS-B glutaminase activity might also reflect a subtle difference in the nature of the N-protonation step for papain and AS-B catalyzed amide hydrolysis. In papain, current evidence suggests that C—N bond cleavage and N-protonation are concerned, that is, that there is general acid catalysis of amine loss from the tetrahedral intermediate involving His-159, which is ideally placed to perform such a function (Kamphuis et al., 1984). Given the apparent absence of a similar catalytic histidine residue in the AS-B GAT-domain, N-protonation might proceed via specific acid catalysis in which the proton is provided by solvent, possibly being facilitated by the N-terminal α-amino group of Cys-1. The KIE associated with N-protonation would be large and inverse (Hermes et al., 1985) thereby acting to reduce the observed $^{15}(V/K)_5$ for C—N bond cleavage.

The molecular basis of the ability of ATP, in the absence of aspartic acid, to stimulate turnover of glutamine to glutamate was also investigated using these heavy-atom KIEs (Stoker et al., 1996). Neither $^{13}(V/K)$ nor $^{15}(V/K)_5$ was affected significantly by the presence of ATP during AS-B catalyzed glutamine hydrolysis, suggesting that the rate-limiting step for this reaction occurs after C-N bond breaking. Thus ATP might exert its effect on the glutaminase activity by altering the conformation of AS-B such that glutamate release is accelerated. This result is consistent with similar proposals for GPA (Kim et al., 1995).

D. REEVALUATING THE KINETIC MECHANISM OF ASPARAGINE SYNTHETASES

As noted earlier, important evidence for the intermediacy of ammonia in nitrogen transfer has been provided by studies of the kinetic mechanism of murine and bovine asparagine synthetases (Milman et al., 1980; Markin et al., 1981). These experiments suggest that glutamine binds to the free enzyme and undergoes hydrolysis with subsequent binding of ATP and aspartic acid to an E.NH$_3$ complex. Such a kinetic mechanism, however, raises a series of important molecular issues. First, it is conceptually difficult to reconcile glutamate release and subsequent substrate binding with a model in which asparagine synthetase undergoes a conformational change to exclude water and sequester ammonia. Second, these observations are distinctive to glutamine-dependent asparagine synthetases in that all other Class II amidotransferases, for which the kinetic mechanism has been determined, bind glutamine after the other substrates involved in the synthetase activity (Golinelli-Pimpaneau et al., 1989; Vanoni et al., 1991b). Third, such a mechanism should exhibit identical ^{15}N KIEs given than C—N bond cleavage would be the first irreversible step in both glutaminase and synthetase activity. The apparent differences in complexity and in many details of the proposed kinetic mechanisms of murine and bovine AS are also surprising, albeit that these enzymes are obtained from different sources.

Some resolution of these issues has been provided by studies on the kinetic mechanism of AS-B (Boehlein et al., unpublished results) using steady-state kinetics (Segel, 1975) and isotope trapping experiments (Rose, 1980). We note that the interpretation of substrate and product inhibition experiments is complicated, however, by the fact that glutamine can, in principle, not only bind to free enzyme, but also to the E.ATP complex since ATP stimulates AS-B glutaminase activity in the absence of aspartic acid (Boehlein et al., 1994b). Isotope trapping experiments using ^{3}H-labeled glutamine are consistent with a mechanism in which catalytically competent E.ATP and E.Asp.ATP complexes can be formed in the absence of glutamine (Table 4), and ATP probably binds to free enzyme as the initial step in the overall synthetase reaction. The dissociation constant, K_d, of 12 μM for the E.ATP complex has been determined using chemical modification experiments employing FSBA (Boehlein et al., 1997c).

TABLE 4

Isotope Trapping Data for the Glutamine-Dependent Synthetase Activity of WT AS-B at 37°C, pH 8[a]

(A) Formation of ^{3}H-AMP Using a Pulse Solution Containing ^{3}H-ATP in the Presence or Absence of Other Substrates

Pulse Substrates	^{3}H-AMP Formed per nmol WT AS-B (nmol)
ATP only	0.29 ± 0.0.039
ATP/Gln	0.29 ± 0.0.008
ATP/Asp	0.44 ± 0.0.024

(B) Formation of ^{14}C-Asn Using a Pulse Solution Containing ^{14}C-Asp in the Presence or Absence of Other Substrates

Pulse Substrates	^{14}C-Asn Formed per nmol WT AS-B (nmol)
Asp only	0.040 ± 0.0.019
Asp/Gln	0.040 ± 0.0.029
ATP/Asp	0.28 ± 0.0.011

[a] Boehlein et al., unpublished observations.

Steady-state kinetics and experiments employing radiolabeled ATP also show that glutamate is not released from the AS-B GAT-domain prior to the binding of either aspartate or ATP during asparagine formation. Furthermore, AS-B cannot catalyze the hydrolysis of ATP to AMP and PP_i in the absence of aspartic acid (Habibzadegah-Tari, unpublished observations), even when glutamine is present in solution. Finally, in common with many other glutamine-dependent amidotransferases (Zalkin, 1993), AS-B does not catalyze ATP/PP_i exchange.

This series of observations are consistent with a simplified kinetic mechanism in which glutamine binding and PP_i release are coupled so that glutamine and β-aspartyl-AMP are present simultaneously on the enzyme. Therefore, in contrast to previous studies (Milman et al., 1980; Markin et al., 1981), nitrogen transfer that is mediated by covalent intermediates cannot be explicitly ruled out for the bacterial enzyme on the basis of the kinetic mechanism.

It seems unlikely that the kinetic mechanism of bacterial and mammalian asparagine synthetases is different. While the early kinetic studies were carried out using AS preparations of substantially less purity than the AS-B obtained using modern expression systems, the underlying basis for the apparent mechanistic discrepancy probably lies in the experimental conditions under which the substrate and product inhibition of the mammalian enzymes were performed. In particular, the patterns observed in double-reciprocal plots for the substrate inhibition of asparagine synthesis by AS-B depend on glutamine concentration. Determination of the initial velocity of AS-B synthetase activity under conditions in which glutamine is maintained at 1-mM concentration and aspartate levels are varied, gives a pattern of intersecting lines for various fixed concentrations of ATP. If glutamine is maintained at 20 mM, parallel lines are observed in the substrate inhibition plots in identical experiments. Such behavior arises from the ability of glutamine to participate in two, independent enzyme-catalyzed reactions (Segel, 1975), and complicates the interpretation of steady-state measurements.

The kinetic mechanism of the AS-B ammonia-dependent synthetase activity has also been determined using similar approaches (Boehlein et al., unpublished results). Again, the nitrogen source binds to the bacterial enzyme after ATP and aspartate in a manner consistent with that observed for ammonia-dependent AS-A (Cedar and Schwartz, 1969b). The kinetic mechanisms of the mammalian asparagine synthetases should therefore be redetermined using purified recombinant material, given that the kinetic behavior of the bacterial AS-B gives rise to a consistent picture of substrate binding across all members of the Class II amidotransferase family.

E.   AN IMIDE INTERMEDIATE IS NOT FORMED IN THE SYNTHETASE REACTION CATALYZED BY ASPARAGINE SYNTHETASE B

All mechanisms underlying nitrogen transfer that do not involve the participation of enzyme-bound ammonia, involve formation of stable intermediates that might be isolable or directly observable under the correct experimental conditions. Although simple methods for characterizing the participation of tetrahedral intermediate 14 in nitrogen transfer are not readily available, imide 17 was expected to be sufficiently stable for isolation from WT AS-B. Further, 17 should be synthesized by the C1A AS-B mutant as the Cys-1 thiolate is required to catalyze C—N bond cleavage to yield asparagine and the thioacylenzyme **15** if this transfer mechanism is operative [Scheme 5(c)]. The highly functionalized, unsymmetrical imide **17** can be prepared in reasonable amounts as a single enantiomer by chemical synthesis (Scheme 8) (Rosa-Rodriguez et al., submitted for publication), allowing an evaluation of its reactivity and behavior with WT AS-B. Interestingly, although **17** is stable in solution below pH 7.0, it undergoes a rearrangement reaction at higher pH to give the amide **23**, the identity of which has been confirmed by independent synthesis. The availability of authentic imide **17** has allowed the characterization of its behavior in electrospray ion cyclotron resonance (ICR) mass spectroscopy (Marshall and Grosshans, 1991), and the development of sensitive reverse-phase HPLC assays (Boehlein et al., 1997b) for detecting low concentrations of the PITC derivative of **23**, formed stoichiometrically from **17** under standard derivatization conditions (pH 10).

Given the availability of large amounts of recombinant WT AS-B, the imide **17** has been incubated with large amounts of the enzyme at pH 6.5. No evidence was obtained for the ability of **17** to behave as a substrate and form asparagine and glutamate under these condi-

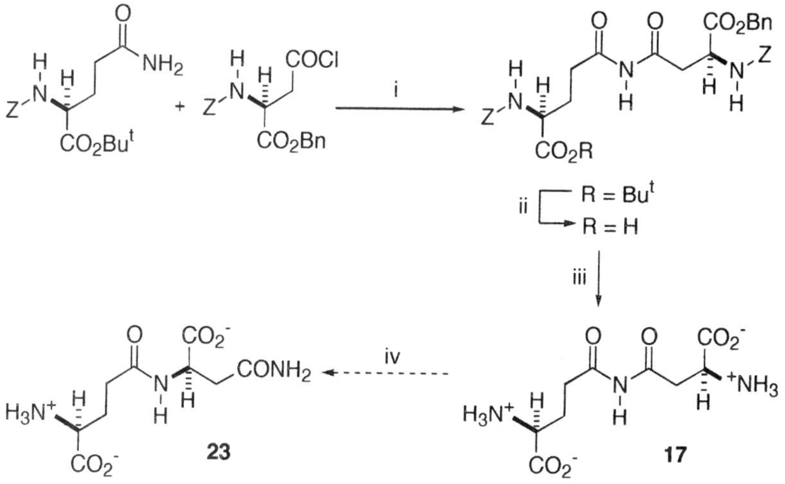

Scheme 8.   Synthesis of putative imide-intermediate 17 from L-glutamine and L-aspartic acid. Z = $PhCH_2OC(=O)$. (i) Pyridine, $CHCl_3$, RT, 32%; (ii) CF_3CO_2H, CH_2Cl_2, 0°C, 94%; (iii) 1,4-Cyclohexadiene, 10% Pd/C. EtOH/THF, reflux, 90%; (iv) Tris buffer, H_2O, pH 8.1. Imide **17** is stable at pH 6.5.

tions. This apparent lack of reactivity does not rule out the involvement of 17 in nitrogen transfer, however, as the correct conformational form for hydrolysis of the imide may not be accessible to the free enzyme. The ability of the C1A AS-B mutant, available in large amounts using our expression system, to synthesize imide 17 at pH 6.5 was therefore investigated. Production of 17 by this AS-B mutant, after incubation with glutamine, ATP and aspartic acid and subsequent denaturation of the protein, is not detectable using either HPLC or ICR analysis. These experiments appear to eliminate the participation of an imide intermediate in nitrogen transfer in bacterial AS.

VI. Conclusions

Despite the lack of a high-resolution structure for any asparagine synthetase, significant progress has been made in understanding the mechanism by which asparagine is synthesized by this enzyme. Residues that mediate aspartate binding and activation have been identified using site-directed mutagenesis and chemical modification experiments. Although the evolutionary relationship of aminoacyl tRNA synthetases and the C-terminal domain of glutamine-dependent amidotransferases is not apparent on the basis of sequence homology, it is likely that these enzymes employ similar functional groups for catalyzing acyl-AMP formation, as expected on the basis of mechanistic imperative (Babbit et al., 1996).

The key issue of the molecular mechanism by which nitrogen is transferred from the GAT-domain to the synthetase active site of the enzyme, however, remains unclear in the absence of crystallographic data. Although a mechanistic relationship has been established between the chemistry catalyzed by the AS-B GAT-domain and papain, KIE measurements indicate that the transition state for the first irreversible step in the synthetase and glutaminase activities of the enzyme may be different. Further, in contrast to previous reports, the kinetic mechanism of AS-B does not rule out a nitrogen-transfer process that is not mediated by ammonia. There are also good chemical arguments for the involvement of tetrahedral intermediate 14 in nitrogen transfer. The picture has been complicated, however, by the exciting discovery of a putative channel in the crystal structure of carbamoyl phosphate synthetase (CPS) (Thoden et al.,

1997). This channel is proposed to allow movement of sequestered ammonia from the glutamine-hydrolyzing site in the small subunit of CPS to a synthetase site in the large subunit. The existence of a similar channel in AS would remove both of the key chemical problems that have been raised against ammonia-mediated nitrogen transfer, and the KIE might then reflect conformational differences in the enzyme during the glutaminase and synthetase reactions. On the other hand, the relevance of the structure and mechanism of CPS, which is composed of two subunits, to Class II amidotransferases, in which the two domains are formed from a single polypeptide chain, has yet to be established. This issue will only be resolved using structural data on AS-B and studies of the mechanism using rapid-quench kinetics techniques.

Whatever the nitrogen-transfer mechanism, it is interesting to speculate on the evolutionary issues in using glutamine as a nitrogen source in asparagine synthesis. On the basis of the palimpsest model, it has been argued that pyridoxal phosphate (PLP) is a relatively recent cofactor (Benner et al., 1993). In the event that asparagine synthetases, and other Class II amidotransferases, represent examples of ancient enzymes, PLP-dependent nitrogen transfer would not have been possible. Release of nitrogen by amide hydrolysis, which requires only a single cysteine residue, is an elegant solution to this problem.

Acknowledgments

We would like to acknowledge the extensive efforts of many co-workers who have been involved in developing the mechanistic understanding of asparagine synthetases in our laboratories. Particular thanks go to Dr. Susan Boehlein who has been an invaluable participant in, and contributor to, all aspects of the studies on *Escherichia coli* AS-B. Our work on asparagine synthetases has also benefitted from many discussions with Dr. V. Jo Davisson (Purdue), Dr. Bernard Badet (CNRS, Gif-sur-Yvette) and Dr. Jon D. Stewart (Florida). We also appreciate the patience of Dr. Dan Purich during the preparation of this chapter. Funding for studies on asparagine synthetase has been provided by the National Institutes of Health (CA 28725).

References

Abell, L. M. and Villafranca, J. J., *Biochemistry*, **30**, 1413–1418 (1991).

Andrulis, I. L., Argonza, R., and Cairney, A. E. L., *Som. Cell. Molec. Genet.*, **16**, 59–65 (1990).

Andrulis, I. L., Duff, C., Evans-Blackler, S., Worton, R., and Siminovitch, L., *Mol. Cell Biol.*, **3**, 391–398 (1983).

Andrulis, I. M., Chen, J., and Ray, P. N., *Mol. Cell. Biol.*, **7**, 2435–2443 (1987).

Arad, D., Langridge, R., and Kollman, P. A., *J. Am. Chem. Soc.*, **112**, 491–502 (1990).

Arfin, S. M., *Biochim. Biophys. Acta*, **136**, 233–244 (1967).

Babbitt, P. C., Hasson, M. S., Wedekind, J. E., Palmer, D. R., Barrett, W. C., Reed, G. H., Rayment, I., Ringe, D., Kenyon, G. L., and Gerlt, J. A., *Biochemistry*, **35**, 16489–16501 (1996).

Badet-Denisot, M. A. and Badet, B., *Arch. Biochem. Biophys.*, **292**, 475–478 (1992).

Badet-Denisot, M. A., Leriche, C., Massiere, F., and Badet, B., *Bioorg. Med. Chem. Lett.*, **5**, 815–820 (1995).

Badet-Denisot, M. A., René, L., and Badet, B., *Bull. Soc. Chim. Fr.*, **130**, 249–255 (1993).

Baldwin, E. T., Bhat, T. N., Gulnik, S., Liu, B., Topol, I. A., Kiso, Y., Mimoto, T., Mitsuya, H., and Erickson, J. W., *Structure*, **3**, 581–590 (1995).

Baldwin, J. E., Moloney, M. G., and North, M., *Tetrahedron*, **45**, 6319–6325 (1989).

Benner, S. A., Cohen, M. A., Gonnet, G. H., Berkowitz, D. B., and Johnsson, K. P., In *The RNA World*, Gesterland, R. and Atkins, J., Eds., Cold Spring Harbor Press, (1993), Cold Spring Harbor, NY, pp. 27–70.

Boehlein, S. K., Rosa-Rodriguez, J. G., Schuster, S. M., and Richards, N. G. J., *J. Am. Chem. Soc.*, **119**, 5785–5791 (1997b).

Boehlein, S. K., Richards, N. G. J., and Schuster, S. M., *J. Biol. Chem.*, **269**, 7450–7457 (1994a).

Boehlein, S. K., Richards, N. G. J., and Schuster, S. M., *J. Biol. Chem.*, **269**, 26789–26795 (1994b).

Boehlein, S. K., Schuster, S. M., and Richards, N. G. J., *Biochemistry*, **35**, 3031–3037 (1996).

Boehlein, S. K., Walworth, E. S., Richards, N. G. J., and Schuster, S. M., *J. Biol. Chem.*, **272**, 12384–12392 (1997a).

Boehlein, S. K., Walworth, E. S., and Schuster, S. M., *Biochemistry*, **36**, 10168–10177 (1997c).

Bork, P. and Koonin, E. V., *Proteins Struct. Funct. Genet.*, **20**, 347–355 (1994).

Bork, P. and Kroonin, E. V., *Curr. Op. Struct. Biol.*, **6**, 366–376 (1996).

Brannigan, J. A., Dodson, G., Duggleby, H. J., Moody, P. C. E., Smith, J. L., Tomchick, D. R., and Murzin, A. G., *Nature (London)*, **378**, 416–419 (1995).

Brisson, J.-R., Carey, P. R., and Storer, A. C., *J. Biol. Chem.*, **261**, 9087–9089 (1986).

Brocklehurst, K., Willenbrock, and F., Salih, E., In *Hydrolytic Enzymes*, Neuberger, A. and Brocklehurst, K., Eds., Elsevier Biomedical, Amsterdam, (1987) pp. 39–158.

Broome, J. D., *J. Exp. Med.*, **127**, 1055–1072 (1968).

Brysk, M. M. and Ressler, C., *J. Biol. Chem.*, **245**, 1156–1160 (1970).

Buchanan, J. M., *Adv. Enzymol. Relat. Areas Mol. Biol.*, **39**, 91–183 (1973).

Bult, C. J., White, O., Olsen, G. J., Zhou, L., Fleischmann, R. D., Sutton, G. G., Blake, J. A., FitzGerald, L. M., Clayton, R. A., Gocayne, J. D., Kerlavage, A. R., Dougherty, B. A., Tomb, J. F., Adams, M. D., Reich, C. I., Overbeek, R., Kirkness, E. F., Weinstock, K. G., Merrick, J. M., Glodek, A., Scott, J. L., Geoghagen, N. S. M., Weidman, J. F., Fuhrmann, J. L., Nguyen, D., Utterback, T. R., Kelley, J. M., Paterson, J. D., Sadow, P. W., Hanna, M. C., Cotton, M. D., Roberts, K. M., Hurst, M. A., Kaine, B. P., Borodovsky, M., Klenk, H. P., Fraser, C. M., Smith, H. O., Woese, C. R., and Venter, J. C., *Science*, **273**, 1058–1073 (1996).

Burchall, J. J., Reichelt, E. C., and Wolin, M. J. *J. Biol. Chem.*, **239**, 1794–1798 (1964).

Carter, C. W., *Annu. Rev. Biochem.*, **62**, 715–748 (1993).

Castric, P. A., Farnden, K. J., and Conn, E. E. *Arch. Biochem. Biophys.*, **152**, 62–69 (1972).

Cedar, H. and Schwartz, J. H., *J. Biol. Chem.*, **244**, 4112–4121 (1969a).

Cedar, H. and Schwartz, J. H., *J. Biol. Chem.*, **244**, 4122–4127 (1969b).

Chakrabati, R., Chakrabati, D., Souba, W., and Schuster, S. M., *J. Biol. Chem.*, **268**, 1298–1303 (1993).

Challis, B. C. and Challis, J. A., In *The Chemistry of Amides* Zabicky, J., Ed., Interscience London, (1970) pp. 731–857.

Chaparian, M. G. and Evans, D. R., *J. Biol. Chem.*, **266**, 3387–3395 (1991).

Chevalier, C., Beorgeois, E., Just, D., and Raymond, P., *Plant J.*, **9**, 1–11 (1996).

Cleland, W. W. and Kreevoy, M. M., *Science*, **264**, 1887–1890 (1994).

Colman, R. F., In *The Enzymes XIX*, 3rd ed, Sigman, D. S. and Boyer, P. D., Eds., Academic San Diego, CA, (1990) pp. 283–321.

Cooney, D. A., Driscoll, J. S., Milman, H. A., Jayaram, H. N., and Davis, R. D., *Cancer Treat. Rep.*, **60**, 1493–1557 (1976).

Cooney, D. A., Jones, M. T., Milman, H. A., Young, D. M., and Jayaram, H. N., *Int. J. Biochem.*, **11**, 519–539 (1980).

Dahnke, T., Shi, Z., Hoggao, Y., Jiang, T. T., and Tsai, M. D., *Biochemistry*, **31**, 6318–6328 (1992).

Davidson, D. and Skovronek, H., *J. Am. Chem. Soc.*, **80**, 376–379 (1958).

Devereux, J. R., Haeberli, P., and Smithies, O., *Nucleic Acids Res.*, **12**, 387–395 (1984).

Dewar, M. J. S., Zoebisch, E. G., Healy, E. F., and Stewart, J. J. P., *J. Am. Chem. Soc.*, **107**, 3902–3909 (1985).

Di Giulio, M., *J. Mol. Evol.*, **37**, 5–10 (1993).

Doolittle, R. F., *Trends Biochem. Sci.*, **19**, 15–18 (1994).

Drenth, J., Jansonius, J., Koekoek, R., and Wolthers, B. G., In *The Enzymes*, 3rd ed., Sigman, D. S. and Boyer, P. D., Eds., Academic San Diego; CA, (1971) pp. 485–499.

Dufour, E., Ménard, R., and Storer, A. C., *Biochemistry*, **34**, 9136–9143 (1995b).

Dufour, E., Storer, A. C., Ménard, *Biochemistry*, **34**, 16382–16388 (1995a).

Duggleby, H. J., Tolley, S. P., Hill, C. P., Dodson, E. J., Dodson, G., and Moody, P. C., *Nature (London)*, **373**, 264–268 (1995).

Ertel, I. J., Nesbit, M. G., Hammon, D., Weiner, J., and Sather, H., *Cancer Res.*, **39**, 3893–3896 (1979).

Feng, D. F. and Doolittle, R. F., *J. Mol. Evol.*, **25**, 351–360 (1987).

Foote, J., Lauritzen, A. M., and Lipscomb, W. N., *J. Biol. Chem.*, **260**, 9624–9629 (1985).

Gerlt, J. A. and Gassman, P. G., *J. Am. Chem. Soc.*, **115**, 11552–11568 (1993).

Ghosh, G., Brunie, S., and Schulman, L.H., *J. Biol. Chem.*, **266**, 17136–17141 (1991b).

Ghosh, G., Pelka, H., Schulman, and L., Brunie, S., *Biochemistry*, **30**, 9569–9575 (1991a).

Gibbs. R. A., Taylor, S., and Benkovic, S. J., *Science*, **258**, 803–805 (1992).

Golinelli-Pimpaneau, B., Le Goffic, and F., Badet, B., *J. Am. Chem. Soc.*, **111**, 3029–3034 (1989).

Gong, S. S. and Basilico, C., *Nucleic Acids Res.*, **18**, 3509–3513 (1990).

Gong, S. S., Guerrini, L., and Basilico, C., *Mol. Cell. Biol.*, **11**, 6059–6066 (1991).

Greer, J., *Proteins Struct. Funct. Genet.*, **7**, 314–334 (1990).

Groll, M., Ditzel, L., Lowe, J., Stock, D., Bochtler, M., Bartunik, and H. D., Huber, R., *Nature (London)*, **386**, 463–471 (1997).

Guerrini, L., Gong, S. S., Mangasarian, and K., Basilico, C., *Mol. Cell. Biol.*, **13**, 3203–3212 (1993).

Guida, W. C., Bohacek, R. S., and Erion, M. D., *J. Comput. Chem.*, **13**, 214–218 (1992).

Handschumacher, R. E., Bates, C. J., Chang, P. K., Andrews, A. T., and Fischer, G. A., *Science*, **161**, 62–63 (1968).

Hanzlik, R. S., Zygmunt, J., and Moon, J. B., *Biochim. Biophys. Acta*, **1035**, 62–70 (1990).

Harris, C. M. and Harris, T. M., *J. Am. Chem. Soc.*, **104**, 4293–4295 (1982).

Heng H. H. Q., Shi X.-M., Scherer S. W., Andrulis I. L., and Tsui L.-C., *Cytogenet Cell Genet.*, **66**, 135–138 (1994).

Hermes, J. D., Weiss, P. M., and Cleland, W. W., *Biochemistry*, **24**, 2959–2967 (1985).

Hersh, E. M., *Transplantation*, **12**, 368–376 (1971).

Hinchman, S. K., Henikoff, S. A., and Schuster, S. M., *J. Biol. Chem.*, **267**, 144–149 (1992).

Hinchman, S. K., and Schuster, S. M., *Protein Eng.*, **5**, 279–283 (1992).

Hol, W. G. J., van Duijnen, P. T., and Berendsen, H. C., *Nature (London)*, **73**, 443–446 (1978).

Hongo, S., Motosugu, F., Shioda, S., Nakai, Y., Takeda, M., and Sato, T., *Biochem. Biophys.*, **295**, 120–125 (1992).

Hongo, S., and Sato, T., *Biochem. Biophys. Acta*, **742**, 484–489 (1983).

Horowitz, B. and Meister, A., *J. Biol. Chem.*, **247**, 6708–6719 (1972).

Howard, A. E. and Kollman, P. A., *J. Am. Chem. Soc.*, **110**, 7195–7200 (1988).

Huang, Y.-Z. and Knox, E. W., *Enzyme*, **19**, 314–328 (1975).

Hutson, R. G., Kitoh, T., Moraga-Amador, D., Cosic, S., Schuster, S. M., and Kilberg, M. S., *Am. J. Physiol.*, **272**, C1691–C1699 (1997).

Isupov, M. N., Obmolova, G., Butterworth, S., Badet-Denisot, M. A., Badet, B., Polikarpov, I., Littlechild, J. A., and Teplyakov, A., *Structure*, **4**, 801–810 (1996).

Jayaram, H. N., Cooney, D. A., Ryan, J. A., Neil, G., Dion, R. L., and Bono, V. H., *Cancer Chemother. Rep.* **59**, 481–491 (1975).

Jayaram, H. N., Cooney, D. A., Milman, H. A., Homan, E. R., and Rosenbluth, R. J., *Biochem. Pharmacol.*, **25**, 1571–1582 (1976).

Kamphuis, I. G., Kalk, H. M., Swarte, M. B. A., and Drenth, J., *J. Mol. Biol.*, **179**, 233–256 (1984).

Kaplan, H. A., Welply, J. K., and Lennarz, W. J., *Biochim. Biophys. Acta*, **906**, 161–173 (1987).

Keefer, J. F., Moraga-Amador, D., and Schuster, S. M., *Biochem. Med.*, **34**, 135–150 (1985a).

Keefer, J. F., Moraga-Amador, D., and Schuster, S. M., *Biochem. Pharmacol.*, **34**, 559–565 (1985b).

Kim, J. H., Krahn, J. M., Tomchick, D. R., Smith, J. L., and Zalkin, H., *J. Biol. Chem.*, **271**, 15549–15557 (1996).

Kim, J. H., Wolle, D., Haridas, K., Parry, R. J., Smith, J. L., and Zalkin, H., *J. Biol. Chem.*, **270**, 17394–17399 (1995).

Klamt, A. and Schüürmann, G., *J. Chem. Soc. Perkin Trans.*, **2**, 799–805 (1993).

Kornfeld, R., and Kornfeld, S., *Annu. Rev. Biochem.*, **54**, 631–664 (1985).

Kraut, J., *Annu. Rev. Biochem.*, **46**, 331–358 (1977).

Lam, H. M., Coschigano, K. T., Oliveira, I. C., Melo-Oliveira, R., and Coruzzi, G. M., *Annu. Rev. Plant Physiol.*, **47**, 569–593 (1996).

Lam, H. M., Peng, S. S., and Coruzzi, G. M., *Plant Physiol.*, **106**, 1347–1357 (1994).

Lewis, Jr., C. A. and Wolfenden, R., *Biochemistry*, **16**, 4890–4894 (1977).

Leyh, T. S., Vogt, T. F., and Suo, Y., *J. Biol. Chem.*, **267**, 10405–10410 (1992).

Liang, T. C. and Abeles, R. H., *Arch. Biochem. Biophys.*, **252**, 626–634 (1987).

Lowe, J., Stock, D., Jap, B., Zwickl, P., Baumeister, W., and Huber, R., *Science*, **268**, 533–539 (1995).

Lowe, G. and Williams, A., *Biochem. J.*, **96**, 194–198 (1964).

Lu, Y., and Hill, K. A., *J. Biol. Chem.*, **269**, 12137–12141 (1994).

Luehr, C. A. and Schuster, S. M., *Arch. Biochem. Biophys.*, **237**, 335–346 (1985).

Mackenzie, N. E., Grant, S. K., Scott, A. I., and Malthouse, J. P. G., *Biochemistry*, **25**, 2293–2298 (1986).

Mäntsälä, P. and Zalkin, H., *J. Bacteriol.*, **174**, 1883–1890 (1992).

Markin, R. S., Leuhr, C. A., and Schuster, S. M., *Biochemistry*, **20**, 7226–7232 (1981).

Marshall, A. G. and Grosshans, P. B., *Analyt. Chem.*, **63**, 215A–229A (1991).

Maul, D. M. and Schuster, S. M., *Life Sci.*, **30**, 1051–1057 (1982).

Maul, D. M. and Schuster, S. M., *Arch. Biochem. Biophys.*, **251**, 577–584 (1986a).

Maul, D. M. and Schuster, S. M., *Arch. Biochem. Biophys.*, **251**, 585–593 (1986b).

Mei, B. and Zalkin, H., *Biol. Chem.*, **264**, 16613–16619 (1989).

Ménard, R., Plouffe, C., Laflamme, P., Vernet, T., Tessier, D. C., Thomas, D. Y., and Storer, A. C., *Biochemistry*, **34**, 464–471 (1995).

Milman, H. A. and Cooney, D. A., *Biochem. J.*, **181**, 51–59 (1979).

Milman, H. A., Cooney, D. A., and Huang, C. Y., *J. Biol. Chem.*, **255**, 1862–1866 (1980).

Mohamadi, F., Richards, N. G. J., Guida, W. C., Liskamp, R. M. J., Lipton, M. A., Caufield, C. E., Chang, G., Hendrickson, T. F., and Still, W. C., *J. Comput. Chem.*, **11**, 440–467 (1990).

Mokotoff, M., Bagaglio, J. F., and Parikh, B. S., *J. Med. Chem.*, **18**, 354–358 (1975).

Montgomery, J. A., Niwas, S., Rose, J. D., Secrist, J. A., Babu, Y. S., Bugg, C. E., Erion, M. D., Guida, W. C., and Ealick, S. E., *J. Med. Chem.*, **36**, 1847–1854 (1993).

Moon, J. B., Coleman, R. S., and Hanzlik, R. P., *J. Am. Chem. Soc.*, **108**, 1350–1351 (1986).

Musil, D., Zucic, D., Turk, D., Engh, R. A., Mayr, I., Huber, R., Popovic, T., Turk, V., Towatari, T., Katunuma, X., and Bode, W., *EMBO J.*, **10**, 2321–2330 (1991).

Northrop, D. B., In Cleland, W. W., O'Leary, M. H., Northrop, D. B., Eds., *Isotope Effects on Enzyme Catalyzed Reactions*, University Park Press, Baltimore, 1977, pp. 122–148.

Oinonen, C., Tikkanen, R., Rouvinen, J., and Peltonen, L., *Nature Struct. Biol.*, **2**, 1105–1108 (1995).

O'Leary, M. H., *Adv. Enzymol.*, **64**, 83–104 (1980).

O'Leary, M. H., *Annu. Rev. Biochem.*, **58**, 377–401 (1989).

O'Leary, M. H. and Kluetz, M. D., *J. Am. Chem. Soc.*, **92**, 6089–6090 (1970).

O'Leary, M. H. and Kluetz, M. D., *J. Am. Chem. Soc.*, **94**, 3585–3589 (1972).

O'Leary, M. H. and Marlier, J. F., *J. Am. Chem. Soc.*, **101**, 3300–3306 (1979).

O'Leary, M. H., Urberg, M., and Young, A. P., *Biochemistry*, **13**, 2077–2081 (1974).

Parr, I. B., Boehlein, S. K., Dribben, A. B., Schuster, S. M., and Richards, N. G. J., *J. Med. Chem.*, **39**, 2367–2378 (1996).

Patterson, M. K. and Orr, G. R., *J. Biol. Chem.*, **243**, 376–380 (1968).

Peräkylä, M. and Kollman, P. A., *J. Am. Chem. Soc.*, **119**, 1189–1196 (1997).

Perona, J. J., Hedstrom, L., Wagner, R. L., Rutter, W. J., Craik, C. S., and Fletterick, R. J., *Biochemistry*, **33**, 3252–3259 (1994).

Pfeiffer, N. E., Mehlhaff, P. M., Wylie, D. E., and Schuster, S. M., *J. Biol. Chem.*, **261**, 1914–1919 (1986).

Pfeiffer, N. E., Mehlhaff, P. M., Wylie, D. E., and Schuster, S. M., *J. Biol. Chem.*, **262**, 11565–11570 (1987).

Phillips, M. A., Hedstrom, L., and Rutter, W. J., *Protein Sci.*, **1**, 517–521 (1992).

Ramos, F. and Waime, J. M., *Eur. J. Biochem.*, **94**, 409–417 (1979).

Ratner, S., *Adv. Enzymol. Related Areas Mol. Biol.*, **39**, 1–90 (1973).

Ravel, J. M., Norton, S. J., Humphreys, J. S., and Shive, W., *J. Biol. Chem.*, **237**, 2845–2849 (1962).

Reitzer, L. J. and Magasanik, B., *J. Bacteriol.*, **151**, 1299–1313 (1982).

Ressler, C. and Ratzkin, H., *J. Org. Chem.*, **26**, 3356–3360 (1961).

Richards, N. G. J. and Schuster, S. M., *FEBS Lett.*, **313**, 98–102 (1992).

Rose, I. A., *Methods Enzymol.*, **64**, 47–59 (1980).

Roux, B. and Walsh, C. T., *Biochemistry*, **31**, 6904–6910 (1992).

Ryan, W. L. and Dworak, T. E., *Cancer Res.*, **30**, 1206–1209 (1970).

Ryan, W. L. and Sorenson, G. L., *Science*, **167**, 1512–1513 (1970).

Sanz, G. F., Sanz, M. A., Rafecas, F. J., Martinez, J. A., Martin-Aragon, G., and Marty, M. L., *Cancer Treat. Rep.*, **70**, 1321–1324 (1986).

Schulz, G. E., *Curr. Op. Struct. Biol.*, **2**, 61–67 (1992).

Schuster, S. M., *In Biochemistry of Metabolic Processes*, Lennon, D. L. F., Stratman, F. W., and Zahlten, R. N., Eds., Elsevier Biomedical, New York, (1982) pp. 323–336.

Scofield, M. A., Lewis, W. S., and Schuster, S. M., *J. Biol. Chem.*, **265**, 12895–12902 (1990).

Segel, I. H., *Enzyme Kinetics*, Wiley-Interscience, New York; (1975).

Sheng, S., Moraga-Amador, D. A., Van Heeke, G., Allison, R. D., Richards, N. G. J., and Schuster, S. M., *J. Biol. Chem.*, **268**, 16771–16780 (1993).

Silva, A. M., Cachau, R. E., Sham, H. L., and Erickson, J. W., *J. Mol. Biol.*, **255**, 321–346 (1996).

Smith, J. L., *Biochem. Soc. Trans.*, **23**, 894–898 (1995).

Smith, J. L., Zaluzec, E. J., Wery, J.-P., Niu, L., Switzer, R. L., Zalkin, H., and Satow, Y., *Science*, **264**, 1427–1433 (1994).

Stoker, P. W., O'Leary, M. H., Boehlein, S. K., Schuster, S. M., and Richards, N. G. J., *Biochemistry*, **35**, 3024–3030 (1996).

Storer, A. C. and Ménard, R., *Methods Enzymol.*, **244**, 486–500 (1994).

Sur, P., Fernandez, D. J., Kute, T. E., and Capizzi, R. L., *Cancer Res.*, **47**, 1313–1318 (1987).

Surh, L. C., Morris, S. M., O'Brien, W. E., and Beaudet, A. L., *Nucleic Acids Res.*, **16**, 9532–9532 (1988).

Tesmer, J. J. G., Klem, T. J., Deras, M. L., Davisson, V. J., and Smith, J. L., *Nature Struct. Biol.*, **3**, 74–86 (1996).

Thoden, J. B., Holden, H. M., Wesenberg, G., Raushel, F. M., and Rayment, I., *Biochemistry*, **36**, 6305–6316 (1997).

Thornberry, N. A. and Molineaux, S. M., *Protein Sci.*, **4**, 3–12 (1995).

Tsai, F. Y. and Coruzzi, G. M., *EMBO J.*, **9**, 323–332 (1990).

Tso, J. Y., Zalkin, H., van Cleemput, M., Yanofsky, C., and Smith, J. M., *J. Biol. Chem.*, **257**, 3525–3531 (1982).

Uren, J. R., Chang, P. K., and Handschumacher, R. E., *Biochem. Pharmacol.*, **26**, 1405–1410 (1977).

Vadlamudi, S., Krishna, B., Subbareddy, V. V., and Goldin, A., *Cancer Res.*, **33**, 2014–2019 (1973).

Van Heeke, G. and Schuster, S. M., *J. Biol. Chem.*, **264**, 19475–19477 (1989).

Van Heeke, G. and Schuster, S. M., *Protein Eng.*, **3**, 739–744 (1990).

Vanoni, M. A., Edmondson, D. E., Rescigno, M., Zanetti, G., and Curti, B., *Biochemistry*, **30**, 11478–11484 (1991a).

Vanoni, M. A., Nuzzi, L., Rescigno, M., Zanetti, G., and Curti, B., *Eur. J. Biochem.*, **202**, 181–189 (1991b).

Villafranca, J. J., *Curr. Op. Structural Biol.*, **1**, 821–825 (1991).

Walker, J. E., Saraste, M., Runswick, M. J., and Gay, N. J., *EMBO J.*, **1**, 945–951 (1982).

Waterhouse, R. N., Smyth, A. J., Massoneau, A., Prosser, I. M., and Clarkson, D. T., *Plant Mol. Biol.*, **30**, 833–897 (1996).

Wright, T. H., *Crit. Rev. Biochem. Mol. Biol.*, **26**, 1–52 (1991).

Yamashita, M. M., Almassey, R. J., Janson, C. A., Cascio, D., and Eisenberg, D., *J. Biol. Chem.*, **264**, 17681–17690 (1989).

Yoshida, K., Seki, S., Fujimura, M., Miwa, Y., and Fujita, Y., *DNA Res.*, **2**, 61–69 (1995).

Zalkin, H., *Adv. Enzymol. Relat. Areas Mol. Biol.*, **66**, 203–309 (1993).

MECHANISMS OF CYSTEINE S-CONJUGATE β-LYASES

By ARTHUR J. L. COOPER, *Departments of Biochemistry and of Neurology and Neuroscience, Cornell University Medical College, New York 10021* and *Burke Medical Research Institute, Cornell University Medical College, White Plains, NY 10605*

CONTENTS

Advances in Enzymology and Related Areas of Molecular Biology, Volume 72: Amino Acid Metabolism, Part A, Edited by Daniel L. Purich
ISBN 0-471-24643-3 ©1998 John Wiley & Sons, Inc.

I. Introduction

Mercapturates [S-(N-acetyl)-L-cysteine conjugates] were first discovered over a century ago (Baumann and Preusse, 1879; Baumann, 1883) and were soon recognized to play a role in the detoxification of potentially harmful xenobiotics. The mercapturate pathway is now also recognized to play an important role in the metabolism of a number of endogenous substances, such as estrogens and leukotrienes (Nicholson, 1993). In 1959 the cysteinyl portion of mercapturates was shown to be derived from glutathione (Barnes, 1959; Bray et al., 1959). Glutathione is found in almost all bacterial, plant, and mammalian cells in high concentrations (0.5–12 mM): (For representative reviews on glutathione see Meister and Anderson, 1983; Meister, 1984; 1988; Tanaguchi et al., 1989; Dolphin et al., 1989; Cooper, 1997.) The mercapturate pathway begins with the formation of a thioether bond between glutathione and electrophilic xenobiotic (or endogenous substance) in a reaction catalyzed, or enhanced, by various glutathione S-transferases. [For a comprehensive volume on the glutathione S-transferases see Tew et al. (1993).] The resulting glutathione S-conjugate is subsequently cleaved to the corresponding cysteinylglycine S-conjugate and then to the cysteine S-conjugate by the consecutive actions of γ-glutamyltranspeptidase and cysteinylglycine dipeptidase, respectively. Finally, the cysteine S-conjugate is N-acetylated to the corresponding mercapturate. The mercapturate is generally more water soluble than is the parent compound and is readily excreted. Reactions involved in the mercapturate pathway and associated transformations are shown in Figure 1. (For representative reviews on the mercapturate pathway see Boyland and Chasseaud, 1969; Jakoby, 1978; Chasseaud, 1979; Wood, 1970; Ketterer et al., 1983; Reed and Meredith, 1984; Reed, 1986; Stevens and Jones, 1989; Cooper and Tate, 1997.) Although the mercapturate pathway generally plays a beneficial role, it is also apparent that the pathway can serve as a branch point for some deleterious reactions. In this case, the pathway leads to bioactivation (toxification) rather than to detoxification. Many cysteine S-conjugates formed from the corresponding glutathione S-conjugates through the mercapturate pathway are converted to ammonia, pyruvate and an SH-containing fragment by the action of cysteine S-conjugate β-lyases (Eq. 1) in a reaction that competes with the acetylation reaction. In some instances, the eliminated fragment is relatively stable and not especially toxic. In other cases, the eliminated fragment is quite reactive and toxic, particularly to the kidney. These reactions are discussed in more detail below.

$$RSH_2CH(NH_2)CO_2H + H_2O \rightarrow CH_3C(O)CO_2H + RSH + NH_3$$
$$(1)$$

The purpose of this chapter is to describe the discovery of the cysteine *S*-conjugate β-lyases and the realization that in mammals they are generally pyridoxal 5'-phosphate (PLP)-containing enzymes that do not normally catalyze a β-elimination reaction but that are "coerced" into carrying out such a reaction by the strong electron-withdrawing properies of the group attached at the sulfur. This chapter also describes the mechanisms involved in the various cysteine *S*-conjugate β-lyase reactions and the potential of the cysteine *S*-conjugates as targets for useful prodrugs.

II. Discovery

A. β-LYASE REACTIONS THAT LEAD TO THE FORMATION OF STABLE THIOLS AND THIOMETHYL PRODUCTS

The cysteine *S*-conjugate β-lyase reaction generates a product containing a thiol that is ultimately derived from the sulfur of glutathione (Fig. 1, consecutive reactions 1 → 2 → 3 → 6). One of the earliest tracer experiments to show that the sulfur of glutathione is incorporated into a xenobiotic metabolite was carried out by Colucci and Buyske (1965), who showed that benzothiazolyl 2-sulfonamide is converted in several mammalian species to *S*-(2-benzothiazolyl)-glutathione and then to the corresponding cysteine *S*-conjugate and mercapturate. Benzothiazolyl mercaptan (mostly in the form of its glucuronide) was also formed. The biotransformation of benzothiazolyl 2-sulfonamide to the product benzothiazolyl mercaptan was shown to involve glutathione, and the sulfur of the mercaptan group was shown to replace the sulfonamide group. The thiol-containing metabolite derived from benzothiazolyl 2-sulfonamide is relatively stable.

Some xenobiotics (e.g., phenacitin and acetaminophen) are known to be metabolized in part to thiomethyl compounds *in vivo*. At first it was assumed that thiomethylation reactions involve forma-

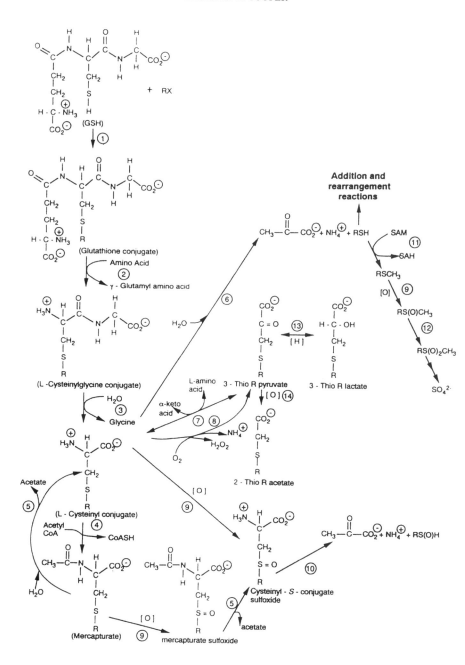

tion of a reactive ester followed by nucleophilic attack by the sulfur of methionine; the resulting sulfonium ion would then decompose to the thiomethyl compound and homocysteine. However, Chatfield and Hunter (1973) showed that the conversion of 2-acetamido-4-chloromethylthiazole to 2-acetamido-4-methylthiomethylthiazole in the rat most likely involved the mercapturate pathway and a methylation reaction. The sequence of events can be traced in Figure 1 (1 → 2 → 3 → 6 → 11). Further evidence for such a thiomethylation pathway was obtained by Tateishi and co-workers (1978a) who investigated the thiomethylation of bromazepam in rats and identified the corresponding mercapturate and 6'-methylthiobromazepam in the bile. The authors also provided additional evidence for the thiomethylation pathway by showing that the thioether bonds of the cysteine S-conjugates of 2,4-dinitrobenzene and bromobenzene are readily cleaved by an enzyme present in rat liver cytosol. Incubation of S-(2,4-dinitrophenyl)-L-cysteine with the enzyme preparation resulted in formation of pyruvate, ammonia, and relatively stable 2,4-dinitrobenzenethiol (Tateishi et al., 1978b). The thiol product was methylated with S-adenosylmethionine in a reaction catalyzed by a microsomal thiomethyl transferase. Tateishi et al. (1978b) were the first to name enzymes that catalyze β-elimination reactions with cysteine S-conjugates as cysteine conjugate β-lyases.

Figure 1. Formation of mercapturates and some related reactions. (1) Glutathione S-transferases, (2) γ-glutamyltranspeptidase, (3) cysteinylglycine dipeptidase, (4) cysteine S-conjugate N-acetyl transferase, (5) deacylases, (6) cysteine S-conjugate β-lyases, (7) glutamine transaminase K/kynurenine pyruvate aminotransferase, (8) L-amino acid oxidase (in the rat the B form of L-α-hydroxy acid oxidase), (9) S-oxidase, (10) cysteine S-conjugate sulfoxide β-lyase (related to the alliinase reaction), (11) S-methyl transferase, (12) S-sulfoxide oxidase. Glutathione = GSH, S-adenosylmethionine = SAM; S-adenosylhomocysteine = SAH. In this diagram, the first step is shown as a nucleophilic displacement (substitution), but examples of glutathione S-transferase-catalyzed Michael addition to a polarized double (or triple) bond or to nucleophilic opening of an oxirane ring (epoxide) are also known. An example of an addition reaction is the attack of glutathione on dichloroacetylene (Fig. 2). The diagram is a composite of reactions that occur in many parts of the body and is meant to illustrate some of the key pathways that have been uncovered. The metabolic fate of the halogenated xenobiotic (RX) depends on the nature of R, species, age, sex, nutritional status. [From Cooper (1994) with permission.]

B. β-LYASE REACTIONS THAT GIVE RISE TO UNSTABLE THIOLS

Early this century it was noted that cattle fed soybean meal contaminated with trichloroethylene developed aplastic anemia (Stockman, 1916). More than 40 years later, the toxic metabolite of trichloroethylene was shown to be the cysteine S-conjugate S-(1,2-
dichlorovinyl)-L-cysteine (DCVC) (McKinney et al., 1959; Schultze
et al., 1959) (Fig. 2). DCVC and related cysteine S-conjugates are
now known to be nephrotoxic to all experimental animals studied
(Terracini and Parker, 1965; Jaffe et al., 1984; Koechel et al., 1991).
(For representative reviews that deal with the biotransformations of
DCVC, see, Lash and Anders, 1986; Stevens and Jones, 1989; Koob
and Dekant, 1991; Elfarra, 1993; Cooper, 1994; Dekant et al., 1994.)
Clues as to the mechanism whereby DCVC is toxic came from tracer
studies showing that reactive sulfur-containing metabolites are generated from this conjugate. Early pioneering work of Schultze and
co-worker led to the important finding that metabolism of DCVC
involves a "C—S" lyase (Anderson and Schultze, 1965) and that
metabolism of ^{35}S-labeled DCVC results in incorporation of a labeled
sulfur fragment into protein and DNA (Saari and Schulze, 1965;
Bhattacharya and Schultze, 1967, 1971a,b, 1972, 1973a,b).

Trichloroethylene is now known to be metabolized in part through
the mercapturate pathway. In this pathway, metabolism begins with
the conversion of trichloroethylene to the corresponding glutathione
S-conjugate in a reaction catalyzed by microsomal glutathione S-

--→

Figure 2.   Bioactivation of dichloroacetylene and trichloroethylene. In the first step
microsomal glutathione S-transferase catalyzes the formation of S-(1,2-dichlorovinyl)-
glutathione (DCVG) either via direct addition (with dichloroacetylene) or via an elimination reaction (with trichloroethylene). The DCVG is converted to S-(1,2-dichloro-
vinyl)-L-cysteinylglycine and then to DCVC by the consecutive action of γ-glutamyl-
transpeptidase and dipeptidase. The DCVC is converted to pyruvate, ammonia and
1,2-mercaptoethylene by cysteine S-conjugate β-lyases. The thiol-containing fragment has not been isolated or trapped. However, much evidence suggests that it forms
a thioketene that thioacetylates macromolecules (Koob and Dekant, 1991; Dekant et
al., 1994). Although both trichloroethylene and dichloroacetylene give rise to the
same glutathione S-conjugate, dichloroacetylene is much more acutely toxic. Dichloroacetylene is a much better substrate of the microsomal glutathione S-transferase
than is trichloroacetylene; in fact it is the best known substrate (Koob and Dekant,
1991; Dekant et al., 1994). [From Cooper (1994) with permission.]

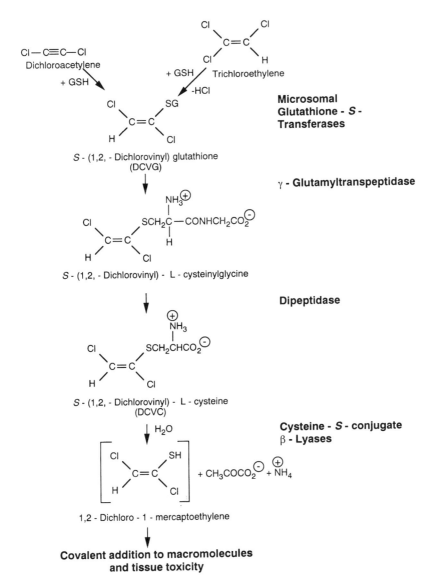

S - (1,2, - Dichlorovinyl) glutathione (DCVG)

S - (1,2, - Dichlorovinyl) - L - cysteinylglycine

S - (1,2, - Dichlorovinyl) - L - cysteine (DCVC)

1,2 - Dichloro - 1 - mercaptoethylene

Covalent addition to macromolecules and tissue toxicity

transferase. The corresponding cysteine *S*-conjugate (DCVC) formed from the glutathione *S*-conjugate is a substrate for the cysteine *S*-conjugate β-lyase reaction (Fig. 1; sequential reactions 1 → 2 → 3 → 6; Fig. 2) (Dekant et al., 1990; 1994; Koob and Dekant, 1991). The expected thiol-containing elimination product (1,2-dichloro-1-mercaptoethylene), however, is highly unstable and cannot be trapped. Dekant et al. (1990, 1991, 1994) and Koob and Dekant (1991) have elegantly shown that the most probable breakdown product of 1,2-dichloro-1-mercaptoethylene is a thioketene arising by elimination of HCl (Fig. 3). Thioketenes are known to be thioacylating agents that react rapidly with nitrogen nucleophiles (Raasch, 1970, 1972; Adiwidjaja et al., 1991). In addition to DCVC, other nephrotoxic unsaturated chlorinated cysteine *S*-conjugates are also probably converted to thioketenes via the cysteine *S*-conjugate β-lyase reaction followed by spontaneous elimination of HCl. Examples include *S*-(pentachlorobutadienyl)-L-cysteine (formed from the metabolism of hexachloro-1,3-butadiene) and *S*-(1,2,2-trichlorovinyl)-L-cysteine (formed from the metabolism of tetrachloroethylene) (Dekant et al., 1994).

α-Fluoroethanethiols are produced by the action of cysteine *S*-conjugate β-lyases on cysteine *S*-conjugates derived from fluorinated ethanes. Examples of such conjugates include *S*-(2-chloro-1,1,2-trifluoroethyl)-L-cysteine and *S*-(1,1,2,2-tetrafluoroethyl)-L-cysteine (TFEC) (Green and Odum, 1985; Dekant et al., 1987; Dekant et al.,

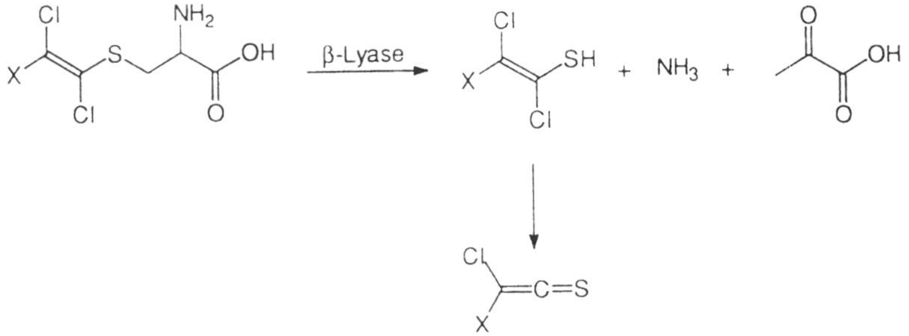

Figure 3.   Cysteine *S*-conjugate β-lyase-catalyzed conversion of *S*-(haloakenyl)-L-cysteine conjugates to halothioalkenes followed by spontaneous conversion to thioketenes. X = H, DCVC; X = Cl, *S*-(1,2,2-trichlorovinyl)-L-cysteine; X = C_2Cl_3, *S*-(pentachlorobutadienyl)-L-cysteine. [From Dekant et al. (1994) with permission.]

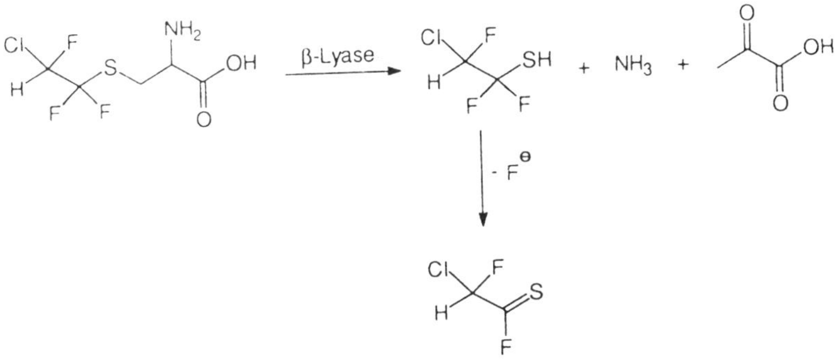

Figure 4.   Cysteine S-conjugate β-lyase-catalyzed conversion of S-(1-chloro-1,2,3-trifluoroethyl)-L-cysteine conjugate to an α-fluoroethanethiol followed by spontaneous elimination of fluoride to yield the thioacylfluoride. [From Dekant et al. (1994) with permission.]

1994). Unlike the unsaturated thiol-containing fragment generated from the β-lyase reaction on DCVC, the saturated fragment released from the fluoroalkylcysteine S-conjugate (α-fluoroethanethiol) can be trapped in model systems that mimic the cysteine S-conjugate β-lyase reaction. The α-fluoroethanethiol, however, rapidly eliminates HF to yield the thioacyl fluoride (Fig. 4). Thioacyl fluorides, like thioketenes, are strong thioacylating agents (Dekant et al., 1987, 1991, 1994; Anders et al., 1993; Müller et al., 1991).

III.   Toxicity of Cysteine S-Conjugates

Cysteine S-conjugates of alkyl, aromatic, heteroaromatic, and alkylaromatic compounds are substrates of various cysteine S-conjugate β-lyases (Bakke and Gustafsson, 1984; Dekant et al., 1994). The eliminated thiol-containing fragments are generally relatively stable. They may be excreted unchanged. However, the fragments may also be methylated via the thiomethyl shunt with S-adenosylmethionine as the methyl donor (Jakoby and Stevens, 1984; Jakoby et al., 1984) or be converted *in vivo* to mixed disulfides (especially with glutathione), sulfenic acids, sulfinic acids, sulfenic/sulfinic acid-mixed disulfides, or S-glucuronides. The thiomethylated compounds may in turn be converted to sulfoxides and sulfones by flavin-dependent monooxygenases (Ziegler, 1984) or be converted to S-glucuronides. The sulfoxides and sulfones may be further metabolized to sulfate (and CO_2) (Barnsley, 1964; Horner and Kuchinskas, 1959; Sklan and Barnsley, 1968; James and Needham, 1973; Turnbull et al., 1978; Mitchel et al., 1984). These reaction are depicted in Figure 1, reactions $6 \rightarrow 11 \rightarrow 9 \rightarrow 12 \rightarrow\rightarrow$ sulfate). In another pathway, the cysteine S-conjugate may be converted to the corresponding α-keto acid (e.g., by the action of a transaminase or L-amino acid oxidase), which in turn may be oxidatively decarboxylated to a thioacetate or be reduced to an α-hydroxy acid (a thiolactate) (Fig. 1, reactions 7,8 $\rightarrow$ 13). The α-keto acid, thioacetate, and thiolactate may in turn be S-oxidized. Alternatively, the cysteine S-conjugate may undergo S oxidation to the corresponding sulfoxide, which is itself a substrate of the cysteine S-conjugate β-lyase reaction yielding pyruvate, ammonia, and a sulfenic acid (Fig. 1, Reactions $9 \rightarrow 10$) (Sausen and Elfarra, 1990; Sausen et al., 1993; Lash et al., 1994). Finally, the cysteine S-conjugate may be converted to the corresponding methyl sulfoxide (Fig. 1, Reactions $6 \rightarrow 11 \rightarrow 9$). This compound is reacted with glutathione to yield the original glutathione S-conjugate and methane sulfenic acid (Eq. 2), which is oxidized to CO_2 and sulfate (Bakke, 1986).

$$RS(O)CH_3 + GSH \rightarrow HS(O)CH_3 + GSR \qquad (2)$$

As pointed out by Dekant et al. (1994), the possible toxic effects of the relatively stable thiol-containing fragments generated from the cysteine S-conjugate β-lyase reaction have not been studied in detail, but limited studies suggest that they are not especially cytotoxic or mutagenic in the Ames test. This situation contrasts with the unstable thiol-containing elimination products, which are usually toxic.

As noted above, DCVC (an α-chlorovinyl cysteine S-conjugate) causes aplastic anemia in cattle. This toxic manifestation is unique to cattle and also appears to be unique to DCVC among cysteine S-conjugates tested (Lock et al., 1996). As also noted, DCVC is nephrotoxic to all species investigated, including cattle. In addition to DCVC, the cysteine S-conjugates formed from many other halogenated alkenes are nephrotoxic. Examples include hexachloro-1,3-butadiene (noted above), tetrafluoroethylene, 1,1-dichloro-2,2-difluoroethylene, 1,1-dibromo-2,2-difluoroethylene, 2-chloro-1,1,2-

trifluoroethylene, tetrachloroethylene (noted above), hexafluoropro-
pylene, and 1,1,2-trichloro-3,3,3-trifluoro-1-propylene. In addition
to these examples of cysteine S-conjugates that possess a carbon–
carbon double bond, the cysteine S-conjugates formed from some
fluorinated ethanes are also nephrotoxic. Examples include TFEC,
S-(2-bromo-2-chloro-1,1-difluoroethyl)-L-cysteine, S-(2-bromo-
-1,1,2-trifluoroethyl)-L-cysteine, and S-(2,2-dibromo-1,1,-difluoro-
ethyl)-L-cysteine. [See Cooper (1994) and Dekant et al. (1994) for
original references.] Finally, the haloalkyne dichloroacetylene is
strongly neurotoxic, nephrotoxic and nephrocarcinogenic (Reichert
et al., 1976, 1984; Koob and Dekant, 1991). In this case, metabolism
results in formation of the same cysteine S-conjugate as that formed
from trichloroethylene (i.e., DCVC) (Fig. 2).

Trichloroethylene (and dichloroacetylene as noted above), tetrachloroethylene, and hexachlorobutadiene are all nephrocarcinogenic
in experimental animals (Koob and Dekant, 1991). The corresponding cysteine S-conjugates [i.e., DCVC, S-(1,1,2,-trichlorovinyl)-L-
cysteine, and S-(pentachlorobutadienyl)-L-cysteine, respectively]
are strongly mutagenic in the Ames test without the addition of an
exogenous activating system. The mutagenicity of DCVC in the
Ames test is presumably a consequence of the fact that *Staphylococcus typhimurium* contains relatively high levels of cysteine S-conjugate β-lyase activity (Dekant et al., 1986). The accumulated evidence
suggests that the nephrocarcinogenic haloalkenes exert their properties through formation of the corresponding glutathione S-conjugate
followed by conversion to the cysteine S-conjugate. The latter is
cleaved by the action of β-lyases to give a reactive thiol-containing
fragment (Dekant et al., 1986, 1994; Vamvakas et al., 1988, 1989).
This reactive thiol, as noted above, can add to DNA (see also Müller
et al., 1994). Evidence has been presented that the thioketene may
bring about point mutations, alter gene expression by inducing Ca^{2+}-
dependent DNA double-strand breaks and increase poly(ADP-ribo-
syl)ation of nuclear proteins (Jaffe et al., 1985; see also the discussion
by Dekant et al., 1994). DCVC also induces the protooncogenes *c-
fos* and *c-myc* in LLC-PK1 cells (Vamvakas and Koster, 1993). [For
an excellent discussion, see Dekant et al. (1994).] Interestingly, haloethyl cysteine S-conjugates containing at least one bromine constituent [e.g., S-(2-bromo-1,1,2-trifluoroethyl)-L-cysteine and S-(2,2,-
dibromo-1,1,-difluoroethyl)-L-cysteine] are weakly mutagenic,

whereas haloethyl cysteine S-conjugates lacking a bromine constituent are not mutagenic (Finkelstein et al., 1994). The accumulated findings may be rationalized by the fact that, as mentioned above, metabolism of DCVC by the cysteine S-conjugate β-lyase pathway produces a thiol fragment that reacts with DNA. Similar work has shown that the fragment released from the cysteine S-conjugate β-lyase reaction on S-(pentachlorobutadienyl)-L-cysteine also binds to DNA (especially mitochondrial DNA) (Schrenk and Dekant, 1989). On the other hand, thioacylfluorides generated from the metabolism of S-(2-chloro-1,1,2-trifluoroethyl)-L-cysteine and TFEC do not produce detectable DNA adducts and are not nephrocarcinogenic although they are nephrotoxic (Müller et al., 1991).

As noted above, both thioketenes and thioacyl fluorides are strong thioacylating agents. Amino groups and hydroxyl groups of proteins and lipoproteins are susceptible to thioacylation (Dekant et al., 1994). Such covalent attachment is associated with nephrotoxicity and, in those conjugates that give rise to thioketenes, mutagenicity (Darnerud et al., 1988; Hayden and Stevens, 1990; Koob and Dekant, 1991). The reactive fragments generated from toxic cysteine S-conjugates apparently destroy renal epithelial cells in part by a combination of covalent binding to macromolecules, depletion of non-protein thiols and lipid peroxidation (Chen et al., 1990). The S_3 region of the kidney tubules is especially vulnerable to the toxic action of cysteine S-conjugates (Terracini and Parker, 1965; Nash et al., 1984; Jaffe et al., 1984; Odum and Green, 1984; Green and Odum, 1985; Dohn et al., 1985; Koechel et al., 1991). Within kidney cells, mitochondria are damaged, in part through perturbation of Ca^{2+} homeostasis, modification of membrane proteins and changes in pyridine nucleotide redox balance (Jones et al., 1986; Lash and Anders, 1986, 1987; Chen et al., 1990; Hayden and Stevens, 1990; Dekant et al., 1994). Bruschi et al. (1993), noted that after rats are exposed to TFEC five proteins in kidney mitochondria are heavily tagged. Two of the more heavily tagged proteins are the heat shock proteins HSP60- and HSP70-like protein. In addition, a less heavily tagged protein was identified as mitochondrial aspartate aminotransferase (mitAspAT) (Bruschi et al., 1993). Anders et al. (1987) showed that DCVC inhibits oxygen consumption in isolated kidney cells. Lash et al. (1986) and Stevens et al. (1988) showed that DCVC inhibits state 3 respiration in isolated rat kidney mitochondria and that this inhibitory effect is due to the action of cysteine *S*-conjugate β-lyases. Stonard and Parker (1971) showed that a metabolite of DCVC inhibits pyruvate- and α-ketoglutarate dehydrogenase complexes in isolated rat liver mitochondria. Kato et al. (1996) showed that incubation of mitAspAT and cytosolic aspartate aminotransferase (cytAspAT) (two important components of the malate–aspartate shuttle) with DCVC (or TFEC) in the presence of α-ketoglutarate results in irreversible inactivation (see also below). Thus, the evidence suggests that energy metabolism in mitochondria may be compromised upon release of reactive fragments from toxic cysteine *S*-conjugates. Indeed, ATP levels were decreased in whole kidney and in the S_2 region of kidney tubules of rats treated with hexachloro-1,3-butadiene (Kim et al., 1997). Unexpectedly, ATP levels were slightly increased in the S_3 region, but the findings may be explained by a possible loss of functions that require ATP (Kim et al., 1997). (For two relatively recent volumes dealing extensively with the chemistry, enzymology and toxicity of cysteine *S*-conjugates see Anders et al., 1993, 1994.)

Human kidney contains cysteine *S*-conjugate β-lyase activity albeit with a somewhat lower specific activity than that of rat and bovine kidney (Lash et al., 1990a). As noted above, trichloroethylene is carcinogenic in experimental animals. Long-term exposure to this compound induces carcinomas in rat kidney proximal tubules (Koob and Dekant, 1991). Inasmuch as human kidneys contain cysteine *S*-conjugate β-lyase and human renal cancers commonly originate from the proximal tubules it is possible that humans may also be susceptible to renal carcinomas induced by environmental trichloroethylene and similar halogenated compounds (Cooper and Anders, 1990). Henschler et al. (1995) recently reported that cardboard workers exposed to a high level of trichloroethylene had a greatly elevated incidence of kidney cancers. However, this report has been criticized (Swaen, 1995; Bloemen and Tomenson, 1995). Nevertheless, the facts that (a) long-term exposure to certain halogenated alkenes induces renal cancers in laboratory animals, (b) the corresponding glutathione and cysteine *S*-conjugates are nephrotoxic, and (c) humans possess renal cysteine *S*-conjugate β-lyase activity, are cause for concern. Several halogenated alkenes (e.g., trichloroethylene and perchloroethylene) are in widespread use. Trichloroethylene is used as a solvent, degreasing agent, and in the synthesis of various chlorinated compounds. Perchloroethylene (Perc) is used in the dry cleaning industry. Due to their widespread use and improper disposal, trichloroethylene and perchloroethylene are sometimes major contaminants of ground water (e.g., Roush, 1995). Because of the widespread use of trichloroethylene and perchloroethylene, and long-term exposure of a segment of the population to a potential renal carcinogen it is important to understand the mechanisms whereby these compounds are bioactivated in the kidney. In addition, there are other reasons for concern. Several reports have been published suggesting that under certain circumstances trichlorethylene may be neurotoxic to humans (e.g. Buxton and Hayward, 1967; Schaumburg, 1993). Trichloroethylene and dichloracetylene have been detected in some submarine and spacecraft atmospheres (Saunders, 1967). Dichloroacetylene, like trichloroethylene, is metabolized to DCVC *in vivo*. However, dichloroacetylene is a much better substrate of microsomal glutathione S-transferase than is trichloroethylene (Koob and Dekant, 1991). This fact explains why dichloroacetylene is acutely nephrotoxic whereas trichloroethylene is not (Kanhai et al., 1989). Trichloroethylene can be converted to dichloroacetylene under alkaline conditions, and dichloroacetylene appears to be the causative agent in the reports of brain damage in humans exposed to trichloroethylene (Greim et al., 1984). It is important to understand the role of cysteine S-conjugate β-lyases in the neurotoxic response to halogenated xenobiotics such a dichloroacetylene.

IV. Relationship of Cysteine S-Conjugate β-Lyases to Other Enzymes That Catalyze β-Elimination Reactions

A number of enzymes are known that catalyze a β-elimination reaction with S-substituted cysteines [$RSCH_2CH(NH_2)CO_2H$] to yield pyruvate [$CH_3C(O)CO_2H$], ammonia, and RSH. The reaction is analogous to the cysteine S-conjugate β-lyase reaction (Eq. 1). An example is β-cystathionase—an enzyme that catalyzes an important step in the pathway leading to formation of methionine in microorganisms and plants. This enzyme catalyzes the conversion of cystathionine [$R = CH_2CH_2CH(NH_2)CO_2H$] to pyruvate, ammonia, and homocysteine (Wijesundera and Woods, 1962; Flavin and Slaughter, 1964). In another example, γ-cystathionase, which normally catalyzes a γ-elimination reaction, can catalyze a β-elimination reaction if presented with an appropriate substrate, such as cystine [R = $SCH_2CH(NH_2)CO_2H$], djenkolic acid [R = $CH_2SCH_2CH(NH_2)$-CO_2H, and lanthionine [R = $CH_2CH(NH_2)CO_2H$]. With cystine the eliminated thiol-containing fragment is thiocysteine [$HSSCH_2$-$CH(NH_2)CO_2H$] (see Braunstein and Goryachenkova, 1984). In addition, alliinase, an enzyme present in onions and garlic, catalyzes a β-elimination reaction with relatively simple alkoxide analogues of cysteine [$RS(O)CH_2CH(NH_2)CO_2H$] to yield pyruvate, ammonia, and a sulfenic acid [$RS(O)H$]. Cysteine and *S*-methylcysteine are not substrates (Schwimmer and Mazelis, 1963; Braunstein and Goryachenkova, 1984). A cysteine *S*-conjugate β-lyase from enteric bacteria and another from human liver are able to catalyze the β cleavage of cysteine *S*-conjugate sulfoxides (see below). The mechanism of the reaction catalyzed by these enzymes is obviously related to that of alliinase. In addition, some cysteine *S*-conjugate β-lyases isolated from enteric bacteria can catalyze β-elimination reactions with cysteine *S*-conjugates [$RSCH_2CH(NH_2)CO_2$] containing (a) simple *S*-alkyl groups [e.g., *S*-methyl-L-cysteine, R = CH_3]; (b) more complex halogenated groups [e.g., DCVC, R = $(Cl)C\!\!=\!\!C(Cl)H$]; and (c) an amino acid moiety [e.g., cystathionine, R = $CH_2CH_2CH(NH_2)$-CO_2H]. Clearly, these cysteine *S*-conjugate β-lyases from enteric bacteria have considerable substrate overlap not only with alliinase, but also with β-cystathionase and γ-cystathionase. However, for the mammalian cysteine *S*-conjugate β-lyases thus far studied simple *S*-alkyl-containing cysteine *S*-conjugates are not substrates. For β elimination to be catalyzed by these enzymes the cysteine *S*-conjugate must contain a sulfur at the oxidation state of a sulfoxide $S(O)R$ or the SR portion must be a good leaving group (i.e., be activated, usually by a halogen, double bond, or an aromatic group). This requirement may be due to the fact that most mammalian cysteine *S*-conjugate β-lyases studied in detail so far are PLP-containing enzymes that normally catalyze another type of reaction. These enzymes appear to be "coerced" into catalyzing the β-elimination reaction only by virtue of the strong electron-withdrawing properties of the SR [or $S(O)R$] portion of the conjugate (see below). The "unnatural" nature of the catalyzed reaction is shown by the fact that many of the mammalian cysteine *S*-conjugate β-lyases are inactivated during the process of catalyzing the elimination reaction. Following elimination of the sulfhydryl-containing fragment, amino-acrylate is released from the active site. The aminoacrylate (an eneamine) rearranges to α-iminopropionate, which spontaneously hydrolyzes to pyruvate and ammonia. However, in some cases the generated aminoacrylate can cause enzyme inactivation by alkylating a nearby residue or, more likely, by attacking the cofactor (see below).

V. Cysteine S-Conjugate β-Lyases in Bacteria and Fungi

Cysteine S-conjugate β-lyase activity is present in fungi (Hafsah et al., 1987; Shimomura et al., 1992) and enteric bacteria (Larsen, 1985; Saari and Schultze, 1965; Larsen and Stevens, 1986; Larsen and Bakke, 1983; Larsen et al., 1983; Bakke and Gustafsson, 1984; Tomisawa et al., 1984; Bernström et al., 1989). In fact, gut bacteria can catalyze the biotransformations of certain xenobiotics (Goldman, 1978) including glutathione S-conjugates (Bakke et al., 1981). The fact that cysteine S-conjugate β-lyase activity is widespread among gut bacteria and that a major portion of cysteine S-conjugates are excreted in the bile suggests that gut flora may be important for the incorporation of sulfur into some xenobiotics (Larsen and Stevens, 1986). Moreover, the bacterial cysteine S-conjugate β-lyases seem to have a wider substrate specificity than do the mammalian enzymes. As noted above, many sulfhydryl-containing fragments released from cysteine S-conjugates are exceptionally reactive. These fragments are likely to exert their toxic effects locally. Whether metabolism by gut bacteria of cysteine S-conjugates to reactive fragments can damage the gut remains to be established. Additionally, whether more stable eliminated fragments can be released from the gut to other regions thereby exerting toxic effects in other organs remains to be determined (Larsen and Stevens, 1986).

The PLP-dependent cysteine S-conjugate β-lyases from *Fusobacterium necrophorum* (Larsen et al., 1983) and *F. varium* (Tomisawa et al., 1984) have been partially purified and characterized. The enzyme from *F. necrophorum* has a M_r of 228,000 and catalyzes the cleavage of the cysteine S-conjugate of propachlor to 2-mercapto N-isopropylacetanilide, pyruvate, and ammonia. The enzyme also catalyzes the cleavage of S-(benzothiazolyl)-L-cysteine and 1,2-dihydroxy-1-hydroxy-2-cysteinylnaphthalene. The enzyme from *F. va-*

rium has a M_r of 70,000 and is also active with aromatic cysteine *S*-conjugates. The enzyme exhibits some activity toward relatively simple alkyl cysteine *S*-conjugates. A cysteine *S*-conjugate β-lyase has been highly purified from *Eubacterium limosum* (Larsen and Stevens, 1986). The enzyme has a M_r of 75,000, is composed of two identical subunits, and is active with aromatic cysteine *S*-conjugates. The *E. limosum* enzyme is not activated by addition of PLP, but is inactivated by addition of carbonyl reagents, suggesting that the cofactor is tightly bound. Interestingly, the enzyme is active with the relatively simple alkyl cysteine *S*-conjugate *S*-ethyl-L-cysteine and especially active with the *S*-oxide of propachlor (Larsen and Stevens, 1986). The fragment eliminated from the *S*-oxide is a sulfenic acid [RS(O)H] (Fig. 1, Reaction 10) which, as noted above, is a better leaving group than is RSH. The *F. necrophorum* and *F. varium* enzymes do not catalyze a β-lyase reaction with cystathionine. However, the *E. limosum* cysteine *S*-conjugate β-lyase is quite active with cystathionine, djenkolic acid and cystine, suggesting that this enzyme is closely related to β- and γ-cystathionases. Bernström et al. (1989) showed that rat fecal contents and the purified *E. limosum* enzyme catalyze the conversion of the cysteine *S*-conjugate leukotriene-E$_4$ to 5-hydroxy-6-mercapto-7,9-*trans*-11,14-*cis*-eicosatetraenoic acid. The purified *E. limosum* enzyme also catalyzes a β-elimination reaction with β-chloroalanine to yield pyuvate (via aminoacrylate), HCl, and ammonia. The enzyme is not inactivated during the turnover of β-chloroalanine to pyruvate. Evidently, the bacterial cysteine *S*-conjugate β-lyases are a disparate group of enzymes and, like most mammalian cysteine *S*-conjugate β-lyases, may catalyze cysteine *S*-conjugate β-lyase reactions secondary to more important, unrelated metabolic reactions.

VI.   Mammalian Cysteine *S*-Conjugate β-Lyases

A.   HUMAN LIVER CYSTEINE *S*-CONJUGATE β-LYASE

Human liver contains a PLP-dependent cysteine *S*-conjugate β-lyase that is active with *S*-arylcysteines but not with *S*-alkylcysteines (Tomisawa et al., 1986). The enzyme, which catalyzes the stoichiometric formation of pyruvate, ammonia, and *p*-bromophenylmercaptan from *S*-(*p*-bromophenyl)-L-cysteine has a pH optimum of 8.5 and has a native M_r of 88,000. The enzyme is nine times more active with S-phenylcysteine sulfoxide than with S-phenylcysteine—again attesting to the excellent leaving group properties of a sulfenic acid.

B. RAT LIVER CYSTEINE S-CONJUGATE β-LYASE/KYNURENINASE

Rat liver possesses an enzyme capable of cleaving cysteine S-conjugates (e.g. DCVC, S-(2,4-dinitrophenyl)-L-cysteine, S-benzothiazolyl-L-cysteine) to pyruvate, ammonia, and a sulfhydryl-containing fragment (Tateishi et al., 1978b; Stevens and Jakoby, 1983; Stevens, 1985a). The enzyme contains tightly bound PLP, has a M_r of about 100,000, is composed of two identical subunits, and is identical with kynureninase (Stevens, 1985a). The enzyme also catalyzes a β-elimination reaction with β-chloroalanine. However, kynureninase is slowly inactivated by both DCVC and β-chloroalanine in a pseudo-first-order process. In both cases, partitioning between events that lead to pyruvate formation and events that lead to inactivation is about 600 (Stevens, 1985a). Interestingly, kynureninase of *Pseudomonas marginalis* also catalyzes a β-elimination reaction with β-chloroalanine [and with S-(o-nitrophenyl)-L-cysteine] and is also similarly inactivated after several hundred turnover events (Kishore, 1984). Kynureninase plays a role in the conversion of tryptophan to quinolinate, which in turn is a precursor of the important cofactor nicotinamide adenine dinucleotide (NAD$^+$). Although some of the NAD$^+$ requirement may be met by ingestion of the cofactor precursors nicotinate/nicotinamide in the diet, it is possible that inactivation of kynureninase by cysteine S-conjugates may result in lowered NAD$^+$ levels. Whether a loss of kynureninase contributes to organ toxicity of cysteine S-conjugates is not clear (Stevens, 1985a; Cooper, 1994).

Presumably the "normal" function of kynureninase is to catalyze the hydrolytic cleavage of kynurenine (or hydroxykynurenine) to anthranilate (or hydroxyanthranilate) and alanine. Pyuvate is not a product of this reaction. Although the kynureninase reaction involves cleavage at the β position of the amino acid substrate this is not a true β-elimination reaction. Loss of the α-proton does not occur during turnover and the β carbon of the eneamine intermediate has carbanion character (Bild and Morris, 1984) (Fig. 5). After hydrolytic cleavage and protonation of the carbanionic cofactor intermediate,

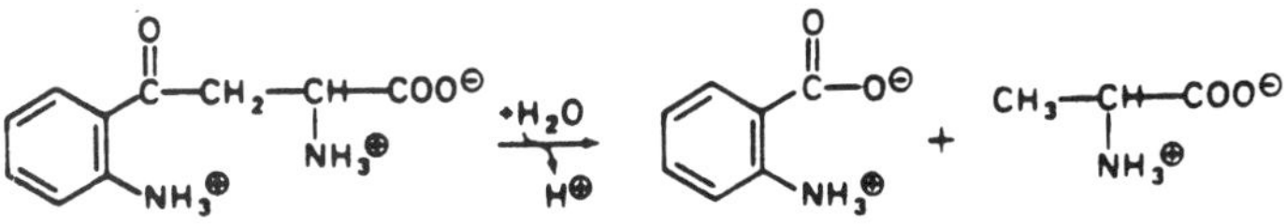

Figure 5. Differences in the mechanisms of the hydrolytic cleavage of kynurenine and β-elimination reaction catalyzed by kynureninase. Equation 3, the "normal" kynureninase reaction; Eq. 4, the overall β-lyase reaction; Eq. 5, resonance hybrid structures showing the partial carbanion nature of the eneamine intermediate resulting from kynurenine cleavage (the kynureninase reaction); Eq. 6, resonance hybrid structures showing partial carbocation character of the eneamine intermediate resulting from β elimination (cysteine S-conjugate β-lyase reaction). [From Stevens (1985a) with permission.]

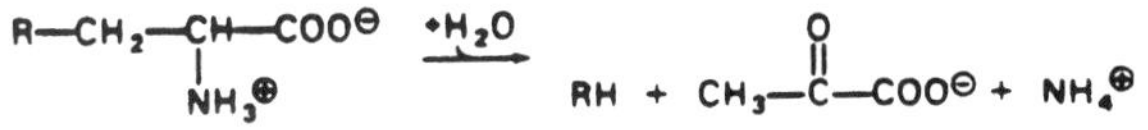

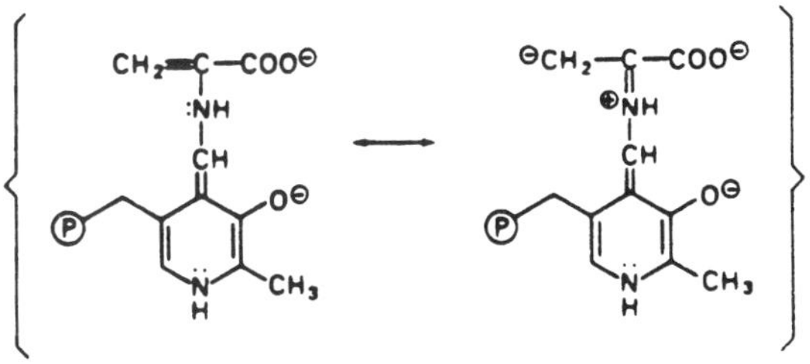

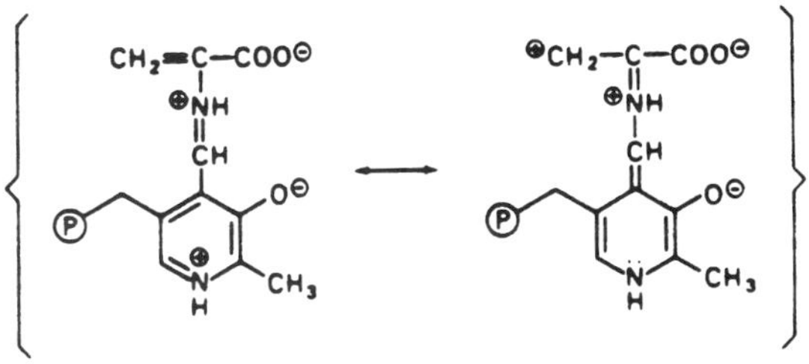

the resulting intermediate is alanine bound in Schiff's base linkage to PLP cofactor. This intermediate is subsequently hydrolyzed to free alanine and free PLP cofactor (Stevens, 1985a). On the other hand, if an amino acid with a good leaving group in the β position (e.g., DCVC, β-chloroalanine) binds at the active site, electron flow from a quinonoid intermediate toward the leaving group is facilitated, resulting in β elimination of aminoacrylate. The aminoacrylate undergoes non-enzymatic conversion to pyruvate and ammonia, or, in every few hundred events, binds to the enzyme thereby inactivating it. In the β-elimination reaction, the enzyme abstracts a proton from the α position. This abstraction results in an eneamine with carbocation character at the β carbon facilitating the elimination reaction (Snell and Di Mari, 1970; Stevens, 1985a).

At this point, it is worth noting that aminoacrylate can be generated from substituted alanines and cysteines with good leaving groups by the catalytic action of a number of PLP-containing enzymes. Examples of enzymes that catalyze the generation of aminoacrylate from haloalanines include pig heart cytAspAT (Morino and Okamoto, 1973), pig heart alanine aminotransferase (Golichowski and Jenkins, 1978), E. coli alanine racemase (Wang and Walsh, 1978), Alcaligenes feacalis L-aspartate β-decarboxylase (Relyea et al., 1974), and bacterial D-amino acid aminotransferase (Soper and Manning, 1978). In each of these examples, the enzymes are also inactivated by the release of the aminoacrylate. In the case of AspAT, it is probable that the aminoacrylate-induced inactivation is due to the formation of a pyruvate-aldol product with the PLP cofactor (Ueno et al., 1982). In the case of L-aspartate β-decarboxylase inactivation appears to be due to covalent modification of an active site glutamate residue (Relyea et al., 1974).

In addition to β-chloroalanine, cytAspAT can catalyze β elimination from a number of other β-substituted amino acids such as serine O-sulfate (Cavallini et al., 1973; Ueno et al., 1982) and erythro β-chloro-DL-glutamate (Manning et al., 1968). The enzyme is not inactivated by the substituted glutamate, presumably because the released eneamine cannot orientate at the active site to form an aldol adduct with the PLP cofactor. On the other hand, the enzyme is inactivated by serine O-sulfate, presumably because the product formed from this amino acid, like that formed from β-chloro-L-alanine, is aminoacrylate which is small enough to orientate in such a way as to form a product with PLP at the active site (Ueno et al., 1982). Interestingly, it has recently been shown that cytAspAT can catalyze a slow β-elimination reaction with DCVC and TFEC. At the same time the enzyme is inactivated, presumably via formation of aminoacrylate and attack on the PLP cofactor, or possibly by attack of the enzyme-generated thioacylating fragment on a susceptible active site residue (Kato et al., 1996).

C. RAT KIDNEY CYSTEINE *S*-CONJUGATE β-LYASE/CYTOSOLIC GLUTAMINE TRANSAMINASE K

Stevens (1985b) showed that the cytosolic and mitochondrial fractions of rat kidney possess lyase(s) distinct from kynureninase. Subsequently, Stevens et al. (1986b) showed that a major cysteine *S*-conjugate β-lyase of rat kidney cytosol is identical to cytosolic glutamine transaminase K (cytGTK). Interestingly, in order to efficiently catalyze β elimination with DCVC the enzyme requires the presence of an α-keto acid, L-amino acid oxidase (Stevens et al., 1986b) or PLP (Abraham et al., 1995a). The L-amino acid oxidase catalyzes the conversion of DCVC to the corresponding α-keto acid which is a substrate of cytGTK. The β-lyase reaction with DCVC and TFEC is especially well supported by the addition of phenylpyruvate or α-keto-γ-methiolbutyrate—the two best α-keto acid substrates for the transaminase reactions catalyzed by cytGTK (see below).

The strict requirement for an α-keto acid or PLP to promote the cysteine *S*-conjugate β-lyase reaction can be rationalized in terms of well-characterized PLP biochemistry (Fig. 6). Thus, the cysteine *S*-conjugate binds at the active site of cytGTK to eventually form a quinonoid intermediate (Fig. 6; quinonoid I). In the "normal" transamination reaction catalyzed by the enzyme, this quinonoid intermediate is exclusively converted to pyridoxamine 5'-phosphate (PMP) plus α-keto acid (Fig. 6, electron flow a). However, if the amino acid bound as a quinonoid contains a good leaving group in the β position electron flow may be redirected toward the leaving group (Fig. 6, electron flow b). Following the β-elimination reaction, aminoacrylate is released and the original PLP form of the enzyme is regenerated to allow a new catalytic cycle to be repeated. However, if transamination occurs then the enzyme is converted to the PMP form and cannot productively bind another amino acid. For

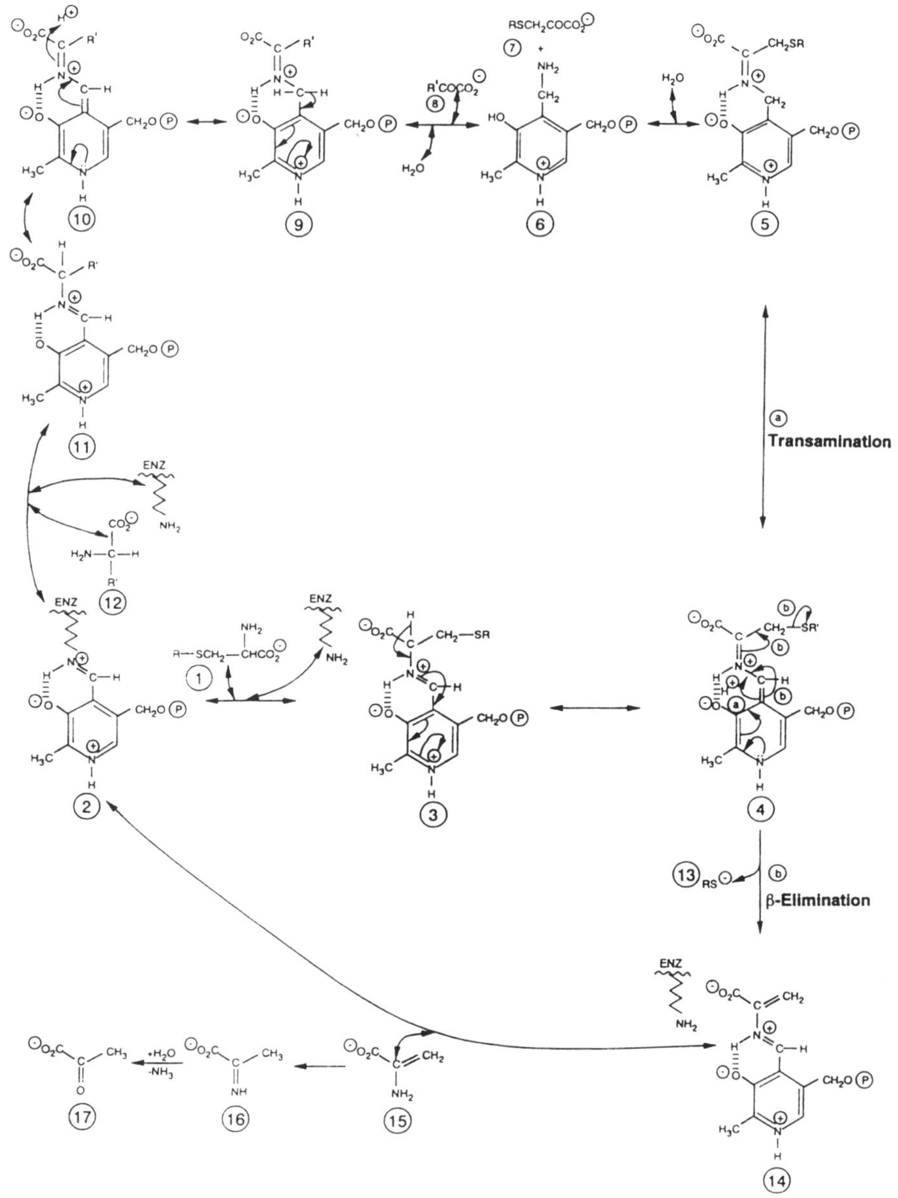
Transamination
β-Elimination
ENZ
H_2O
CH_2O P
$RSCH_2COCO_2^-$
$+H_2O$
$-NH_3$

the β-lyase reaction to continue, an α-keto acid or PLP must be present to convert the PMP form of the enzyme to the PLP form (Stevens et al., 1986b: Stevens and Jones, 1989; Cooper, 1994). The partitioning between transamination and β elimination with cysteine S-conjugates depends very much on the electron-withdrawing properties of the group attached to the sulfur. For example, the simplest alkyl cysteine S-conjugate, namely, S-methyl-L-cysteine, is an excellent substrate in the transaminase reaction catalyzed by cytGTK (Cooper and Meister, 1981). However, CH_3S^- is a relatively poor leaving group and there is no evidence that cytGTK catalyzes an elimination reaction with S-methyl-L-cysteine. On the other hand, DCVC and TFEC contain good leaving groups attached at the sulfur. With these substrates, cytGTK-catalyzed partitioning between transamination and β elimination is about 5:1–10:1 (Stevens et al., 1986b, Abraham et al., 1995a,b).

cytGTK has an M_r of 91,600, contains tightly bound PLP, and is composed of two identical subunits. The amino acid sequence has recently been deduced from the cDNA (Abraham and Cooper, 1996). The enzyme exhibits greater than 90% identity with cytosolic kynurenine pyruvate aminotransferase although it is slightly smaller (see below). Activity staining showed the enzyme to be present in the rat kidney only in the proximal tubules. Within the tubules the activity is present in the S_1, S_2, and S_3 regions (Kim et al., 1997). Thus, inasmuch as nephrotoxic cysteine S-conjugates exert their toxicity mostly toward the S_3 segment it is quite likely that cytGTK plays a role in the toxification process. The situation with the brain is less clear. In the rat, the specific activity of GTK in the brain is relatively low and the activity is mostly represented by a mitochondrial form

←————————————————————————————————————

Figure 6. Proposed mechanism for the competing transamination and β-lyase reactions catalyzed by cytosolic glutamine transaminase K (cytGTK). (**1**) Cysteine S-conjugate substrate; (**2**) PLP bound at the active site through Schiff base linkage to the ε-amino group of a protein lysine; (**3**) external aldimine I; (**4**) quinonoid intermediate I; (**5**) ketimine I; (**6**) enzyme-bound PMP; (**7**) α-keto acid I; (**8**) α-keto acid II; (**9**) ketimine II; (**10**) quinonoid II; (**11**) external aldimine II; (**12**) amino acid substrate II; (**13**) eliminated thiol-containing fragment; (**14**) external aldimine III; (**15**) aminoacrylate; (**16**) 2-iminobutyrate; (**17**) pyruvate. [From Cooper (1994) with permission.]

that has low cysteine S-conjugate β-lyase activity (Cooper, 1988; Abraham et al., 1995b). However, the choroid plexus contains cytGTK of relatively high specific activity (Cooper et al., 1993). Whether cytGTK contributes to the neurotoxicity of dichloroacetylene remains to be established.

cytGTK has a broad amino acid and α-keto acid substrate specificity. In general, amino acids with the structure $X(CH_2)_nCH(NH_2)$-CO_2H, where X is a noncharged group and $n = 1$ or 2 are excellent substrates (Cooper, 1994; Cooper and Anders, 1990). Thus, the enzyme is active with glutamine, phenylalanine, methionine, S-methylcysteine, and their corresponding α-keto acids. The standard assay for measuring GTK-type activity utilizes a reaction mixture containing L-phenylalanine and α-keto-γ-methiolbutyrate.

D. RAT KIDNEY CYSTEINE S-CONJUGATE β-LYASE/KYNURENINE PYRUVATE AMINOTRANSFERASE

In 1993, Perry et al. reported the amino acid sequence of a rat kidney cytosolic cysteine S-conjugate β-lyase deduced for the cDNA. The authors showed that the monomer has a M_r of 47,800 and contains a conserved PLP binding site. Shortly after the report of Perry et al., Mosca et al. (1994) reported the deduced amino acid sequence for rat kidney kynurenine pyruvate aminotransferase. The enzyme expressed in COS-1 cells exhibited both GTK activity (i.e., ability to catalyze the characteristic transamination reaction between L-phenylalanine and α-keto-γ-methiolbutyrate) and kynurenine aminotransferase activity. The amino acid sequence deduced by Mosca et al. (1994) for kynurenine pyruvate aminotransferase was identical to that published previously by Perry et al. (1993) except for a Val for Ile at residue 177 and an Ala for Arg at residue 107. Mosca et al. (1994) assumed that because their expressed enzyme exhibited appreciable GTK activity and because the kidney cysteine S-conjugate β-lyase characterized by Perry et al. (1993) also possessed GTK activity the three enzymes (i.e., cytGTK, kynurenine pyruvate aminotransferase and cytosolic kidney cysteine S-conjugate β-lyase) are identical. However, the situation is more complicated. Certainly, kynurenine aminotransferase possesses appreciable cysteine S-conjugate β-lyase and GTK activities. However, as noted above work from our laboratory (Abraham and Cooper, 1996) has resulted in the characterization and cloning of a cytGTK that is slightly smaller than kynurenine pyruvate aminotransferase. The cytGTK cloned by Abraham and Cooper (1996) had appreciable cysteine S-conjugate β-lyase activity and some kynurenine pyruvate aminotransferase activity (Abraham and Cooper, 1996). Evidently, the two enzymes exhibit considerable overlap in substrate specificity. The various reports also highlight the difficulties of working with enzymes with overlapping specificities. Overlapping specificities may have led to contradictory reports in the literature. Perry et al. (1993) stated that the message for kidney cysteine S-conjugate β-lyase could only be detected in the kidney cytosol and not in the liver cytosol, yet the liver contains appreciable GTK activity ($\sim$30% the specific activity of that in the kidney) (Cooper, 1988; Jones et al., 1988). The same group (MacFarlane et al., 1988) reported that cysteine S-conjugate β-lyase/cytGTK is present only in the S_3 region of the nephron, whereas Jones et al. (1988) reported cytGTK to be present in the S_1, S_2, and S_3 regions. We have confirmed the findings of Jones et al. by direct assay of individual segments of the rat proximal tubules (Kim et al., 1997). Moreover, we have also detect mRNA for cytGTK in rat liver and other organs (Abraham and Cooper, 1996).

E.   RAT KIDNEY HIGH M_r CYSTEINE S-CONJUGATE β-LYASE

This enzyme was discovered when homogenates of rat kidney were subjected to nondenaturing polyacrylamide gel electrophoresis followed by activity staining with DCVC (Abraham and Cooper, 1991). The enzyme was estimated to have a M_r of about 330,000. The enzyme has some weak GTK-type activity and hence full cysteine S-conjugate β-lyase activity is dependent on the presence of added α-keto acid (e.g., α-keto-γ-methiolbutyrate) or PLP. With DCVC as substrate it was estimated that the high M_r enzyme contributes to about 50% of the cysteine S-conjugate β-lyase activity in rat kidney cytosol. The high M_r lyase has been partially purified from the cytosolic fraction of rat kidney and separated from the low M_r forms (cytGTK/kynurenine pyruvate aminotransferase) (Abraham et al., 1995a). The high M_r enzyme has a broader specificity toward potential β-lyase substrates than does the low M_r enzyme. Like cytGTK, the high M_r enzyme is active with both DCVC and TFEC. However, unlike cytGTK the high M_r lyase is also active with leukotriene E_4

and with 5'-S-cysteinyldopamine. Pyruvate and ammonia were identified as reaction products, but in neither case was the nature of the
sulfhydryl-containing product identified. Evidently, the active site
of the high M_r lyase is able to accomodate relatively large amino
acid substrates and may have some similarities with the *E. limosum*
enzyme mentioned above.

At present, the relationship of the high M_r enzyme to other PLP-
containing enzymes is not known, but some evidence suggests that
this enzyme may be important in the bioactivation of nephrotoxic
cysteine S-conjugates. Thus, we have shown that the enzyme is well
represented in the rat proximal tubules, especially in the S_2 and S_3
segments (Kim et al., 1997). Moreover, the enzyme is present in
both the cytosolic and mitochondrial fractions in the rat kidney
(Abraham et al., 1995a,b) and, as noted above, mitochondria are
especially vulnerable to the toxic effects of cysteine S-conjugates.
Rat kidney mitochondria possess GTK activity (Cooper and Meister,
1981; Stevens et al., 1988). However, GTK purified from rat kidney
mitochondria has almost no cysteine S-conjugate β-lyase activity
(Abraham et al., 1995a). Finally, activity of the high M_r lyase in both
the cytosol and mitochondria is markedly stimulated by physiological levels of α-ketoglutarate. Whether there are distinct cytosolic
and mitochondrial isoforms of the high M_r lyase remains to be determined. In future studies it will be interesting to determine the relationship of the high M_r lyase(s) to other PLP enzymes and to determine its role in the biological response to nephrotoxic cysteine
S-conjugates.

F. POSSIBLE MECHANISMS FOR THE CONTROL OF CYSTEINE
S-CONJUGATE β-LYASES *IN VIVO*

The activity of the cysteine S-conjugate β-lyases may be regulated
in vivo (Stevens and Jones, 1989). For example, the liver enzyme
(kynureninase) may be inactivated by the product aminoacrylate, in
which case restoration of activity may require new protein synthesis.
However, it is also possible that some naturally occurring thiols may
protect against inactivation by aminoacrylate (Cavallini et al., 1973).
Some PLP-containing enzymes such as AspAT may be inactivated
by attack of aminoacrylate on cofactor, in which case restoration of
activity will depend to some extent on the availability of new cofac-

tor. In the case of the low M_r cysteine S-conjugate β-lyases of kidney (kynurenine pyruvate aminotransferase, cytGTK), the β-elimination reaction may be regulated by competition with a transamination reaction. In this case, the reaction will depend in part on the availability of suitable α-keto acid substrates (or PLP) that are capable of converting the enzyme to a form that can catalyze a β-elimination reaction with cysteine S-conjugates. This situation is reminiscent of the regulation of the activity of some PLP-containing enzymes by transamination as a side reaction. For example, bacterial aspartate β-decarboxylase undergoes an occasional transamination reaction to yield the inactive PMP form of the enzyme. Activity is restored by addition of pyruvate, which undergoes transamination at the active site to yield the active PLP form of the enzyme (Novogrodsky et al., 1963). The activity of mammalian glutamate decarboxylase also appears to be similarly regulated by a transamination cycle (Spink and Martin, 1983).

Finally, it has been shown that a rat kidney cysteine S-conjugate β-lyase activity (= kynurenine pyruvate aminotransferase) can be induced in rats by injection of a single nonnephrotoxic dose of N-acetyl-S-(1,2,3,4,4-pentachloro-1,3-butadienyl)-L-cysteine (MacFarlane et al., 1993). The enzyme activity and mRNA were both elevated (MacFarlane et al., 1993). On the other hand, repeated injections of toxic doses resulted in decreased enzyme activity in the kidney (MacFarlane et al., 1993). We have shown that 24 h after a single injection of hexachlorobutadiene the specific activity of small M_r lyases (= cytGT, kynurenine pyruvate aminotransferase) are increased in rat kidney (Kim et al., 1997). However, the specific activity in the S_3 segment of the proximal tubules, as judged by activity staining, is greatly diminished by this treatment. This finding suggests that although there is increased synthesis of small M_r lyases after acute exposure to S-(1,2,3,4,4-pentachloro-1,3-butadienyl)-L-cysteine, this increase is negated by leakage of cytosolic contents from the tubule (Kim et al., 1997).

Much work has established the importance of PLP and α-keto acids in regulating the toxic response toward cysteine S-conjugates. Inactivation of cysteine S-conjugate β-lyase activity by PLP antagonists such as aminooxyacetate (Dohn et al., 1985; Stevens et al., 1986a; Elfarra et al., 1986a; Wallin et al., 1987; Commandeur et al., 1988) and thiosemicarbazide (Jones et al., 1986) reduces the toxicity of cysteine S-conjugates under a variety of conditions. Interestingly, S-(1,2-dichlorovinyl)-D-cysteine is almost as toxic as the L-enantiomer (Wolfgang et al., 1989). Rat kidney possesses a large amount of D-amino acid oxidase (Meister et al., 1960), which will convert this D-isomer to the corresponding α-keto acid. Transamination of this α-keto acid in a reaction catalyzed by cytGTK will generate the nephrotoxic L-isomer. The α-methyl analogues of DCVC (Elfarra et al., 1986b) and S-(1-chloro-1,2,2-trifluoroethyl)-L-cysteine (Dohn et al., 1985) are nontoxic. These compounds cannot form a quinonoid intermediate with the PLP cofactor (Fig. 6, Structure 4) and hence cannot undergo β elimination. Finally, α-keto acids exacerbate the toxic effects of cysteine S-conjugates (Elfarra et al., 1987; Lash et al., 1986; Stevens et al., 1986a, 1988). The above discussion suggests that regulation of the various cysteine S-conjugate β-lyases *in vivo* is likely to be complex.

VII.   Cysteine S-Conjugate β-Lyases as Targets for Prodrugs Targeted to the Kidney

Kidney cancers in the United States are relatively rare, accounting for about 2% of all cancers (Henderson et al., 1991). Nevertheless, kidney cancer is serious because metastatic renal cell carcinomas are resistant to chemotherapy. In humans, most renal carcinomas appear to arise from the proximal tubules, the region most susceptible to damage by cysteine S-conjugates in experimental animals. [See Cooper (1994) for original references.] As noted above, cysteine S-conjugate β-lyase activity is present in human kidney. Thus, a link may exist between human renal cancer and exposure to certain halogenated compounds in the environment. As also noted above, such a link has been claimed (Henschler et al., 1995) and disputed (Swaen, 1995; Bloemen and Tomenson, 1995). Nevertheless, assuming that cysteine S-conjugate β-lyases in the human kidney, as in the rat kidney, are prominently located to the proximal tubules, then this location suggests a useful therapy based on the delivery of prodrugs to the kidney.

Hwang and Elfarra (1989, 1991) synthesized the cysteine S-conjugate of the anticancer compound 6-mercaptopurine as a kidney-directed prodrug. This cysteine S-conjugate is converted to pyruvate, ammonia, and a stable thiol (i.e., 6-mercaptopurine). Because the eliminated sulfhydryl-containing fragment is not especially reactive, the eliminated drug is not a renal toxin (Hwang and Elfarra, 1989, 1991; Elfarra and Hwang, 1990; 1993). Thirty minutes after administration of S-(6-purinyl)-L-cysteine to rats the concentrations of 6-mercaptopurine and its metabolites were 90- and 2.5-fold higher in kidney than in plasma and liver, respectively. Cytosolic and mitochondrial fractions of rat kidney were about equally effective in converting the prodrug to 6-mercaptopurine as were the cytosolic fractions of rat kidney and liver. Rat kidney mitochondria were, however, more effective than liver mitochondria.

The homocysteine conjugate of 6-mercaptopurine [i.e., S-(6-purinyl)-L-homocysteine] is also an effective prodrug (Elfarra and Hwang, 1993). To appreciate this idea some background chemistry and additional discussion of the specificity of cytGTK (= small M_r cysteine S-conjugate β-lyase) will be given. γ-Elimination of 6-mercaptopurine from the homocysteine S-conjugate could theoretically occur via a γ-cystathionase-type reaction with generation of α-ketobutyrate, ammonia and 6-mercaptopurine (XH = eliminated thiol-containing fragment) (Eq. 7).

$$XCH_2CH_2CH(NH_2)CO_2H + H_2O$$
$$\rightarrow XH + CH_3CH_2C(O)CO_2H + NH_3 \tag{7}$$

An alternative mechanism involves activation of the β C—H bond by conversion of the amino acid to the corresponding imino acid by L-amino acid oxidase. The activated bond facilitates a non-enzymatic β,γ-elimination reaction with the formation of vinylglyoxylate (2-oxobutenoate) (Eqs. 8,9).

$$XCH_2CH_2CH(NH_2)CO_2H + O_2$$
$$\rightarrow XCH_2CH_2C(=NH)CO_2H + H_2O_2 \tag{8}$$

$$XCH_2CH_2C(=NH)CO_2H + H_2O$$
$$\rightarrow XH + CH_2=CHC(O)CO_2H + NH_3 \tag{9}$$

228   ARTHUR J. L. COOPER

In a related reaction, elimination may occur from the α-keto acid (generated via the action of a transaminase or dehydrogenase) (Eq. 10).

$$XCH_2CH_2C(O)CO_2H \rightarrow XH + CH_2{=}CHC(O)CO_2H \quad (10)$$

The formation of 6-mercaptopurine from its homocysteine conjugate in rat kidney probably occurs via transamination in a reaction catalyzed by cytGTK followed by nonenzymatic decomposition of the corresponding α-keto acid to vinylglyoxylate and 6-mercaptopurine (Eq. 10). Thus, α-ketobutyrate is not a product in the elimination reaction and α-keto-γ-methiolbutyrate stimulated the reaction (Elfarra and Hwang, 1993). The various reaction pathways to vinylglyoxylate are summarized in Figure 7. Examples of L-amino acids that contain a good leaving group in the γ position and that undergo spontaneous β,γ-elimination after conversion to the corresponding α-imino or α-keto acid include S-adenosylmethionine (Stoner and Eisenberg, 1975), methionine sulfoximine (Cooper et al., 1976), homocysteine (Cooper and Meister, 1985), canavanine (Hollander et al., 1989; Cooper et al., 1989), S-methylmethionine (Rhodes et al., 1997), and homocysteine S-conjugates (Elfarra et al., 1986b). The vinylglyoxylate generated in these reactions is very reactive, but can be readily trapped with a suitable nucleophile such as 2-mercaptoethanol (Cooper et al., 1976; Lash et al., 1990b).

S-(1,2-Dichlorovinyl)-L-homocysteine (DCVHC) is even more toxic than is DCVC (Elfarra et al., 1986b, Lash et al., 1986, Lash and Anders, 1987). As noted above, DCVC is strongly mutagenic in the Ames test but DCVHC is less so. Aminooxyacetate blocks the

---

Figure 7. Elimination and γ-addition reactions associated with α-oxidation/transamination of amino acids that possess a good leaving group (X) in the γ position. 1, γ-substituted amino acid; 2, 4-substituted 2-oxobutyrate; 3, four-substituted 2-iminobutyrate; 4, vinylglyoxylate (2-oxobutenoate); 5, 2-imino-3-butenoate; 6, addition product of 4 with nucleophile (Nu:⁻). Abbreviations: AAO-L- (or D-) amino acid oxidase; PLP, pyridoxal 5'-phosphate; PMP-pyridoxamine 5'-phosphate. Pyridoxal 5'-phosphate is the cofactor for glutamine transaminase K/cysteine S-conjugate β-lyase, which participates in the transamination of some γ-substituted amino acids (see the text). [From Cooper et al. (1989) with permission.]

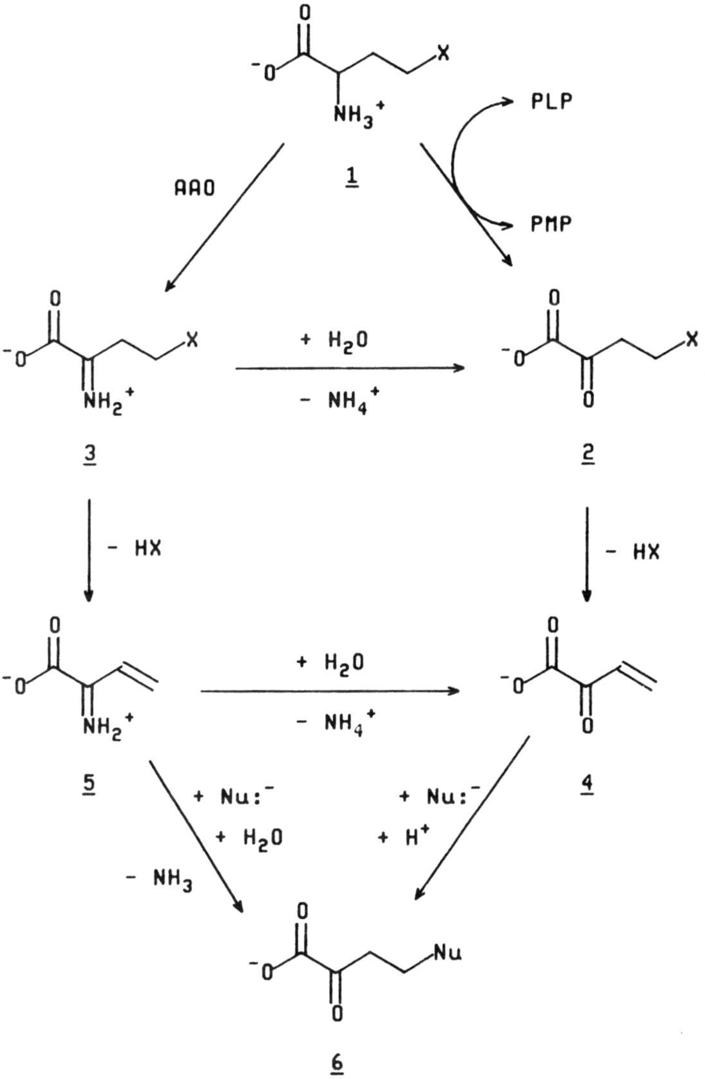
O
⁻O
X
NH₃⁺
1
PLP
PMP
AAO
O
⁻O
X
NH₂⁺
3
+ H₂O
− NH₄⁺
O
⁻O
X
O
2
− HX
− HX
O
⁻O
NH₂⁺
5
+ H₂O
− NH₄⁺
O
⁻O
O
4
+ Nu:⁻
+ H₂O
− NH₃
+ Nu:⁻
+ H⁺
O
⁻O
Nu
O
6

mutagenicity of both compounds (Dekant et al., 1989). The α-methyl analogue of DCVHC is not toxic. Evidence was presented that the toxicity of DCVHC is not due to a γ-cystathionase-type reaction, but rather to a transamination reaction, followed by a spontaneous non-enzymatic elimination reaction. It was shown that S-(benzothiazolyl)-L-homocysteine, methionine sulfoximine, and canavanine are toxic to isolated rat kidney cells (Lash et al., 1990b). Inasmuch as the SH fragment released from S-(benzothiazolyl)-L-homocysteine is nontoxic, it seems reasonable to assume that the toxicity in each case is due to metabolism to highly toxic vinylglyoxylate. A Micheal acceptor related to vinylglyoxylate—methyl vinyl ketone—is also toxic. Thus, the greater toxicity of DCVHC compared with DCVC is probably due to the formation of two toxic species from the former following enzymatic transformation catalyzed by cytGTK/cysteine S-conjugate β-lyase (i.e., vinylglyoxylate and 1-mercapto-1,2-dichloroethylene). With DCVC only one toxic species is generated (i.e., 1-mercapto-1,2-dichloroethylene).

In conclusion, both the cysteine- and homocysteine S-conjugates of 6-mercaptopurine are effective kidney-directed prodrugs. Both are substrates for glutamine transaminase K/cysteine S-conjugate β-lyase. Despite the fact that the mechanism is quite different in the two cases, interaction with the enzyme gives rise in both cases to an eliminated fragment (6-mercaptopurine). In the future, it may be possible to exploit the β- and γ-elimination reactions associated with cysteine- and homocysteine S-conjugates to design additional kidney-directed prodrugs.

VIII. Conclusions

Because of the possible link between human disease and exposure to toxins in the environment the cysteine S-conjugate β-lyases should be further studied. However, in addition to the epidemiological aspects, the enzymes deserve to be studied from other perspectives. In the future, we can probably look forward to the discovery of more PLP-containing enzymes capable of catalyzing cysteine S-conjugate β-lyase reactions. It was once thought that a given PLP-containing enzyme can catalyze only a single type of reaction. However, several examples are now known where a PLP-containing enzyme can catalyze at least two types of reaction. Such diversity may play a role in regulation of enzyme activity. It will be interesting to compare the sequences that allow enzymes that normally catalyze disparate reactions (e.g., kynureninase, cytGTK) to catalyze a β-elimination reaction with cysteine S-conjugates. Finally, experiments designed to determine the identity of the high M_r lyase and its relationship to other PLP enzymes should be especially rewarding.

Acknowledgments

I thank Dr. M. W Anders and Dr. J. L Stevens for their help and many useful discussions.

Abbreviations

AspAT	Aspartate aminotransferase
cyt	Cytosolic
DCVC	S-(1,2-Dichlorovinyl)-L-cysteine
DCVG	S-(1,2-Dichlorovinyl)glutathione
DCVHC	S-(1,2-Dichlorovinyl)-L-homocysteine
GTK	Glutamine transaminase K
mit	Mitochondrial
PLP	Pyridoxal 5'-phosphate
PMP	Pyridoxamine 5'-phosphate
TFEC	S-(1,1,2,2-Tetrafluoroethyl)-L-cysteine

References

Abraham, D. G. and Cooper, A. J. L., *Anal. Biochem.*, **197**, 421–427 (1991).

Abraham, D. G. and Cooper, A. J. L., *Arch. Biochem.*, **335**, 311–320 (1996).

Abraham, D. G., Patel, P., and Cooper, A. J. L., *J. Biol. Chem.*, **270**, 180–188 (1995a).

Abraham, D. G., Thomas, R. J., and Cooper, A. J. L., *Mol. Pharmacol.*, **48**, 855–860 (1995b).

Adiwidjaja, G., Kirsch, C., Pedersen, F., Schaumann, E., Schmerse, G. C., and Senning, A., *Chem. Ber.*, **124**, 1485–1487 (1991).

Anders, M. W. and Dekant, W., Eds., *Conjugation-Dependent Carcinogenicity and Toxicity of Foreign Compounds: Advances in Phamacology*, Vol. 27 Academic, San Diego, 1994.

Anders, M. W., Dekant, W., Henschler, D., Oberleithner, H., and Silbernagl, S. Eds., *Renal Disposition and Nephrotoxicity of Xenobiotics*, Academic, San Diego (1993).

Anders, M. W., Elfarra, A. A., and Lash, L. H., *Arch. Toxicol.*, **60**, 103–108 (1987).

Anderson, P. M. and Schultze, M. O., *Arch. Biochem. Biophys.*, **111**, 593–602 (1965).

Bakke, J. E., *Xenobiotic Conjugation Chemistry*, American Chemical Society, Symposium Series 199, 1986, pp. 388–396.

Bakke, J. E. and Gustafsson, J.-A., *Trends Pharmacol. Sci. Dec.*, 517–521 (1984).

Bakke, J. E., Larsen, G. L., Aschbacher, P. W., Rafter, J. J., Gustafsson, J.-Å., and Gustafsson, B. E., In *Sulfur in Pesticide Action and Metabolism*, Rosen J. D., Magee P. S., Casida J. E. (eds): American Chemical Society, Washington, DC, 1981, pp. 165–178.

Barnes, M. M., *Biochem. J.*, **71**, 680–690 (1959).

Barnsley, E. A., *Biochim. Biophys. Acta*, **90**, 24–36 (1964).

Baumann, E., *Hoppe-Seylers Physiol. Chem.*, **84**, 190–197 (1883).

Baumann, E. and Preusse, C., *Ber. Dtsch.*, **12**, 806–810 (1879).

Bernström, K., Larsen, G. L., and Hammerström, S., *Arch. Biochem. Biophys.*, **275**, 531–539 (1989).

Bhattacharya, R. K. and Schulze, M. O., *Comp. Biochem. Physiol.*, **22**, 723–735 (1967).

Bhattacharya, R. K. and Schulze, M. O., *Arch. Biochem. Biophys.*, **145**, 575–582 (1971a).

Bhattacharya, R. K. and Schulze, M. O., *Biochem. Biophys. Res. Commun.*, **45**, 1526–1532 (1971b).

Bhattacharya, R. K. and Schulze, M. O., *Arch. Biochem. Biophys.*, **153**, 105–115 (1972).

Bhattacharya, R. K. and Schulze, M. O., *Biochem. Biophys. Res. Commun.*, **54**, 172–181 (1973a).

Bhattacharya, R. K. and Schulze, M. O., *Biochem. Biophys. Res. Commun.*, **54**, 538–543 (1973b).

Bild, G. S. and Morris, J. C., *Arch. Biochem. Biophys.*, **235**, 41–47 (1984).

Bloemen, L. J. and Tomenson, J., *Arch. Toxicol.*, **70**, 131–133 (1995).

Boyland, E. and Chasseaud, L. F., *Adv. Enzymol. Related Areas Mol. Biol.*, **46**, 383–414 (1969).

Braunstein, A. E. and Goryachenkova, E. V., *Adv. Enzymol. Relat. Areas Mol. Biol.*, **56**, 1–89 (1984).

Bray, H. G., Franklin, T. J., and James, S. P., *Biochem. J.*, **71**, 690–696 (1959).

Bruschi, S. A., West, K. A., Crabb, J. W., Gupta, R. S., and Stevens, J. L., *J. Biol. Chem.*, **268**, 23157–23161 (1993).

Buxton, P. H. and Hayward, M., *J. Neurol. Neurosurg. Psychiatry*, **30**, 511–518 (1967).

Cavallini, D. E., Federici, G., Bossa, F., and Granata, F., *Eur. J. Biochem.* **39**, 301–304 (1973).

Chasseaud, L. F., *Adv. Cancer Res.*, **29**, 175–274 (1979).

Chatfield, D. H. and Hunter, W. H., *Biochem. J.*, **134**, 879–884 (1973).

Chen, Q., Jones, T. W., Brown, P. C., and Stevens, J. L., *J. Biol. Chem.*, **265**, 21603–21611 (1990).

Colucci, D. F. and Buyske, D. A., *Biochem. Pharmacol.*, **14**, 457–466 (1965).

Commandeur J. N. M., Brakenhoff J. P. G., De Kanter, F. J. J., and Vermeulen, N. P. E., *Biochem. Pharmacol.*, **37**, 4495–4504 (1988).

Cooper, A. J. L., In *Glutamine and Glutamate in Mammals*, Kvamme, E. ed., CRC Press, Boca Raton, FL, (1988), pp. 33–52.

Cooper, A. J. L., *Adv. Pharmacol.*, **27**, 71–113 (1994).

Cooper, A. J. L., In *The Molecular and Genetic Basis of Neurological Disease*, 2nd ed., Rosenberg, R. N., Prusiner, S. B., DiMauro, S., and Barchi, R. L. eds., Butterworth-Heinemann, Boston, MA, 1997, pp. 1195–1230.

Cooper, A. J. L., Abraham, D. G., Gelbard, A. S., Lai, J. C. K., and Petito, C. K., *J. Neurochem.*, **61**, 1731–1741 (1993).

Cooper, A. J. L. and Anders, M. W., *Proc. NY Acad. Sci.*, **585**, 118–127 (1990).

Cooper, A. J. L., Hollander, M. M., and Anders, M. W., *Biochem. Pharmacol.*, **38**, 3895–3901 (1989).

Cooper, A. J. L. and Meister, A., *Comp. Biochem. Physiol.*, **69B**, 137–145 (1981).

Cooper, A. J. L. and Meister, A., *Arch. Biochem. Biophys.*, **239**, 556–566 (1985).

Cooper, A. J. L., Stephani, R. A., and Meister, A., *J. Biol. Chem.*, **251**, 6674–6682 (1976).

Cooper, A. J. L. and Tate, S. S., In *Comprehensive Toxicology: Volume 3. Biotransformations*, Sipes, G., McQueen, C. A., and Gandolfi, A. J., Eds., Elsevier, Oxford, 1997, pp. 329–363.

Darnerud, P. O., Brandt, I., Feil, V. J., and Bakke, J. E., *Toxicol. Appl. Pharmacol.*, **95**, 423–434 (1988).

Dekant, W., Koob, M., and Henschler, D., *Chem.-Biol. Interact.*, **73**, 89–101 (1990).

Dekant, W., Lash, L. H., and Anders, M. W., *Proc. Natl. Acad. Sci. USA*, **84**, 7443–7447 (1987).

Dekant, W., Urban, G., Görsman, C., and Anders, M. W., *J. Am. Chem. Soc.*, **113**, 5120–5122 (1991).

Dekant, W., Vamvakas, S., Berthold, K., Schmidt, S., Wild, D., and Henschler, D., *Chem.-Biol. Interact.*, **60**, 31–45 (1986).

Dekant, W., Vamvakas, S., and Anders, M. W., *Drug Metabol. Rev.*, **20**, 43–83 (1989).

Dekant, W., Vamvakas, S., and Anders, M. W., *Adv. Pharmacol.*, **27**, 1114–162 (1994).

Dohn, D. R., Leininger, J. R., Lash, L. H., Quebbemann, A. J., and Anders, M. W., *J. Pharmacol. Exp. Ther.*, **235**, 851–857 (1985).

Dolphin, D., Avramovic, O., and Poulson, R., *Glutathione: Chemical, Biochemical and Medical Aspects*, Wiley, New York, 1989.

Elfarra, A. A., In *Toxicology of the Kidney*, 2nd ed., Hook, J. B. and Goldstein R. S., Eds., Raven, New York, 1993, pp. 387–4141.

Elfarra, A. A. and Hwang, I. Y., *Drug Metab. Dispos.*, **18**, 917–922 (1990).

Elfarra, A. A. and Hwang, I. Y., *Drug Metab. Dispos.*, **21**, 841–845 (1993).

Elfarra, A. A., Jakobson, I., and Anders, M. W., *Biochem. Pharmacol.*, **35**, 283–288 (1986a).

Elfarra, A. A., Lash, L. H., and Anders, M. W., *Proc. Natl. Acad. Sci. USA*, **83**, 2667–2671 (1986b).

Elfarra, A. A., Lash, L. H., and Anders, M. W., *Mol. Pharmacol.*, **31**, 208–217 (1987).

Finkelstein, M. B., Vamvakas, S., Bittner, D., and Anders, M. W., *Chem. Res. Toxicol.*, **7**, 157–163 (1994).

Flavin, M., and Slaughter, C., *J. Biol. Chem.*, **239**, 2212–2219 (1964).

Goldman, P., *Annu. Rev. Pharmacol. Toxicol.*, **18**, 523–539 (1978).

Golichowski, A. and Jenkins, W. T., *Arch. Biochem. Biophys.*, **189**, 109–114 (1978).

Green, T. and Odum, J., *Chem.-Biol. Interact.*, **54**, 15–31 (1985).

Greim, H., Wolff, T., Höffler, M., and Lahaniatis, E., *Arch. Toxicol.*, **56**, 74–77 (1984).

Hafsah, Z., Tahara, S., and Mizutani, J., *J. Pesticide Res.*, **12**, 617–623 (1987).

Hayden, P. J. and Stevens, J. L., *Mol. Pharmacol.*, **37**, 468–476 (1990).

Henderson, B. E., Ross, R. K., and Pike, M. C., *Science*, **254**, 1131–1138 (1991).

Henschler, D., Vamvakas, S., Lammer, M., Dekant, W., Kraus, B., Thomas, B., and Ulm, K., *Arch. Toxicol.*, **69**, 291–299 (1995).

Hollander, M. M., Reiter, A. J., Horner, W. H., and Cooper, A. J. L., *Arch. Biochem. Biophys.*, **270**, 698–713 (1989).

Horner, W. H. and Kuchinskas, E. J., *J. Biol. Chem.*, **234**, 2935–2937 (1959).

Hwang, I. Y. and Elfarra, A. A., *J. Pharmacol. Exp. Ther.*, **251**, 448–454 (1989).

Hwang, I. Y. and Elfarra, A. A., *J. Pharmacol. Exp. Ther.*, **258**, 171–177 (1991).

Jaffe, D. R., Gandolfi, A. J., and Nagle, R. B., *J. Appl. Toxicol.*, **4**, 315–319 (1984).

Jaffe, D. R., Hassall, C. D., Gandolfi, A. J., and Brendel, K., *Toxicology*, **35**, 25–33 (1985).

Jakoby, W. B., *Adv. Enzymol. Relat. Anas. Mol. Biol.*, **46**, 383–414 (1978).

Jakoby, W. B. and Stevens, J. *Biochem. Soc. Trans.*, **12**, 33–35 (1984).

Jakoby, W. B., and Stevens, J. L., Duffel, M. W., and Wiesiger, R. A., *Rev. Biochem. Toxicol.*, **6**, 97–115 (1984).

James, S. P. and Needham, D., *Xenobiotica*, **3**, 207–218 (1973).

Jones, T. W., Qin, C., Schaeffer, V. H., and Stevens, J. L., *Mol. Pharmacol.*, **34**, 621–627 (1988).

Jones, T. W., Wallin, A., Thor, H., Gerdes, R. G., Ormstad, K., and Orrenius, S., *Arch. Biochem. Biophys.*, **251**, 504–513 (1986).

Kanhai, W., Dekant, W., and Henschler, D., *Chem. Res. Toxicol.*, **2**, 51–56 (1989).

Kato, Y., Asano, Y., and Cooper, A. J. L., *Dev. Neurosci.*, **18**, 505–514 (1996).

Ketterer, B., Coles, B., and Meyer, D. J., *Environ. Health Perspect.*, **49**, 59–69 (1983).

Kim, H. S., Cha, S. H., Abraham, D. G., Cooper, A. J. L., and Endou, H., *Arch. Toxicol.*, **71**, 131–141 (1997).

Kishore, G. M., *J. Biol. Chem.*, **259**, 10669–10674 (1984).

Koechel, D. A., Krejci, M. E., and Ridgewell, R. E., *Fundamental Appl. Toxicol.*, **17**, 17–33 (1991).

Koob, M., and Dekant, W., *Chem.-Biol. Interact.*, **77**, 107–136 (1991).

Larsen, G. L., *Xenobiotica*, **15**, 199–209 (1985).

Larsen, G. L., and Bakke, J. E., *Xenobiotica*, **13**, 115–126 (1983).

Larsen, G. L., Larsen, J. D., and Gustafsson, J.-E., *Xenobiotica*, **13**, 689–700 (1983).

Larsen, G. L., and Stevens, J. L., *Mol. Pharmacol.*, **29**, 97–103 (1986).

Lash, L. H. and Anders, M. W., *Comments Toxicol.*, **1**, 87–107 (1986).

Lash, L. H., and Anders, M. W., *Mol. Pharmacol.*, **32**, 549–556 (1987).

Lash, L. H., Elfarra, A. A., and Anders, M. W., *J. Biol. Chem.*, **261**, 5930–5935 (1986).

Lash, L. H., Elfarra, A. A., Rakiewicz-Nemeth, D., and Anders, M. W., *Arch. Biochem. Biophys.*, **276**, 322–330 (1990b).

Lash, L. H., Nelson, R. M., Van Dyke, R., and Anders, M. W., *Drug Metab. Dispos.*, **18**, 50–54 (1990a).

Lash, L. H., Sausen, P. J., Duescher, R. J., Cooley, A. J., and Elfarra, A. A., *J. Pharmacol. Exp. Ther.*, **269**, 374–383 (1994).

Lock, E. A., Sani, Y., Moore, R. B., Finkelstein, M. B., Anders, M. W., and Seawright, A. A., *Arch. Toxicol.*, **70**, 607–619 (1996).

MacFarlane, M., Foster, J. R., Gibson, G. G., King, L. J., and Lock, E. A., *Toxicol. Appl. Pharmacol.*, **98**, 185–195 (1988).

MacFarlane, M., Schofield, M., Parker, N., Roelandt, L., David, M., Lock, E. A., King, L. J., Goldfarb, P. S., and Gibson, G. G., *Toxicology*, **77**, 133–144 (1993).

Manning, J. M., Khomutov, R. M., and Fasella, P., *Eur. J. Biochem.*, **5**, 199–208 (1968).

McKinney, L. L., Picken, J. C., Jr., Weakley, F. B., Eldridge, A. C., Campbell, R. E., Cowan, J. C., and Biester, H. E., *J. Am. Chem. Soc.*, **81**, 909–915 (1959).

Meister, A., *Fed. Proc.*, **43**, 3031–3042 (1984).

Meister, A., *TIBS*, **13**, 185–188 (1988).

Meister, A. and Anderson, M. E., *Annu. Rev. Biochem.*, **52**, 711–760 (1983).

Meister, A., Wellner, D., and Scott, S. J., *J. Natl. Cancer Inst. (US)*, **24**, 31–49 (1960).

Mitchell, S. C., Smith, R. L., Waring, R. H., and Aldington, G. F., *Xenobiotica*, **14**, 767–779 (1984).

Mosca, M., Cozzi, L., Breton, J., Speciale, C., Okuno, R., Schwartcz, R., and Benatti, L., *FEBS Lett.*, **353**, 21–24 (1994).

Morino, Y. and Okamoto, M., *Biochem. Biophys. Res. Commun.*, **50**, 1061–1067 (1973).

Müller, D. A., Birner, G., Henschler, D., and Dekant, W., *IARC Sci. Publ. Ser.*, **115**, 423–428 (1994).

Müller, D. A., Urban, G., and Dekant, W., *Chem.-Biol. Interact.*, 77, 159–172 (1991).

Nash, J. A., King, L. J., Lock, E. A., and Green, T., *Toxicol. Appl. Pharmacol.*, 73, 124–137 (1984).

Nicholson, D. W., In *Structure and Function of Glutathione Transferases*, Tew, K. D., Pickett, C. B., Mantle, T. J., Mannervik, B., and Hayes, J. D., Eds., CRC, Boca Raton, FL, 1993, pp. 47–62.

Novogrodsky, A., Nishimura, J. S., and Meister, A., *J. Biol. Chem.*, 238, PC1903–PC1905 (1963).

Odum, J. and Green, T., *Toxicol. Appl. Pharmacol.*, 76, 306–318 (1984).

Perry, S. J., Schofield, M. A., MacFarlane, M., Lock, E. A., King, L. J., Gibson, G. G., and Goldfarb, P. S., *Mol. Pharmacol.*, 43, 660–665 (1993).

Raasch, M. S., *J. Org. Chem.*, 35, 3470–3483 (1970).

Raasch, M. S., *J. Org. Chem.*, 37, 1347–1355 (1972).

Reed, D. J., *Int. J. Radiat. Oncol. Biol. Biophys.*, 12, 1457–1461 (1986).

Reed, D. J., and Meredith, M. J., In *Drugs and Nutrients: The interactive effects*, Roe, D. A., and Campbell, T. C., Eds., Marcel-Dekker, New York, 1984, pp. 179–225.

Reichert, D., Spengler, U., Romen, W., and Henschler, D., *Carcinogenesis (London)*, 5, 1411–1420 (1984).

Reichert, D., Liebalt, G., and Henschler, D., *Arch. Toxicol.*, 37, 23–38 (1976).

Relyea, N. M., Tate, S. S., and Meister, A., *J. Biol. Chem.*, 249, 1519–1524 (1974).

Rhodes, D., Gage, D. A., Cooper, A. J. L., and Hanson, A. D., *Plant Physiol.*, 115, 1541–1548 (1997).

Roush, W., *Science*, 269, 473 (1995).

Saari, J. C. and Schultze, M. O., *Arch. Biochem. Biophys.*, 109, 595–602 (1965).

Saunders, R. A., *Arch. Environ. Health*, 14, 380–384 (1967).

Sausen, P. J., Duescher, R. J., and Elfarra, A. A., *Mol. Pharmacol.*, 43, 388–396 (1993).

Sausen, P. J., and Elfarra, A. A., *J. Biol. Chem.*, 265, 6139–6145 (1990).

Schaumburg, H. H., In *Disorders of the Central Nervous System*, 2nd Ed., Asbury, A. K., McKahnn, G. M., and MacDonald, W. I., Eds., Saunders, Philadelphia, 1993, pp. 1238–1249.

Schrenk, D. and Dekant, W., *Carcinogenesis (London)*, 10, 1139–1141 (1989).

Schultze, M. O., Klubes, P., Perman, V., Mizuno, N. S., Bates, F. W., and Sautter, J. H., *Blood*, 14, 1015–1025 (1959).

Schwimmer, S. and Mazelis, M., *Arch. Biochem. Biophys.*, 100, 66–73 (1963).

Shimomura, N., Honma, M., Chiba, S., Tahara, S., and Mizutani, J., *Biosci. Biotech. Biochem.*, 56, 963–964 (1992).

Sklan, N. M. and Barnsley, E. A., *Biochem. J.*, 107, 217–223 (1968).

Snell, E. E. and Di Mari, S. J., In *The Enzymes*. 3rd ed., Vol. 2, Boyer, P. D., Ed., Academic, New York, 1970, pp. 335–370.

Soper, T. S. and Manning, J. M., *Biochemistry*, **17**, 3377–3384 (1978).

Spink, D. C. and Martin, D., In *Glutamine, Glutamate and GABA in the Central Nervous System*, Hertz, L., Kvamme, E., McGeer, E. G., and Shousboe, A., Eds., Liss, 1983 New York, pp. 129–143.

Stevens, J. L., *J. Biol. Chem.*, **260**, 7945–7950 (1985a).

Stevens, J. L., *Biochem. Biophys. Res. Commun.*, **129**, 499–504 (1985b).

Stevens, J. L. and Jakoby, W. B., *Mol. Pharmacol.*, **23**, 761–765 (1983).

Stevens, J. L. and Jones, D. P., In *Glutathione: Chemical, Biochemical and Medical Aspects, Part B*, Dolphin, D., Poulson, R., and Avramovic, O. Eds., Wiley, New York, 1989, pp. 45–84.

Stevens, J. L., Hayden, P., and Taylor, S., *J. Biol. Chem.*, **261**, 3325–3332 (1986a).

Stevens, J. L., Robbins, J. D., and Byrd, R. A., *J. Biol. Chem.*, **261**, 15529–15537 (1986b).

Stevens, J. L., Ayoubi, N., and Robbins, J. D., *J. Biol. Chem.*, **263**, 3395–3401 (1988).

Stockman, S., *J. Comparat. Pathol.*, **29**, 95–107 (1916).

Stonard, M. D. and Parker, V. H., *Biochem. Pharmacol.*, **20**, 2417–2427 (1971).

Stoner, G. L. and Eisenberg, M. A., *J. Biol. Chem.*, **250**, 4029–4036 (1975).

Swaen, G. M. H., *Arch. Toxicol.*, **70**, 127–128 (1995).

Tanaguchi, N., Higashi, Y., Sakamoto, Y., and Meister, A., *Glutathione Centennial: Molecular Perspective and Clinical Implications* Academic, New York, 1989.

Tateishi, M., Suzuki, S., and Shimuzu, S., *Biochem. Pharmacol.*, **27**, 809–810 (1978a).

Tateishi, M., Suzuki, S. and Shimuzu, S., *J. Biol. Chem.*, **253**, 8854–8859 (1978b).

Terracini, B. and Parker, V. H., *Food Cosmet Toxicol.*, **3**, 67–74 (1965).

Tew, K. D., Pickett, C. B., Mantle, T. J., Mannervik, B., and Hayes, J. D., *Structure and Function of Glutathione Transferases*, CRC Press, Boca Raton, FL, (1993).

Tomisawa, H., Ichihara, S., Fukazawa, H., Ichimoto, N., Tateishi, M., and Yamamoto, I., *Biochem. J.*, **235**, 569–575 (1986).

Tomisawa, H., Susuki, S., Ichihara, S., Fukazawa, H., and Tateishi, M., *J. Biol. Chem.*, **259**, 2588–2593 (1984).

Turnbull, L. B., Teng, L., Kinzie, J. M., Pitts, J. E., Pinchbeck, F. M. and Bruce, R. B., *Xenobiotica*, **8**, 621–628 (1978).

Ueno, H., Likos, J. J., and Metzler, D. E., *Biochemistry*, **21**, 4387–4393 (1982).

Vamvakas, S., Herkenhoff, M., Dekant, W., and Henschler, D., *J. Biochem. Toxicol.*, **4**, 21–27 (1989).

Vamvakas, S., Kordowich, F. J., Dekant, W., Neudecker, T., and Henschler, D., *Carcinogenesis (London)*, **9**, 907–910 (1988).

Vamvakas, S., and Köster, U., *Cell. Biol. Toxicol.*, **9**, 1–13 (1993).

Wallin, A., Jones, T. W., Vercesi, A. E., Cotgreave, I., Ormstadt, K., and Orrenius, S., *Arch. Biochem. Biophys.*, **258**, 365–372 (1987).

Wang, E., and Walsh, C., *Biochemistry*, **17**, 13–1321 (1978).

Wijesundera, S., and Woods, D. D., *J. Gen. Microbiol.*, **29**, 353–366 (1962).

Wolfgang, G. H. I., Gandolfi, A. J., Stevens J. L., and Brendel, K., *Toxicology*, **58**, 33–42 (1989).

Wood, J. L., In Fishman WA (ed): "Metabolic Conjugation and Metabolic Hydrolysis." New York: Academic Press, pp 261–291 (1970).

Ziegler, D. M., In Mitchell J. R., Horning M. G. (Eds.), New York: Raven Press, pp 33–53 (1984).

γ-GLUTAMYL TRANSPEPTIDASE: CATALYTIC MECHANISM AND GENE EXPRESSION

By NAOYUKI TANIGUCHI
and YOSHITAKA IKEDA, *The Department of Biochemistry, Osaka University Medical School, Suita, Osaka 565, Japan*

CONTENTS

Advances in Enzymology and Related Areas of Molecular Biology, Volume 72: Amino Acid Metabolism, Part A, Edited by Daniel L. Purich
ISBN 0-471-24643-3 ©1998 John Wiley & Sons, Inc.

I. Introduction

γ-Glutamyl transpeptidase (5-L-glutamyl)-peptide/amino acid 5-glutamyltransferase, (EC 2. 3. 2. 2), which is also known as γ-glutamyltransferase, is a key enzyme in glutathione metabolism and plays a central role in the γ-glutamyl cycle involving the synthesis and degradation of glutathione (Meister and Larsson, 1995; Meister and Tate, 1976; Meister and Anderson, 1983). In mammalian cells, the enzyme exists as a membrane-bound protein, and is anchored to the extracellular surface of the plasma membrane. γ-Glutamyl transpeptidase activity is most abundant in the kidney of most mammals. Pancreas, epididymis, and seminal vesicle also contain relatively high levels of activity. In general, activity is most pronounced in cells that have secretory or absorptive functions. γ-Glutamyl transpeptidase hydrolyzes a γ-glutamyl bond of glutathione, its S-substituted derivatives, and γ-glutamyl compounds. In the degradation of glutathione, the cysteinylglycine resulting from cleavage of γ-glutamyl linkage of glutathione is further hydrolyzed into cysteine and glycine by another membrane-bound hydrolytic enzyme, dipeptidase. γ-Glutamyl transpeptidase catalyzes the hydrolysis of glutathione as well as the transfer of the γ-glutamyl moiety of glutathione and other γ-glutamyl compounds to amino acids and dipeptides. It is thought that the reactions catalyzed by the enzyme involve the formation of a γ-glutamyl-enzyme intermediate, in which the γ-glutamyl moiety derived from the substrates is covalently bound to the catalytic site of the enzyme (Tate and Meister, 1981). The linkage between the γ-glutamyl group and the enzyme in the intermediate is disrupted by the attack of water or α-amino groups of amino acid and peptides. Cleavage of the enzyme species by water leads to hydrolysis, and reaction with amino acids or dipeptides results in a transfer to give respective γ-glutamyl compounds.

Most cells that depend on aerobic metabolism are susceptible to damage by reactive oxygen species and most have various mechanisms for protection from oxidative stress. The production of glutathione is one of the most important mechanisms for eliminating oxygen species and peroxides, since this unique tripeptide is a major intracellular reducing agent in a variety of cells, both prokaryotic and eukaryotic (Meister, 1992). Therefore, the depletion of glutathione would be expected to lead to oxidative damage in tissues (Jain et al., 1991). In addition to its significant protective role against oxidation, the conjugation of glutathione with other chemical compounds facilitates their removal from the body via excretion into the urine. Thus it is also involved in the detoxification of xenobiotics or in drug metabolism. The maintenance of glutathione is, therefore, closely associated with cellular defense against the environment. While glutathione is synthesized in the successive reactions catalyzed by γ-glutamylcysteine synthetase and glutathione synthetase in an ATP-dependent manner within cells, the tripeptide is exclusively degraded outside cells by γ-glutamyl transpeptidase because the catalytic domain of this enzyme extends into the extracellular space (Inoue et al., 1977; Meister and Larsson, 1995). The production of glutathione is regulated by the activity of γ-glutamylcysteine synthetase but is also controlled by the intracellular concentrations of the constituent amino acids. In terms of the regulation of glutathione synthesis by its constituents, the synthesis appears to be most significantly affected by the concentration of cysteine, since the intracellular concentration is the lowest for this amino acid. The capability of γ-glutamyl transpeptidase to catalyze the degradation of glutathione increases the availability of cyst(e)ine from extracellular glutathione as the source and further allows cells to recover the amino acid(s) by uptake of γ-glutamyl derivatives of the amino acids through the γ-glutamyl cycle. Thus, γ-glutamyl transpeptidase participates in the regulation of glutathione synthesis. Consequently, the enzyme would be expected to modulate the cellular functions involving glutathione.

Xenobiotics and drugs are conjugated with the sulfhydryl group of glutathione by the action of glutathione-S-transferase isozymes. The resultant glutathione-S-conjugates undergo removal of their γ-glutamyl groups by γ-glutamyl transpeptidase. Subsequent cleavage of the glycine moiety, followed by its N-acetylation leads to the formation of the corresponding mercapturic acids and to their excretion into urine. This type of metabolism of glutathione-S-conjugates is not only directed toward foreign chemical compounds but also occurs for endogeneous substances, as found in analogous reactions involved in conversion of leukotriene C4 to leukotriene D4 (Anderson et al., 1982) and the metabolism of prostaglandins (Cagen et al., 1976). Furthermore, there is accumulating evidence that conjugation with glutathione negatively affects the potency of certain antitumor agents by removal of these agents from cells in a similar manner (Godwin et al., 1992). Therefore, γ-glutamyl transpeptidase might be a component responsible for drug resistance in cancer cells, and could be a target to overcome resistance, as well as γ-glutamyl cysteine synthetase (Godwin et al., 1992).

Expression of γ-glutamyl transpeptidase has long been known to be induced in liver by alcohol (Nishimura et al., 1981; Barouki et al., 1983), and it is also elevated in hepatocarcinogenesis (Fiala and Fiala, 1973; Taniguchi et al., 1983a; Farber, 1984; Beer et al., 1986; Hanigan and Pitot, 1985). The enzyme therefore has been regarded as a tumor marker and has been extensively investigated for application in the diagnosis of cancer (Sawabu et al., 1978; Taniguchi et al., 1985). Elevation of γ-glutamyl transpeptidase activity or protein in serum has been found to be associated with liver carcinogenesis. In chemically induced hepatocellular carcinoma, it was found that the enzyme is strongly expressed in neoplastic nodules, whereas the expression is not apparent in the normal region. Although γ-glutamyl transpeptidase occurring in response to carcinogenesis was earlier regarded as a tumor-specific isozyme, it appears unlikely that the difference associated with tumor is reflected by differences in the polypeptide backbone. The apparent isozymic forms appear to be due to cancer-associated changes of sugar chain moieties. Thus, γ-glutamyl transpeptidase also has clinical significance, in addition to its importance in glutathione metabolism.

In this chapter, we summarize the recent findings on roles of γ-glutamyl transpeptidase in glutathione metabolism, its catalytic mechanism, and gene expression, most of which have been obtained for mammalian γ-glutamyl transpeptidases.

II.   Structural and Topological Features of γ-Glutamyl Transpeptidase

γ-Glutamyl transpeptidase is a membrane-bound and heavily glycosylated protein. The enzyme is a heterodimer, in which a large and a small subunit associate in a noncovalent manner. Several methods of purification have been established and applied to preparation from various mammalian tissues (Meister et al., 1981; Tate and Meister, 1985; Hughey and Curthoys, 1976; Taniguchi, 1974; Taniguchi et al., 1975; Matsuda et al., 1983). In all purification procedures, solubilization of the enzyme from cell membrane are required for preparation of highly purified enzyme, and detergents such as Triton X-100, Lubrol PX, and deoxycholate have been used for extraction of the enzyme. Treatment by these detergents, however, does not cause dissociation of the subunits despite the fact that they are associated by noncovalent interactions. Alternatively, instead of the detergents, proteases such as papain and bromelain also efficiently solubilize the enzyme (Tate and Meister, 1985; Hughey and Curthoys, 1976). The enzymatic properties of both detergent- and protease-solubilized enzymes are indistinguishable. The SDA–PAGE analysis of the protease-treated enzyme indicates a smaller heavy subunit, as compared with that of the detergent-solubilized form (Hughey and Curthoys, 1976). Electrophoresis indicates no apparent difference in the light subunits. It was suggested that the amino terminal hydrophobic segment is necessary for the protein to interact with a lipid bilayer (Hughey et al., 1979). Amino acid sequence analyses of those enzymes showed that treatment with proteases removed approximately 30 amino acid residues of the amino terminal sequence and that the carboxyl terminal amino acids were identified as tyrosine for both enzymes, indicating that the amino terminal region is the domain which anchors it to the cell membrane (Matsuda et al., 1983).

The cDNAs for γ-glutamyl transpeptidases have been cloned from a variety of mammalian species (Coloma and Pitot, 1986; Goodspeed et al., 1989; Laperche et al., 1986; Papandrikopoulou et al., 1989; Sakamuro et al., 1988; Shi et al., 1995; Rajpert-De-Meyts et al., 1988). The structural analyses of the primary structures indicated that the heavy and light subunits are encoded by a common mRNA in this order. It has also been revealed that the amino terminal region of the heavy subunits represents a possible single transmembrane domain, in agreement with the findings relative to the truncation of the heavy subunit in protease-solubilized enzyme. The extreme amino terminal small region followed by the transmembrane domain contains positively charged amino acids, suggesting that the short segment serves as the cytosolic domain of type-II membrane proteins. (von Heijne, 1990). The 5–6 potential sites for N-glycosylation (Asn-X-Thr/Ser) are present in the heavy subunit and 1–2 positions in the light subunit. This would account for the extensive glycosylation and the great heterogeneities with respect to molecular mass found in mammalian enzymes, since sugar chains typically exhibit structural heterogeneity.

III. Biosynthesis and Processing of γ-Glutamyl Transpeptidase

In vitro translation studies and pulse-chase labeling experiments have revealed that γ-glutamyl transpeptidase is translated as a single-chain precursor form, which is then glycosylated. The glycosylated precursor is subsequently processed into two subunits by the processing protease(s), which have not yet been identified (Capraro and Hughey, 1983; Finidori et al., 1984; Barouki et al., 1984; Nash and Tate, 1982; Yokosawa et al., 1983; Nash and Tate, 1984). In rat kidney, this proteolytic processing was found to take place at the brush border membrane (Kuno et al., 1983). This behavior and the topological features are consistent with a typical type-II membrane-bound protein, which has a single transmembrane domain at the amino terminus (von Heijne, 1990). In such types of membrane proteins, the amino-terminal single transmembrane domain functions as an uncleavable signal peptide, which is involved in targeting the protein to the endoplasmic reticulum. The proteins are then translocated into the lumen of endoplasmic reticulum, followed by glycosylation. Thus, it had been generally thought that the transmembrane domain of γ-glutamyl transpeptidase functions, not only to anchor the enzyme to cell membrane, but also as a signal peptide (Tate and Nash, 1987a). Nevertheless, when insect cells were infected with the recombinant baculovirus carrying the cDNA encoding human γ-glutamyl transpeptidase mutant that lacks the transmembrane domain, the infected cells secreted the mutant enzyme into the culture medium (Ikeda et al., 1995a). Furthermore, the mutant enzyme purified from the medium was glycosylated with N-linked oligosaccharides. Its secretion from cells was inhibited by brefeldin A, a fungal metabolite that inhibits the transport of membrane-bound and secretory proteins from the trans-golgi network (Klausner et al., 1992; Lippincott-Schwartz et al., 1991). These observations suggest that γ-glutamyl transpeptidase can be directed to the ER and translocated into the lumen in the absence of the putative signal/anchor domain.

The single-chain precursor form of γ-glutamyl transpeptidase was isolated and characterized (Tate, 1986). The precursor form exhibited about 2% the activity of the heterodimeric mature enzyme, and seemed to be virtually inactive. This study suggests that processing by proteinase(s) into two subunits is required for activation of the enzyme. However, an active single-chain form of the enzyme was discovered in a human hepatocellular carcinoma cell line, HepG2 cells (Tate and Galbraith, 1987b). This cell line produces only the single-chain form of γ-glutamyl transpeptidase. Comparison of the structures of cDNAs for γ-glutamyl transpeptidases from the cells and other human tissues showed that there are no mutations in amino acid sequence (Goodspeed et al., 1989). Furthermore, when RNA, extracted from the cells was subjected to *in vitro* translation with dog pancreas microsome, the heterodimeric form of the enzyme was produced. This finding suggests that the cells lack the processing protease(s) (Tate and Galbraith, 1988). In fact, when the cDNA for the enzyme derived from HepG2 cells was introduced into other cell lines, the γ-glutamyl transpeptidase protein expressed was processed into the heterodimeric form (Visvikis et al., 1991). Even though γ-glutamyl transpeptidase produced by HepG2 cells remains a single chain, the enzyme might prefer an active conformation without processing, possibly due to unknown factors such as much heavier glycosylation.

IV. Role of γ-Glutamyl Transpeptidase in Glutathione Metabolism

γ-Glutamyl transpeptidase catalyzes the initial step of glutathione degradation. Based on the capability to catalyze the transfer of a γ-glutamyl moiety, Meister proposed the γ-glutamyl cycle, which includes glutathione synthesis, the formation of γ-glutamyl amino acids outside cells, and the regeneration of amino acids inside cells (Meister and Tate, 1976; Griffith et al., 1978; Griffith et al., 1979a; Meister and Anderson, 1983; Meister and Larsson, 1995). A series of reactions by this cycle accompanies the uptake of amino acids by cells. Of six enzymes constituting the reaction cycle, only γ-glutamyl transpeptidase functions outside cells, while the other members function exclusively inside cells. Thus, γ-glutamyl transpeptidase is thought to be of critical importance in this cycle in terms of amino acid transport.

McIntyre and Curthoys (1979) suggested that activity of γ-glutamyl transpeptidase is a major contributor to the hydrolysis of glutathione and γ-glutamyl compounds, rather than to the transfer of a γ-glutamyl groups under physiological conditions. It should be emphasized that the hydrolytic activity of the enzyme is a significant *in vivo* function since K_m values for amino acids functioning as acceptor substrates of the γ-glutamyl moiety are relatively high, as compared with their physiological concentrations. Hydrolysis appears to be a major reaction only under the conditions where one type of amino acid is contained as an acceptor at the physiological concentration in the enzyme activity assay. However, the total concentration of amino acids able to serve as acceptors in body fluid seems sufficient for the occurrence of a transpeptidation reaction (Allison and Meister, 1981). The relative contributions of hydrolysis and transpeptidation to degradation of glutathione could vary, dependent on the metabolic state of amino acids.

Many amino acids serve as acceptor substrates of γ-glutamyl transpeptidase. Of these amino acids, cystine is found to be one of the best acceptors of the enzyme (Thompson and Meister, 1975; Thompson and Meister, 1976; Griffith et al., 1981). The K_m value for cystine as an acceptor in transpeptidation is much lower than the others. In the synthesis of glutathione, γ-glutamylcysteine synthetase is known as a regulatory enzyme, whose activity is inhibited by the final product, glutathione, in a feedback manner. In addition to the regulation by γ-glutamylcysteine synthetase, glutathione synthesis is also regulated by the concentrations of amino acids required to produce the tripeptide. The intracellular concentration of cysteine is the lowest in the three amino acids constituting glutathione, indicating that the supply of cysteine is rate limiting in the synthesis of the tripeptide. The kinetic properties of γ-glutamyl transpeptidase with respect to cystine allows for the efficient formation of its γ-glutamyl derivative. The uptake of the γ-glutamyl derivative of cystine into cells might represent one of the pathways to supply cells with cysteine. Therefore, γ-glutamyl transpeptidase would play a significant role in the salvage of the amino acids involved in the synthesis of glutathione, particularly cysteine from an extracellular source.

Animal models, developed by inhibition of γ-glutamyl transpeptidase, have been used in early studies of the role(s) of γ-glutamyl transpeptidase in glutathione metabolism (Anderson and Meister, 1986; Griffith and Meister, 1979b; Griffith and Meister, 1980). When the enzyme is inhibited *in vivo* by inhibitors, the excretion of glutathione and cystine into urine was significantly increased. This provides evidence that γ-glutamyl transpeptidase is involved in renal degradation of glutathione and recovery of constituent amino acids. Although a glutathione transporter(s) was found in various tissues and appears to function in the uptake of extracellular glutathione in its intact form into cells (Kannan et al., 1995; Kannan et al., 1996), the relative importance of the transporter and the γ-glutamyl cycle in the recovery of glutathione is uncertain. However, since inhibition of γ-glutamyl transpeptidase leads to the excretion of a large amount of glutathione, it is likely that the γ-glutamyl cycle significantly contributes to the uptake of glutathione, even if not intact, in the kidney and probably also in some other tissues.

Although the animal model developed by inhibitors of γ-glutamyl transpeptidase has provided information on the roles of the enzyme in glutathione metabolism, the model did not allow one to further describe in detail the effects of a deficiency of γ-glutamyl transpeptidase on the development and/or growth of animals and/or about the metabolic consequence that result in impairment of cellular functions. γ-Glutamyl transpeptidase knocked out mice, established recently by Lieberman et al. (1996), provided evidence for the important role of the enzyme in glutathione metabolism (Lieberman et al., 1996). The mice exhibited growth retardation, were sexually immature, had coats with a gray cast, and developed cataracts even though the mice had appeared normal at birth. The mouse study showed that a deficiency in the enzyme leads, not only to an elevation of glutathione levels in body fluids such as urine and plasma, but also causes a decrease of the levels in tissues. This result clearly indicates the involvement of γ-glutamyl transpeptidase in the regulation of intracellular levels of glutathione. Furthermore, cyst(e)ine levels in plasma were also significantly decreased, suggesting that the enzyme plays a role in the metabolism of both glutathione and cyst(e)ine. Interestingly, the symptoms caused by a deficiency of γ-glutamyl transpeptidase were partially restored by orally administered N-acetylcysteine. This finding is consistent with the function of γ-glutamyl transpeptidase in the γ-glutamyl cycle as described above.

V.   Reactions Catalyzed by γ-Glutamyl Transpeptidase and Substrate Specificity

γ-Glutamyl transpeptidase catalyzes the transfer of a γ-glutamyl moiety of glutathione, its S-substituted derivatives, and γ-glutamyl compounds to a variety of acceptor substrates. The transfer reactions are referred to as transpeptidation (Reaction 1). Many amino acids and dipeptides serve as acceptors producing their respective γ-glutamyl derivatives (Allison, 1985; Meister et al., 1981; Tate and Meister, 1981; Tate and Meister, 1985). Since L-γ-glutamyl donor substrates can act as acceptors at sufficiently high concentrations, the γ-glutamyl moiety of the donor may be transferred to the donor (Abbott et al., 1986). This reaction yields a γ-(γ-glutamyl)-glutamyl compound (autotranspeptidation, Reaction 2). Water also serves as an acceptor of the γ-glutamyl moiety so that the reaction also leads to hydrolysis of the γ-glutamyl bond (Reaction 3).

$$\gamma\text{-Glutamyl}—X + Y \rightarrow X + \gamma\text{-glutamyl}—Y \qquad (1)$$

$$\gamma\text{-Glutamyl}—X + \gamma\text{-glutamyl}—X \rightarrow \qquad (2)$$

$$\gamma\text{-}(\gamma\text{-glutamyl})\text{-glutamyl}—X + X$$

$$\gamma\text{-Glutamyl}—X \rightarrow \text{glutamate} + X \qquad (3)$$

The reactions catalyzed by γ-glutamyl transpeptidase are believed to proceed via a γ-glutamyl enzyme intermediate, as shown by a modified ping–pong mechanism in the kinetic analysis of transpeptidation (Allison, 1985). In the presumed acylated enzyme species (γ-glutamyl enzyme), a γ-glutamyl moiety derived from the donor substrate is covalently bound to the enzyme. The nucleophilic attack of the active-site residue of the enzyme against a γ-carbonyl carbon of the substrate complexed with the enzyme leads to formation of the γ-glutamyl enzyme. After release of the leaving groups of the donor substrates, such as, for example, a cysteinylglycine in glutathione, a subsequent nucleophilic substitution by water, or by α-amino groups of amino acids and dipeptides, result in breakdown of the intermediate and the formation of glutamate or a γ-glutamyl compound.

The substrate binding site of γ-glutamyl transpeptidase may be divided into three subsites corresponding to the constituents of glutathione, and include the γ-glutamyl group binding site as well as the subsites for cysteinyl and glycyl residues (Thompson and Meister, 1977; Thompson and Meister, 1979). While the enzyme displays strict stereospecificity for the L-isomer toward acceptor substrates, both L- and D-isomers of γ-glutamyl compounds are capable of acting act as a donor.

γ-Glutamyl transpeptidase exhibits a broad specificity toward the moiety corresponding to the cysteinyl residue of glutathione in the interaction with a γ-glutamyl donor; some of the S-conjugates of glutathione and γ-glutamyl-p-nitroanilide, which is the donor substrate most commonly used in assays for activity, are known to be more active than glutathione itself (McIntyre and Curthoys, 1979; Tate, 1975). This broad substrate specificity for a γ-glutamyl donor appears to allow the enzyme to metabolize a large number of glutathione-S-conjugates of xenobiotics. On the other hand, it seems likely that the subsite(s) for a leaving group of a donor also serves as the acceptor binding site. γ-Glutamyl transpeptidase appears to prefer the L-isomers of neutral amino acids such as cystine, glutamine, methionine, alanine, and serine for the first residue of the dipeptide and for amino acid acceptors (Thompson and Meister, 1977). Amino acids with branched chains are relatively poor acceptor substrates, and D-isomers of amino acids, L-proline, and α-substituted amino acids are absolutely inactive as acceptors (Tate and Meister, 1974a; Tate and Meister, 1985). In contrast, a determination of relative activities of various dipeptides as acceptor substrates showed that glycine was almost exclusively preferred as the second residue of the dipeptide acceptor (Thompson and Meister, 1977). Since substrate substructures favored by the cysteinyl subsite are different between a donor and an acceptor, a subtle conformational change might be induced in the subsite in conjunction with the formation of a γ-glutamyl enzyme intermediate.

L-γ-Glutamyl compounds give rise to autotranspeptidation by acting as the acceptor in the absence of an acceptor such as amino acids and dipeptides, whereas the D-isomer and α-methyl derivative of γ-glutamyl compounds act as the donor but do not serve as the acceptor. These donor substrates having a D-configulation and α-substitution are useful for measurements of hydrolytic activity (Thompson and Meister, 1976). Alternatively, sufficiently low concentrations of L-γ-glutamyl compounds also allow an assessment of the hydrolysis reaction, as described below.

When low concentrations of L-γ-glutamyl-p-nitroanilide are employed in the absence of an acceptor, only hydrolysis is observed because the K_m values for acceptors, in general, are much higher than those for the γ-glutamyl donors. Under such conditions, autotranspeptidation, where the substrate acts as the donor acting as an acceptor, is negligible. The double reciprocal plot of the reaction at low substrate concentrations gives a linear curve, whereas autotranspeptidation becomes more evident as the concentration increases (Thompson and Meister, 1976; Ikeda et al., 1995). The autotranspeptidation is observed as substrate-activation kinetics, which display a progressive downward curveture in a double reciprocal plot. On the other hand, although the reaction mechanism of transpeptidation by γ-glutamyl transpeptidase is regarded as a ping–pong mechanism, the kinetics do not give a set of parallel curves when kinetic data obtained with varied concentrations of the donor and acceptor are plotted as double reciprocal plots of 1/rate versus 1/[donor substrate]. The behavior of the enzyme on the kinetic analysis is explained by the evidence that the binding of the donor substrate is competitively inhibited by acceptor substrates, amino acids, and dipeptide. The simultaneous occurrence of hydrolysis and autotranspeptidation reactions also complicates the kinetic analysis of transpeptidation by γ-glutamyl transpeptidase. A detailed kinetic study dealing with hydrolysis, transpeptidation, and substrate inhibition has been carried out to assess the enzymatic properties with respect to various substrates by Allison (Allison, 1985). If an acceptor substrate is very active as an acceptor and is used at sufficiently high concentrations to prevent autotranspeptidation and hydrolysis, the kinetic analysis may be more simplified (Ikeda et al., 1995b).

Maleate is known to modulate the activity of γ-glutamyl transpeptidase (Tate and Meister, 1974a; Thompson and Meister, 1979). The prevention of the binding of an acceptor substrate by maleate appears to result in a decrease in the contribution of the transpeptidation reaction. In addition, maleate appears to induce a conformational change in the enzyme, thereby fascilitating the hydrolysis reaction. In fact, it was found that maleate alters the reactivities of γ-glutamyl transpeptidase with iodoacetamide, iodoacetic acid, and irreversible inhibitors (Smith and Meister, 1995a). Iodoacetamide is capable of modifying a single cysteine residue of the light subunit only in the presence of maleate. Thus, in the absence of maleate, the thiol group may be buried but it might be exposed to modifying agents as a result of the conformational change induced with maleate.

VI. Active Site of γ-Glutamyl Transpeptidase

A. THE LIGHT SUBUNIT AS A CATALYTIC SUBUNIT

Chemical modification studies have provided information relative to amino acid residues located in the active site (Elce, 1980; Schasteen et al., 1983; Inoue et al., 1987), and a variety of reversible and irreversible inhibitors of the enzyme and substrate analogues have been used to probe the structure or amino acid residues of the active site of γ-glutamyl transpeptidase (Inoue et al., 1977; Tate and Meister, 1977a; Tate and Ross, 1977b; Tate and Meister, 1978). Affinity-labeling reagents such as azaserine and 6-diazo-5-oxo-L-norleucine bind covalently to the enzyme in a ratio of one mol/mol enzyme, and the binding of these reagents are prevented by the presence of γ-glutamyl donor substrates or the substrate analogues (Inoue et al., 1977; Tate and Meister, 1977; Tate and Ross, 1977). When γ-glutamyl transpeptidase was inactivated by the ^{14}C labeled inhibitors, the ^{14}C activity was found to be associated exclusively with the light subunit. This result suggests that the light subunit contains the amino acid residue(s) which are reactive with respect to these γ-glutamyl donor analogues, and that the subunit participates in the γ-glutamyl group-binding site. Moreover, γ-glutamyl transpeptidase is known to exhibit proteinase activity when the heterodimer is dissociated by sodium dodecylsulfate or urea at the neutral pH (Gardell and Tate, 1979). The dissociated light subunit exhibits proteinase activity and degrades the heavy subunit, as well as bovine serum albumin. It can reasonably be concluded that it contains the residues required to hydrolyze an amide or a peptide bond of a substrate. Together with the findings that affinity labeling agents bind exclusively to the light subunit, it has been concluded that the light subunit plays a role in catalysis, and it is generally the subunit that has been regarded as the catalytic subunit. Since the heterodimeric form of the enzyme does not display such proteinase activity, it is thought that the heavy subunit masks the unusual properties of the light subunit as a proteinase.

B. SUBSTRATE BINDING SITES

Although many chemical modification studies have been carried out in attempts to identify the reactive groups and amino acids located in the active site of γ-glutamyl transpeptidase, specific amino acid residue(s) were not identified in earlier studies. The identification of the residues in the active site is required and necessary for an understanding of the chemical or molecular basis of the enzyme reaction. Stole and Meister (1991) used phenylglyoxal, which is known to react with the guanidino group of arginine, to modify purified rat kidney γ-glutamyl transpeptidase. The enzyme was inactivated on binding this reagent to the enzyme. They identified the residues that were blocked by the radiolabeled phenylglyoxal only in the absence of a substrate analogue. These residues were found to be Arg-110 and Lys-99, both of which are located on the heavy subunit. Interestingly, phenylglyoxal was found to modify lysine as well as arginine. This study suggests that these residues are located in the active site, and therefore may play a role in the binding of substrates. The participation of an arginine residue in substrate binding had also been suggested by Schasteen (Schasteen et al., 1983).

The role of Arg-111 and Lys-99 of rat γ-glutamyl transpeptidase in substrate binding was examined by analysis of mutant human γ-glutamyl transpeptidase in which those equivalent residues are replaced with unionizable amino acids (Ikeda et al., 1993). The mutant enzymes with amino acid substitutions were found to be nearly fully active, clearly indicating that these residues are not involved in either substrate binding residues or as catalytic residues, even though they are located in the active site or very close to the substrate binding site(s). In order to identify arginine residues essential for enzymatic activity, the candidates to mutate were chosen on the basis of the homology between two human enzymes, γ-glutamyl transpeptidase and a γ-glutamyl transpeptidase related enzyme, which was found to catalyze the hydrolysis of glutathione in a manner similar to γ-glutamyl transpeptidase. The cDNA for the related enzyme has been cloned as the protein structurally related but distinct from γ-glutamyl transpeptidase (Heisterkamp et al., 1991). When the human mutant enzymes with replacements of the residues by glutamine were expressed in COS-1 cells under the control of SV40 promoter, the substitution at Arg-107 resulted in sufficient amount of the protein but no measurable activity. This finding suggests that these Arg residues are essential for enzyme activity. When the same residue was replaced by Lys so as to retain a positive charge at the position of residue 107, the mutant enzyme exhibited a small percent of the activity of the wild-type enzyme. Kinetic analysis of this mutant revealed that the K_m of the mutant for L-γ-glutamyl-3-carboxy-4-nitroanilide as a donor substrate was significantly increased. This result suggest that Arg-107 plays a role in binding of the donor substrate through an electrostatic interaction and possible hydrogen bonding between its guanidino group and an α-carboxyl group of a γ-glutamyl donor substrate. The rat arginine residue corresponding to human Arg-107 would not be modified by phenylglyoxal, probably because this residue is buried in the pocket of the substrate binding site, or the modifying agent could not gain access to the residue due to its bulkness. Accordingly, it seems likely that the region containing both Arg-107 and Arg-112 (equivalent to rat Arg-111) participates in the active site, but only Arg-107 in human and its equivalent of rat actually function in the binding of substrates. Therefore, the active site of the enzyme would be formed in the interface of both the heavy and the light subunits, unlike the earlier suggestion that the γ-glutamyl group binding site was contained only in the light subunit.

Iodoacetamide inactivates γ-glutamyl transpeptidase, whereas iodoacetic acid has a similar effect only when maleate is present (Szewczuk and Connell, 1965; Smith and Meister, 1995a). Inactivation of the enzyme by iodoacetamide (and by iodoacetic acid) is prevented by the presence of serine–borate complex, which acts as a transition state analogue for the enzyme. Following the modification of purified rat kidney enzyme by unlabeled iodoacetamide in the presence of serine–borate complex, further modifying the enzyme by iodo[^{14}C]-acetamide resulted in the labeling of the active site residue(s) (Smith and Meister, 1995a). The labeled residue was identified as Asp-422, which is in the light subunit. The requirement of the residue was examined by site-directed mutagenesis (Ikeda et al., 1995b). The human mutant γ-glutamyl transpeptidase in which the equivalent residue (Asp-423 in human) was replaced with Ala residue was produced by a baculovirus-insect cell expression system and then purified. Kinetic analysis of the mutant revealed that K_m of the mutant for the donor substrate, L-γ-glutamyl-p-nitroanilide was greatly increased, whereas V_{max} of hydrolysis was found to be relatively modest. Therefore, it was suggested that the residue (Asp-422 in rat and Asp-423 in human) plays an essential role in binding the donor substrate. Furthermore, the neighboring residue, which is also aspartate (Asp-422 in human enzyme and Asp-421 in the rat equivalent), was also found to be involved in the binding of the donor, because the K_m for the substrate was about 10 times higher without substantial change of V_{max} for the hydrolysis reaction. The studies with chemical modification and mutational analyses have shown that those aspartate residues are required for the substrate binding. Thus, the binding of a γ-glutamyl donor substrate to the enzyme would involve interactions between the enzyme active site residues and the ionic groups of the substrate: one is between the arginine residue of the heavy subunit and an α-carboxyl group of the γ-glutamyl moiety of the substrate and the other is between the aspartate residue(s) of the light subunit and an α-amino group of the substrate. Since the enzyme recognizes both L and D isomers of γ-glutamyl group, these two different interactions between the donor and the active site of the enzyme might be of major importance. If other interactions were possible in addition to these, the enzyme would display a far different and distinct stereospecificity for the optical isomers.

C. CATALYTIC SITE AND ACTIVE SITE CHEMISTRY

The reactions catalyzed by γ-glutamyl transpeptidase, hydrolysis of a γ-glutamyl bond of substrates, and transfer of the group to an acceptor substrate, are thought to involve a common enzyme species. The species, γ-glutamyl enzyme intermediate, results from the cleavage of the γ-glutamyl linkage of the donor in the enzyme–substrate complex and the subsequent formation of γ-glutamyl bond with the enzyme. Nucleophilic substitutions by water and other acceptors, such as amino acids and dipeptides in the enzyme species, lead to hydrolysis and transpeptidation, respectively (Allison, 1985). Thus, the transpeptidation is consistent with the kinetic properties in which the enzyme exhibits a ping–pong mechanism. However, the γ-glutamyl enzyme intermediate has never been isolated. Elce (1980) made the observation that the radioactivity of radiolabeled donor substrate was incorporated in the light subunit of the enzyme when the enzyme was incubated with the radiolabeled substrate and the reaction was then terminated by denaturants. Furthermore, it was found that modification of γ-glutamyl transpeptidase with N-acetylimidazole allowed the formation of a stable γ-glutamyl enzyme (Smith and Meister, 1995a). When the enzyme modified with N-acetylimidazole was incubated with L-γ-glutamyl-p-nitroanilide, the enzyme released about 90 mol of p-nitroaniline/mol of enzyme and then became inactivated. Incubation of the modified enzyme with [^{14}C]glutamine or [^{14}C]glutathione resulted in the incorporation of radioactivity exclusively into the light subunit at the ratio of 1 mol/mol of the light subunit. These results suggested that the catalytic residue(s) directly responsible for formation of γ-glutamyl linkage in the intermediate is located on the light subunit, in agreement with the evidence for proteolytic activity found in the light subunit and for the exclusive modification of the subunit with several affinity labeling agents. Nevertheless, because the stable γ-glutamyl enzyme was rapidly hydrolyzed upon denaturation, the residue to which the labeled moiety attached could not be identified. Although attempts to directly identify the residue undergoing γ-glutamylation has never been successful, the findings that the linkage of the γ-glutamyl enzyme is labile against alkaline conditions, rather than acidic and that hydroxylamine efficiently lysed the linkage, provide the suggestion of an ester linkage between the γ-glutamyl moiety and the enzyme, rather than an amide. A thioester also seems unlikely, even though mammalian γ-glutamyl transpeptidases possess a single free thiol, because replacement of the cysteine residue by alanine or serine resulted in fully active enzyme (Ikeda et al., 1995b). Thus, it has been suggested that a hydroxyl group of the light subunit functions as a nucleophile and is required to form the γ-glutamyl enzyme intermediate.

The involvement of the hydroxyl group of the light subunit of γ-glutamyl transpeptidase is also supported by the evidence that the enzyme is inhibited by phenyl methanesulfonyl fluoride, which is known to be an inhibitor of serine-class hydrolases (Inoue et al., 1987). Specific residues involved in the catalysis of γ-glutamyl transpeptidase were only recently identified. Although there are many analogues for γ-glutamyl donor substrates, only those that are known to attach covalently to the enzyme have been utilized to investigate the catalytic residue(s). These analogues or affinity labeling reagents were found to bind to the light subunit and thought to attach to the enzyme through an ester linkage. It is known that both the L and D isomers of serine, in the presence of borate, effectively inhibit γ-glutamyl transpeptidase (Tate and Meister, 1978). Serine mimics the substructure around an α-carbon of a γ-glutamyl moiety of a donor substrate and thus appears to occupy the subsite for the portion of the substrate so that the amino acid alone weakly inhibits the enzyme in a competitive manner against the donor (Allison, 1985). In the presence of borate, the hydroxyl groups of the active site of the enzyme and the serine would be bridged by the borate (Fig. 1). The resultant tetrahedral borate complex strongly inhibits the enzyme and is thought to serve as a transition state analogue at the catalytic center of the enzyme.

Inhibition by acivicin

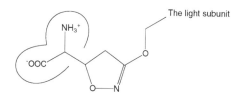

Inhibition by serine-borate complex

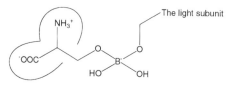

Figure 1.   Inhibition of γ-glutamyl transpeptidase by acivicin and serine–borate complex.

Stole et al. (1990) first identified Thr-523 in rat γ-glutamyl transpeptidase as the hydroxyl-containing residue, which is highly reactive with acivicin [L-(αS, 5S)-α-amino-3-chloro-4, 5-dihydro-5-isoxazoleacetic acid], also known as AT-125), which is one of the most potent irreversible inhibitors of γ-glutamyl transpeptidase, as shown in Figure 1. Treatment of the acivicin-inhibited enzyme with hydroxylamine restores activity and releases the acivicin-derived threo-β-hydroxy-L-γ-glutamyl hydroxamate (Stole et al., 1994). Furthermore, it was shown that acivicin acts as a substrate of the enzyme: γ-Glutamyl transpeptidase hydrolyzes acivicin at very slow rate, yielding threo-β-hydroxy-L-glutamate and hydroxylamine.

In spite of the evidence that Thr-523 is the acivicin-bound residue in rat γ-glutamyl transpeptidase, amino acid sequence of pig enzyme, which was deduced from the structure of the cDNA, indicated that the pig enzyme contains alanine in place of the threonine residue (Fig. 2), although the homology of the entire amino acid sequences between rat and pig was found to be approximately 90% (Papandriko-poulou et al., 1989). This finding clearly shows that the hydroxyl of the threonine residue is not associated with activity. In addition, pig γ-glutamyl transpeptidase is also inactivated by acivicin as well as the rat enzyme (Smith et al., 1995b). In fact, the human mutant enzyme in which Thr-524 (equivalent to rat Thr-523) is replaced with alanine, demonstrated that the threonine residue is not essential for activity and not involved in inactivation by acivicin, since the mutant was shown to be almost fully active and inactivated by acivicin in a manner similar to the wild-type human enzyme. Thus, other residues responsible for attachment of acivicin in human and pig enzymes were identified with [^{14}C]acivicin adducts of the enzymes purified from kidneys. Consequently, they are found to be Ser-406 and Ser-405 in human and pig enzymes, respectively (Fig. 2). Nevertheless, activity assay and acivicin inactivation of the human mutants with replacement of Ser-406 by alanine and with the double replacement of Ser-406 and Thr-524 by alanine revealed that neither of those amino acid residues plays an essential role in enzyme activity and inhibition by acivicin. The studies identifying the acivicin-bound residues and the mutational analyses have suggested that acivicin would initially react with the unidentified hydroxyl group and give rise to transesterification toward other immediate hydroxyl groups such as the aforementioned serine and threonine residues in the acivicin–enzyme complex. It is quite probably that the residue involved in the initial reaction with acivicin might be identical to the catalytic one that is involved in the formation of the γ-glutamyl enzyme intermediate.

A series of mutants of human γ-glutamyl transpeptidase have been prepared in order to investigate which serine residues of the light subunit are required for enzyme activity and for inhibition by acivicin (Ikeda et al., 1995c). The serine residues of the light subunit were selected as candidates on the basis of the sequence homology of the light subunits of γ-glutamyl transpeptidases for a variety of species (Fig. 2), including two bacterial enzymes (*Escherichia coli* and *Pseudomonas*), all of which exhibit both hydrolytic and transpeptidation activities (Suzuki et al., 1989; Ishiye et al. 1993). The mutants were produced by a baculovirus-insect cell expression system and were purified for use in kinetic analyses and inhibition studies with acivicin. Of five replacements of serine residues by alanine, substitution at either Ser-451 or Ser-452 lead to loss of activity and decrease of extent of inhibition by acivicin, whereas the others had essentially

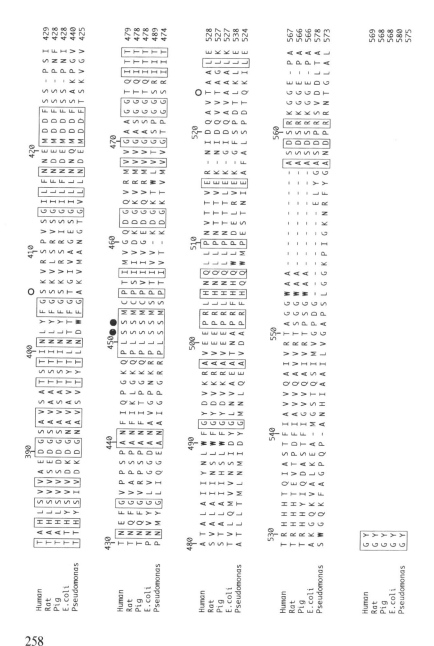

no effect on the enzymatic properties. Furthermore, kinetic analyses of the mutant with the double replacement indicated a great decrease in V_{max} without substantial change of K_m for the substrate, L-γ-glutamyl-p-nitroanilide: The double mutation decreased the V_{max} value for hydrolysis of the substrate to 5×10^{-5} of the wild-type enzyme. The rate of the hydrolysis of the substrate derived from p-nitroaniline by the double mutant (k_{cat} in hydrolysis reaction) is 2000 times faster than in the absence of the enzyme. A similar observation was made for a mutant subtilisin (a serine protease) in which the catalytic serine residues were replaced with alanine (Carter and Wells, 1988). This subtilisin mutant still exhibited a 3000-fold faster rate of hydrolysis of acyl-p-nitroanilide substrate, compared to non-enzymatic hydrolysis. In the mutant with replacement of either of those serine residues, the rate of inactivation by acivicin was dramatically decreased to about 0.5% compared to the wild type. No detactable inhibition was observed in the double mutant. Thus, it was found that the two serine residues, Ser-451 and Ser-452 in human, in the light subunit are of critical importance in both the enzyme catalysis and in the reaction with acivicin.

The wild type of recombinant human γ-glutamyl transpeptidase is inhibited by the serine–borate complex, similar to the rat kidney enzyme. Borate facilitates the inhibition of the enzyme by serine (Fig. 1), as indicated by the dramatic decrease of K_i for serine, probably through formation of a covalent bridge between hydroxyl groups of the enzyme and of serine (Tate and Meister, 1978). The single and double mutantions introduced to Ser-451 and Ser-452 impaired the effect of borate on inhibition by serine. Therefore, these studies with the serine-substituted mutants showed that the serine residues of the light subunit are probably involved in interaction with the borate portion of the serine–borate complex as the transition state analogue. It was thus suggested that the serine residues are located at positions accessible to the γ-carbonyl carbon of a γ-glutamyl

←—————————————————————————————

Figure 2. Amino acid sequence alignment of the light subunits of γ-glutamyl transpeptidases from various species. Boxes indicate the residues conserved among all species. The residue number for human enzyme are given above the sequences. Open circles indicate the acivicin-bound residues identified in rat, human, or pig enzymes. Closed circles show the catalytically important serines identified by site-directed mutagenesis.

moiety of the substrate because the borate portion of the serine–borate complex appears to mimic the tetrahedral intermediate. The experimental evidence described is consistent with the possible roles of the serine residues either as a nucleophile(s) against γ-carbonyl carbon of the substrate or as the residues stabilizing the transition state (Fig. 3), which is analogous to an oxyanion hole discovered in certain hydolases (Robertus et al., 1972). Further study is required for the definite identification of the nucleophile responsible for formation of the γ-glutamyl enzyme intermediate.

In the catalytic mechanism involving a presumed nucleophilic hydroxyl group, a general base might be required to allow the hydroxyl group to react with a substrate. In general, ionic groups such as imidazole, carboxylate, and amino groups are known to potentially function as a base in many enzymes. Since the light subunit of γ-glutamyl transpeptidase exhibits proteolytic activity as described above, all catalytic prerequisits to hydrolyze a peptide bond would be associated with the subunit. The pH profiles of kinetic parameters for the enzyme have been examined in rat and human enzymes using γ-glutamyl derivatives of p-nitroaniline, and indicate that the enzyme activity depends on the base form of an ionic group with a pK_a in the range 6–7 (PetitClerc et al., 1980; Allison, 1985; Ikeda et al., 1996). It is most likely that the ionic group is an imidazole. Furthermore, a determination of enthalpy change (ΔH) for dissociation of the group also supports this suggestion. Among γ-glutamyl transpeptidases from various species, only two histidine are conserved in the light subunits (Ikeda et al., 1996). The corresponding histidine residues are His-383 and His-505 in human enzyme. Thus, kinetic properties and pH profiles of kinetic parameters were investigated in the human mutants with replacements of these histidine residues by alanine, in order to assign the ionic group responsible for activity. The mutants with their single substitutions, however, retained significant activity. About 2% of the activity of the wild type was detected, even in the double mutant. These analyses show that neither of those conserved histidine residues plays an essential catalytic role in γ-glutamyl transpeptidase. The pH-dependence profiles of the double mutant for kinetic parameters are essentially the same as those of the wild type, leading to the conclusion that the conserved histidine residues do not function as acid–base catalysts and that the enzyme activity depends on another histidine or other types of ionic groups. The assignment of the critical ionic group function as the catalyst

As catalytic nucleophiles

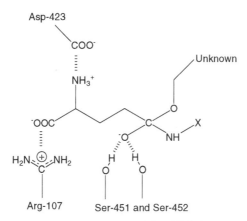

As the residues stabilizing a transition state

Figure 3.   Two possible roles of Ser-451 and Ser-452 in human γ-glutamyl transpeptidase.

remains to be solved in order to elucidate the detailed catalytic mechanism of γ-glutamyl transpeptidase.

As described, the findings on the catalysis of γ-glutamyl transpeptidase have suggested that the presumed hydroxyl group of the light subunit functions as the nucleophlile against the γ-carbonyl carbon of a γ-glutamyl donor substrate, suggesting that specific arginine and aspartate residues of the heavy and light subunit, respectively, would be involved in the binding of the donor substrate. These suggestions, however, need to be confirmed by the structural studies of the enzyme, based on X-ray crystallography. Mammalian γ-glutamyl transpeptidases are heavily glycosylated and are known to exhibit great molecular heterogeneities, probably due to variations of sugar chain structures. The heterogeneities appear to make it difficult to crystalize mammalian enzymes. Although bacterial expression of mammalian γ-glutamyl transpeptidase could allow production of a non-glycosylated enzyme, the level of expression was too poor to obtain sufficient amounts (Angele et al., 1989). Deglycosylation of rat kidney enzyme by glycosidase-cocktail has been attempted as an alternative strategy to obtain a non-glycosylated form of the enzyme (Smith and Meister, 1994). The deglycosylated rat enzyme was found to have an activity comparable to the nontreated one, and thus may allow the preparation of crystals of mammalian γ-glutamyl transpeptidase. Alternatively, since γ-glutamyl transpeptidase from *E. coli* K-12 is the only enzyme that has been crystalized and is subjected to X-ray crystallographic analysis, the bacterial enzyme would be available as a model to discuss structure–function relationships (Kumagai et al., 1993; Sakai et al., 1996). The study has already given an outline of the structure of γ-glutamyl transpeptidase. This outline may provide information on the active site structure and, in turn, may contribute to an understanding of the structural basis of the catalytic properties of the enzyme in the future, if the bacterial enzyme is crystalized in a form complexed with a γ-glutamyl donor or analogue.

VII.   γ-Glutamyl Transpeptidase-Related Enzyme and γ-Glutamyl Transpeptidase Transcripts Unique to Human Lung

Proteins structurally related to γ-glutamyl transpeptidase have been found in human tissues. It has been suggested that these related genes are members of a family of γ-glutamyl transpeptidase.

Heisterkamp et al. (1991) has cloned the cDNA for the protein structurally related to γ-glutamyl transpeptidase from human placenta. Comparison of amino acid sequences of the protein and human γ-glutamyl transpeptidase showed that homology between these proteins is as high as 40% for the entire sequences. When the cDNA was introduced into NIH3T3 cells, the expressed protein displayed hydrolytic activity toward glutathione as well as activity for converting leukotriene C4 into leukotriene D4, as found in γ-glutamyl transpeptidase. Furthermore, acivicin, a potent inhibitor of γ-glutamyl transpeptidase, inhibited the related enzyme. Nevertheless, the protein did not hydrolyze L-γ-glutamyl-p-nitroanilide, which is commonly used for the assay of γ-glutamyl transpeptidase. These unique features indicate that the protein hydrolyzing glutathione is an enzyme similar to, but distinct from, γ-glutamyl transpeptidase and thus the novel enzyme has been named as a γ-glutamyl transpeptidase-related enzyme. It is of interest that there is another enzyme catalyzing the first step of glutathione degradation. Until this novel enzyme was discovered it had been believed that only γ-glutamyl transpeptidase catalyzes the cleavage of the γ-glutamyl bond of glutathione. The related enzyme appears to be heterodimeric, and both heavy and light subunits are encoded in a common mRNA, analogous to γ-glutamyl transpeptidase. In addition, the location of the putative transmembrane domain is also similar to that of γ-glutamyl transpeptidase. Although there are many similarities in structure and function between γ-glutamyl transpeptidase and the related enzyme, no expression of the related enzyme was detected in mouse tissues. It seems unlikely that the related enzyme plays a significant role in the degradation of glutathione *in vivo* because animals do not appear to have an absolute requirement for it. Further study is needed to elucidate the participation of the related enzyme in glutathione metabolism or to understand the biological function of the enzyme.

Wetmore et al. (1993) found the expression of 1.2 kb transcripts of γ-glutamyl transpeptidase (2.4 kb for usual form) in human lung, and cloned cDNAs for the smaller transcripts. In this tissue, the smaller transcripts are more abundant than the 2.4-kb species, unlike other tissues, such as liver and pancreas. Sequence analyses revealed that the transcripts encode the 3' portion of "usual" γ-glutamyl transpeptidase and contain the entire sequence of the light subunit. The clones they obtained have an open reading frame of 675 bp, which is 1032 bp shorter than the usual one, essentially identical to the 3′ 675 bp of the open reading frame of the γ-glutamyl transpeptidase. When the cloned sequences were transcribed and translated *in vitro*, a 24-kDa polypeptide was produced, as expected, from the size of the open reading frame. This fact suggests that the initiation codon of the unique transcripts are functional. However, the expression of the unique human lung clones in COS cells resulted in virtually no expression of γ-glutamyl transpeptidase activity, even if a cDNA encoding the entire sequence of the heavy subunit was cotransfected. The small polypeptide(s) derived from the unique transcripts do not appear to function as the enzyme.

It seems unlikely that these proteins, which are related to γ-glutamyl transpeptidase, participate significantly in glutathione metabolism, but it is possible that they have a distinct biological function(s), even though they exhibit significant structural homology with γ-glutamyl transpeptidase. In particular, if the protein unique to lung has a distinct function, it might indicate that γ-glutamyl transpeptidase also has a biological significance distinct from enzyme activity as an enzyme because the unique one is identical to the portion of γ-glutamyl transpeptidase.

VIII. Gene Structure and Expression

A. γ-GLUTAMYL TRANSPEPTIDASE GENE

γ-Glutamyl transpeptidase is known to be expressed in a tissue-specific manner, and the expression levels of the enzyme are also changed developmentally. The gene structure and mechanism by which γ-glutamyl transpeptidase is expressed has long been of interest in terms of tissue-specific expression and for developmental or differentiation stage-specific expression. Several groups have characterized the structure and organization of mouse and rat γ-glutamyl transpeptidase genes (Rajagopalan et al., 1990; Rajagopalan et al., 1993; Okamoto et al., 1994; Kurauchi et al., 1991; Shi et al., 1995; Carter et al., 1994; Sepulveda et al., 1994; Lieberman et al., 1995). The gene was found to be a single copy in mice and rats, whereas the human gene is a multiple copy (Rajagopalan et al., 1993; Pawlak et al., 1988; Chobert et al., 1990; Figlewics et al., 1993). Rat and mouse γ-glutamyl transpeptidase genes have 12 coding exons interrupted by 11 introns and span about 12 kb (Shi et al., 1995; Rajagopalan et al., 1990; Lieberman et al., 1995). Human γ-glutamyl transpeptidase gene(s) was first assigned to chromosome 22 and was later mapped on other autosomes (Bulle et al., 1987; Rajpert-De-Meyts et al., 1988; Figlewics et al., 1993).

B. MULTIPLE PROMOTERS AND TISSUE-SPECIFIC EXPRESSION

It was found that mouse γ-glutamyl transpeptidase gene is transcribed by use of at least six promoters (Carter et al., 1994; Lieberman et al., 1995). One of the promoters contains a TATA box, but the other contain no obvious TATA box. The transcripts from these different promoters appear to splice to, or are transcribed through, a common noncoding exon followed by coding exons. Thus, the use of multiple promoters would result in the occurrence of a variety of RNAs with different 5' ends, at least six species, although all the different transcripts would encode the same polypeptide of γ-glutamyl transpeptidase. No splice variants or alternative poly(A) addition sites have been identified in the coding region (Lieberman et al., 1995). In mouse adult kidney, the γ-glutamyl transpeptidase gene is transcribed from all six promoters, yielding six different RNA transcripts (Rajagopalan et al., 1993). On the other hand, mRNA transcribed from the third promoter constitutes a major species in mouse fetal liver (Habib et al., 1995). In small intestine, 90% of the transcripts arise from the sixth promoter (Carter et al., 1994).

The rat γ-glutamyl transpeptidase gene is also a single copy, transcribed from multiple promoters, and the organization of rat gene is found to be similar to that of mouse (Fig. 4). To date, four different promoters have been identified on the single copy gene of rat γ-glutamyl transpeptidase, yielding five different mRNAs that encode the same polypeptide (Okamoto et al., 1994). The two promoters most proximal from the initiation codon are utilized in rat kidney, yielding two 2.2-kb mRNA species with alternative 5' ends (Rajagopalan et al., 1990; Chobert et al., 1990; Kurauchi et al., 1991; Lahuna et al., 1992). Northern blot analysis of the kidney poly(A)+ RNA revealed that the ratio of the expression of these two RNAs is fairly constant at different developmental stages. These two mRNAs are not transcribed in normal liver, fetal liver, seminal vesicle, and testis. The most proximal promoter was shown to exhibit a promoter activity in a kidney cell line (LLCPK cells) but not in a hepatoma cell line (HTC cells). Thus, the promoter appears to serve for the tissue-

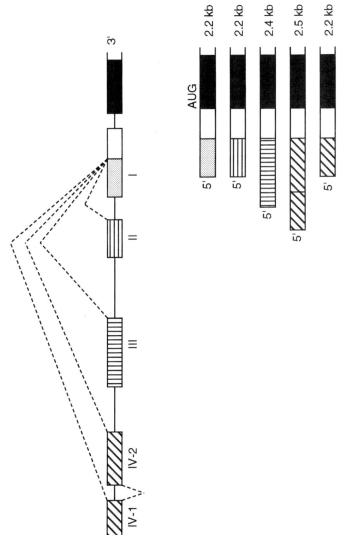

Figure 4. Organization of rat γ-glutamyl transpeptidase gene upstream of the translation start site and transcripts with different 5′ ends from multiple promoters. Boxes I–IV-1 and IV-2 indicate the promoter region of the rat γ-glutamyl transpeptidase gene.

266

specific expression of γ-glutamyl transpeptidase in the kidney. On the other hand, the 2.4-kb mRNA, transcribed from the third promoter upstream of these two promoters, was found in rat fetal liver (Brouillet et al., 1994) and the primary tumors that were induced by diethylnitrosamine (Darbouy et al., 1991). Transcription from the third promoter was observed in hepatoma cells with differentiated phenotypes, such as Fao cells, but not in the undifferentiated state. The fourth promoter has been reported to specifically regulate transcription of γ-glutamyl transpeptidase gene in the small intestine, certain hepatoma cells, and epididymis (Palladino et al., 1994; Okamoto et al., 1994). Expression of this promoter leads to a primary transcript, which subsequently undergoes alternative splicing into two mRNAs of 2.2 and 2.5 kb. Although the promoter exhibits activity in hepatoma cell lines such as HTC and C2Rev7 cells, it is not active in another hepatoma cell line, Fao cells, in normal liver, and even in the fetal liver. Expression of the fourth promoter could be facilitated in undifferentiated cells that have the capacity to differentiate into hepatocytes, biliary cells, or other cell lineages. Thus, HTC cells and C2Rev7 cells could be derived from relatively undifferentiated cells. Because of the divergent specificity of promoters in hepatoma cells, a promoter specific to hepatocarcinogenesis seems unlikely and furthermore, the transcriptional regulation of the gene in hepatoma cells might not necessarily be the same as that in fetal liver. These analyses of promoter specificity in various hepatoma cell lines have shown that the differentiation stages from which hepatoma cells originate might determine which promoter is activated for the transcription of γ-glutamyl transpeptidase gene.

C. γ-GLUTAMYL TRANSPEPTIDASE GENE EXPRESSION IN HUMAN

On the other hand, while human γ-glutamyl transpeptidase gene(s) has not been completely characterized, it was found that the gene is multiple copies (5–7, including pseudogene) in human. Courtay et al. (1994) found seven potential γ-glutamyl transpeptidase genes, and reported that four of these potential genes are expressed in a tissue-restricted manner, in addition to the ubiquitously expressed gene, which would contain known γ-glutamyl transpeptidase-coding sequences. It was reported that a different mRNA of γ-glutamyl transpeptidase is expressed, along with the usual form of the enzyme, in human liver (Pawlak et al., 1989). Deletion of 30 bp and several nucleotide substitutions were found in the mRNA, and the resultant protein lacks a chain of 10 amino acid residues and has 14 amino acid substitutions in the light subunit, some of which result from frame shifting. The γ-glutamyl transpeptidase protein would be inactive because these mutations cause the loss of the residues essential for activity. Furthermore, an alternatively processed mRNA for γ-glutamyl transpeptidase was identified in various human tissues (Pawlak et al., 1990). The mRNA contains a shorter open reading frame because a premature stop codon is introduced due to an alternative splicing. The transcript encompasses most of the heavy subunit but not the light subunit. It is interesting that different transcripts code for the heavy subunit and the light subunit (Wetmore et al., 1993) separately in human tissues. However, the physiological significance of species that do not appear to have enzymatic activities is still unknown. Thus, the expression of a γ-glutamyl transpeptidase gene(s) in human seems more complicated than those of mouse and rat, because multiple genes would be transcribed from multiple promoters and alternative splicing of transcripts occurs, not only in the untranslated region, but also in a coding sequence. In addition, it was shown that the 5′ untranslated region of human γ-glutamyl transpeptidase mRNA serves as a tissue-specific active translational enhancer (Diederich et al., 1993).

Three types of mRNAs with different 5′ noncoding sequences were detected in human tissues (Tsutsumi et al., 1996). These mRNA species correspond to γ-glutamyl transpeptidase mRNAs expressed in fetal liver, HepG2 cells (human hepatocarcinoma cell line), and placenta. Their expression exhibited different tissue distributions. The fetal liver type and HepG2 cell type of mRNAs were found to be expressed in livers from patients with hepatocellular carcinomas. Furthermore, the HepG2 cell type was a major species in the cancerous tissues, whereas both of them were detected in the non-cancerous lesions. Thus, detection of the γ-glutamyl transpeptidase mRNA(s) related to development of hepatocellular carcinoma may contribute to early diagnosis of the malignant disease. However, it is uncertain whether those mRNAs are transcribed from an identical gene by multiple promoters. It is also possible that transcriptions from different copies of the genes lead to occurrence of several type of mRNAs. The elucidation of the complex mechanism regulating the expression in human requires further intensive studies.

D. DNA METHYLATION STATUS AND GENE EXPRESSION

Deoxyribonucleic acid methylation has been proposed as one of the mechanisms regulating gene expression, and correlation between DNA methylation status and gene expression has been shown (Razin and Riggs, 1980; Doerfler, 1983; Cedar, 1988). In the expression of γ-glutamyl transpeptidase, it has been suggested that DNA methylation is involved in tissue-specific expression and development stage-specific expression. Distinct methylation patterns were observed in rat adult kidney and liver, which exhibit the highest activity of γ-glutamyl transpeptidase and very low activity, respectively (Coloma and Garcia-Jimeno, 1991). The gene appears to be fully methylated in the liver, whereas it is hypomethylated in the kidney. Progressive methylation of the gene was also observed in rat liver during development, and the methylation appears to be associated with inactivation of the γ-glutamyl transpeptidase gene in the liver (Baik et al., 1991). However, although the methylation pattern of the gene is altered during hepatocarcinogenesis, the pattern is different from that in fetal liver, which exhibits a relatively high activity of γ-glutamyl transpeptidase. This finding suggests that the mechanism involved in γ-glutamyl transpeptidase expression is different in fetal liver and hepatoma. The two most proximal promoter regions, which are responsible for γ-glutamyl transpeptidase expression in adult rat kidney, are progressively demethylated during kidney development (Baik et al., 1992a). The DNA methylation status in these two promoters seems to determine developmental stage-specific expression of the gene in the kidney. On the other hand, no transcriptions from those two promoters were observed in Fao cells, which originates from hepatic cells, even after the gene was strongly demethylated by treatment with a hypomethylating agent, 5-azacytidine (Baik et al., 1992b). Alternatively, treatment with this agent resulted in expression of the gene by the third promoter. While DNA methylation status is involved, it is not primarily responsible for tissue-specific and developmental stage-specific expression of γ-glutamyl transpeptidase. Both methylation of the gene and tissue-specific transcriptional factors might be necessary for the tissue-specific expression of γ-glutamyl transpeptidase.

IX.   Cancer-Associated Changes of γ-Glutamyl Transpeptidase

It is well known that γ-glutamyl transpeptidase activity is elevated in the liver during chemical hepatocarcinogenesis (Fiala and Fiala, 1973; Taniguchi et al., 1983a; Tateishi et al., 1976; Hanigan and Pitot, 1985; Beer et al., 1986). γ-Glutamyl transpeptidase expressed in hepatoma tissues was initially thought to be a tumor-specific isozyme because differences in physicochemical properties were observed (Sawabu et al., 1978). However, these differences were later found to be due to the alterations in sugar chain moieties and not related to the polypeptide backbone (Tsuchida et al., 1979; Taniguchi et al., 1983b). The finding that the γ-glutamyl transpeptidase gene is a single copy in rat and mouse also indicates that the change associated with carcinogenesis reflects structural differences in oligosaccharides rather than amino acid sequence (Pawlak et al., 1988).

Structures of N-linked oligosaccharides were analyzed in γ-glutamyl transpeptidase purified from rat normal liver and rat ascites hepatoma cells (Yamashita et al., 1983). The analysis showed that the enzyme from hepatoma cells contains a β1-4 linked N-acetylglucosamine residue, which is referred to as a bisecting GlcNAc, whereas the enzyme from the normal tissue does not contain this structure. This study indicated that hepatocarcinogenesis results in, not only induction of γ-glutamyl transpeptidase, but also to alteration of its sugar chain structures. The formation of the unique structure of oligosaccharides is catalyzed by a β1-4 N-acetylglucosaminyltransferase III (Kornfeld and Kornfeld, 1985), and, thus, it was suggested that the glycosyltranferase is also induced during hepatocarcinogenesis.

In fact, although the activity of β1-4 N-acetylglucosaminyltransferase III is nearly undetectable in adult rat normal liver, high activity was found in hepatoma tissues and fetal liver (Nishikawa et al., 1990). In addition, the kidney exhibits the highest N-acetylglucosaminyltransferase III activity as well as γ-glutamyl transpeptidase (Nishikawa et al., 1988). It is of interest that the tissue distribution and the expression pattern of the glycosyltransferase is similar to those of γ-glutamyl transpeptidase. A cDNA for rat N-acetylgluco-saminyltransferase III was cloned (Nishikawa et al., 1992) and a subsequent analysis of the genomic sequences and promoter regions suggested that the glycosyltransferase is also transcribed from multiple promoters, similar to γ-glutamyl transpeptidase (Koyama et al., 1996). Expressions of these two enzymes might involve a similar transcriptional regulation with respect to their tissue-specific and developmental stage-specific expression. Very recently, Sultan (Sultan et al., 1996) reported that Forskolin-treated hepatoma cells induce N-acetylglucosaminyltransferase III and the increase of its product, bisecting N-acetylglucosamine, acts as a negative sorting signal for γ-glutamyl transpeptidase. This finding indicates the bisecting N-acetylyglucosamine may play a role in the protein targeting for membrane-bound enzyme such as γ-glutamyl transpeptidase. In clinical aspects, the detection of γ-glutamyl transpeptidase with altered sugar chains might serve as a marker for the diagnosis of malignant disease.

X. Conclusions

Over the years, extensive studies have been focused on the role(s) of γ-glutamyl transpeptidase in glutathione metabolism. γ-Glutamyl transpeptidase is the only ecto-protein in the enzymes constituting the γ-glutamyl cycle, and therefore is thought to be the most important component of the cycle in terms of the uptake of amino acids and the recovery of cysteine. A recent study, using γ-glutamyl transpeptidase-deficient mice, indicates that a deficiency in the enzyme leads to symptoms that are associated with depletion of tissue glutathione levels, and suggests that glutathione depletion might be due to a decreased availability of cysteine. The study has provided evidence for the role(s) of γ-glutamyl transpeptidase, not only in the recovery of the amino acid but also in regulating glutathione synthesis in spite of extracellular localization of the enzyme. However, the issue of whether γ-glutamyl transpeptidase plays a significant role in transport of other amino acids remains unknown. Since inhibition or deficiency of γ-glutamyl transpeptidase appears to result in the depletion of cellular glutathione, the enzyme might be a component of multidrug resistance in some carcinomas whose resistance depends on glutathione. Therefore, γ-glutamyl transpeptide could be a target to overcome such drug resistance.

Chemical modification studies and mutational analyses of γ-glutamyl transpeptidase have revealed that an arginine residue of the heavy subunit and an aspartic residue(s) of the light subunit serve to bind a substrate. Furthermore, catalytically important serine residues have also been identified in the light subunit, which is regarded as a catalytic subunit. However, a detailed catalytic mechanism of the action of the enzyme is still unclear, despite a considerable amount of investigative work. Structural information on the enzyme will be required to understand the catalytic mechanism and the active site chemistry of γ-glutamyl transpeptidase. The detailed structure of the bacterial enzyme, which has been crystalized and subjected to structural analysis, would provide valuable information on the active site structure. Further studies, for example, definitive identification of a catalytic nucleophile and a general acid–base catalyst, remain to be performed, in order to elucidate the catalytic mechanism.

The expression of γ-glutamyl transpeptidase is regulated in a tissue-specific and a developmental stage-specific manner. A multipromoter system was found to be involved in the regulation of its expression. Analyses of the expression mechanism would lead to understanding, not only of the tissue-specific expression of genes, but also the induction mechanism of proteins during carcinogenesis.

Acknowledgments

We dedicated this paper to the memory of the late Dr. Alton Meister, who was until his sudden death on April 6, 1995, doing collaborative works with us on γ-glutamyl transpeptidase. We also thank Mr. Milton S. Feather and Ms. Taeko Okamura for editing and typing our paper, respectively.

References

Abbott, W. A., Griffith, O. W., and Meister, A., *J. Biol. Chem.*, **261**, 13657–13661 (1986).

Allison, R. D. and Meister, A., *J. Biol. Chem.*, **256**, 2988–2992 (1981).

Allison, R. D., *Methods Enzymol.*, **113**, 419–437 (1985).

Anderson, M. E., Allison, R. D., and Meister, A., *Proc. Natl. Acad. Sci. USA*, **79**, 1088–101 (1982).

Anderson, M. E. and Meister, A., *Proc. Natl. Acad. Sci. USA*, **83**, 5029–5032 (1986).

Angele, C., Wellman, M., C., Guellaen, G., and Siest, G., *Biochem. Biophys. Res. Commun.*, **160**, 1040–1046 (1989).

Baik, J. H., Chikhi, N., Bulle, F., Giuili, G., Guellaen, and G., Siegrist, S., *J. Cell. Physiol.*, **153**, 408–416 (1992b).

Baik, J. H., Griffiths, S., Giuili, G., Manson, M., Siegrist, S., and Guellaen, G., *Carcinogenesis*, **12**, 1035–1039 (1991).

Baik, J. H., Siegrist, S., Giuili, G., Lahuna, O., Bulle, F., and Guellaen, G., *Biochem. J.*, **287**, 691–694 (1992a).

Barouki, R., Chobert, M. N., Finidori, J., Aggerbeck, M., Nalpas, B., and Hanoune, J., *Hepatology*, **3**, 323–329 (1983).

Barouki, R., Finidori, J., Chobert, M. N., Aggerbeck, M., Laperche, Y., and Hanoune, J., *J. Bio. Chem.*, **259**, 7970–7974 (1984).

Beer, D. G., Schwartz, M., Sawada, N., and Pitot, H. C., *Cancer Res.*, **46**, 2435–2441 (1986).

Brouillet, A., Darbouy M., Okamoto, T., Chobert, M. N., Lahuna, O., Garlatti, M., Goodspeed, D., and Laperche, Y., *J. Biol. Chem.*, **269**, 14878–14884 (1994).

Bulle, F., Mattei, M. G., Siegrist, S., Pawlak, A., Passage, E., Chobert, M. N., Laperche, Y., and Guellaen, G., *Hum-Genet.*, **76**, 283–286 (1987).

Cagen, L. M., Fales, H. M., and Pisano, L. J., *J. Biol. Chem.*, **251**, 6550– (1976).

Capraro, M. A., and Hughey, R. P., *FEBS Lett.*, **157**, 139–143 (1983).

Carter, B. Z., Habib, G. M., Sepulveda, A. R., Barrios, R., Wan, D. F., Lebovitz, R. M., and Lieberman, M. W., *J. Biol. Chem.*, **269**, 24581–24585 (1994).

Carter, P. and Wells, J. A., *Nature (London)*, **332**, 564–568 (1988).

Cedar, H., *Cell*, **53**, 3–4 (1988).

Chobert, M. N., Lahuna, O., Lebargy, F., Kurauchi, O., Darbouy, M., Bernaudin, J. F., Guellaen, G., Barouki, R., and Laperche, Y., *J. Biol. Chem.*, **265**, 2352–2357 (1990).

Coloma, J. and Garcia-Jimeno, A., *Biochem. Biophys. Res. Commun.*, **177**, 229–234 (1991).

Coloma, J. and Pitot, H. C., *Nucleic Acids. Res.*, **14**, 1393–1403 (1986).

Courtay, C., Heisterkamp, N., Siest, G., and Groffen, J., *Biochem. J.*, **297**, 503–8 (1994).

Darbouy, M., Chobert M. N., Lahuna, O., Okamoto, T., Bonvalet, J. P., Farman, N., and Laperche, Y., *Am. J. Physiol.*, **261**, C1130–C1137 (1991).

Diederich, M., Wellman, M., Visvikis, A., Puga, A., and Siest, G., *FEBS Lett.*, **332**, 88–92 (1993).

Doerfler, W., *Annu. Rev. Biochem.*, **52**, 93–124 (1983).

Elce, J. S., *Biochem. J.*, **185**, 473–481 (1980).

Farber, E., *Cancer Res.*, **44**, 5463–5474 (1984).

Fiala, S. and Fiala, E. S., *J. Natl. Cancer Inst.*, **51**, 151–158 (1973).

Figlewicz, D. A., Delattre, O., Guellaen, G., Krizus, A., Thomas, G., Zucman, J., and Rouleau, G. A., *Genomics*, **17**, 299–305 (1993).

Finidori, J., Laperche, Y., Haguenauer-Tsapis, R., Barouki, R., Guellaen, G., and Hanoune, J., *J. Biol. Chem.*, **259**, 4687–4690 (1984).

Gardell, S. J. and Tate, S. S., *J. Biol. Chem.*, **254**, 4942–4945 (1979).

Godwin, A. K., Meister, A., O'Dwyer, P. J., Huang, C. S., Hamilton, T. C., and Anderson, M. E., *Proc. Natl. Acad. Sci. USA*, **89**, 3070–3074 (1992).

Goodspeed, D. C., Dunn, T. J., Miller, C. D., and Pitot, H. C., *Gene*, **76**, 1–9 (1989).

Griffith, O. W., Bridges, R. J., and Meister, A., *Proc. Natl. Acad. Sci. USA*, **75**, 5405–5408 (1978).

Griffith, O. W., Bridges, R. J., and Meister, A., *Proc. Natl. Acad. Sci. USA*, **76**, 6319–6322 (1979a).

Griffith, O. W., Bridges, R. J., and Meister, A., *Proc. Natl. Acad. Sci. USA*, **78**, 2777–2781 (1981).

Griffith, O. W. and Meister, A., *Proc. Natl. Acad. Sci. USA*, **76**, 268–272 (1979b).

Griffith, O. W., and Meister, A., *Proc. Natl. Acad. Sci. USA*, **77**, 3384–3387 (1980).

Habib, G. M., Carter, B. Z., Sepulveda, A. R., Shi, Z. Z., Wan, D. F., Lebovitz, R. M., and Lieberman, M. W., *J. Biol. Chem.*, **270**, 13711–13715 (1995).

Hanigan, M. H. and Pitot, H. C., *Carcinogenesis*, **6**, 165–172 (1985).

Heisterkamp, N., Rajpert-De-Meyts, E., Uribe, L., Forman, H. J., and Groffen, J., *Proc. Natl. Acad. Sci. USA*, **88**, 6303–6307 (1991).

Hughey, R. P., Coyle, P. J., and Curthoys, N. P., *J. Biol. Chem.*, **254**, 1124–1128 (1979).

Hughey, R. P. and Curthoys, N., *J. Biol. Chem.*, **251**, 8763–8770 (1976).

Ikeda, Y., Fujii, J., Anderson, M. E., Taniguchi, N., and Meister, A., *J. Biol. Chem.*, **270**, 22223–22228 (1995c).

Ikeda, Y., Fujii, J., and Taniguchi, N., *J. Biol. Chem.*, **268**, 3980–3985 (1993).

Ikeda, Y., Fujii, J., Taniguchi, N., and Meister, A., *Proc. Natl. Acad. Sci. USA*, **92**, 126–130 (1995a).

Ikeda, Y., Fujii, J., Taniguchi, N., and Meister, A., *J. Biol. Chem.*, **270**, 12471–12475 (1995b).

Ikeda, Y., Fujii, J., and Taniguchi, N., *J. Biochem.*, **119**, 1166–1170 (1996).

Inoue, M. Horiuchi, S., and Morino, Y., *Eur. J. Biochem.*, **73**, 335–342 (1977).

Inoue, M. Horiuchi, S., and Morino, Y., *Biochem. Biophys. Res. Commun.*, **82**, 1183–1188 (1987).

Ishiye, M., Yamashita, M., and Niwa, M., *Biotechnol. Prog.*, **9**, 323–331 (1993).

Jain, A., Martensson, J., Stole, E., Auld, P. A., and Meister, A., *Proc. Natl. Acad. Sci. USA*, **88**, 1913–1917 (1991).

Kannan, R., Yi, J. R., Tang, D., Li, Y., Zlokovic, B. V., and Kaplowitz, N., *J. Biol. Chem.*, **271**, 9754–9758 (1996).

Kannan, R., Yi, J. R., Zlokovic, B. V., and Kaplowitz, N., *Invest. Ophthalmol. Vis. Sci.*, **36**, 1785–1792 (1995).

Klausner, R. D., Donaldson, J. G., and Lippincott-Schwartz, J., *J. Cell Biol.*, **116**, 1071–1080 (1992).

Kornfeld, R. and Kornfeld, S., *Annu. Rev. Biochem.*, **54**, 631–664 (1985).

Koyama, N., Miyoshi, E., Ihara, Y., Kang, R., Nishikawa, A., and Taniguchi, N., *Eur. J. Biochem.*, **238**, 853–861 (1996).

Kumagai, H., Nohara, S., Suzuki, H., Hashimoto, W., Yamamoto, K., Sakai, H., Sakabe, K., Fukuyama, K., and Sakabe, N., *J. Mol. Biol.*, **234**, 1259–1262 (1993).

Kuno, T., Matsuda, Y., and Katunuma, N., *Biochem. Biophys. Res. Commun.*, **114**, 889–895 (1983).

Kurauchi, O., Lahuna, O., Darbouy, M., Aggerbeck, M., Chobert, M. N., and Laperche, Y., *Biochemistry*, **30**, 1618–1623 (1991).

Lahuna, O., Brouillet, A., Chobert, M. N., Darbouy, M., Okamoto, T., and Laperche, Y., *Biochemistry*, **31**, 9190–9196 (1992).

Laperche, Y., Bulle, F., Aissani, T., Chobert, M. N., Aggerbeck, M., Hanoune, J., and Guellaen, G., *Proc. Natl. Acad. Sci. USA*, **83**, 937–941 (1986).

Lieberman, M. W., Barrios, R., Carter, B. Z., Habib, G. M., Lebovitz, R. M., Rajagopalan, S., Sepulveda, A. R., Shi, Z. Z., and Wan, D. F., *Am. J. Pathol.*, **147**, 1175–1185 (1995).

Lieberman, M. W., Wiseman, A. L., Shi, Z. Z., Carter, B. Z., Barrios, R., Ou, C. N. Chevez-Barrios, P., Wang, Y., Habib, G. M., Goodman, J. C., Huang, S. L., Lebovitz, R. M., and Matzuk, M. M., *Proc. Natl. Acad. Sci. USA*, **93**, 7923–7926 (1996).

Lippincott-Schwartz, J., Yuan, L. C., Tpper, C., Amherdt, M., Orci, L., and Klausner, R. D., *Cell*, **67**, 601–616 (1991).

Matsuda, Y., Tsuji, A., and Katunuma, N., *J. Biochem.*, **93**, 1427–1433 (1983).

McIntyre, T. M., and Curthoys, N. P., *J. Biol. Chem.*, **254**, 6499–6504 (1979).

Meister, A. and Anderson, M. E., *Annu. Rev. Biochem.*, **52**, 711–760 (1983).

Meister, A., *Biochem. Pharmacol.*, **44**, 1905–1915 (1992).

Meister, A. and Larsson, A., *In The Metabolic and Molecular Bases of Inherited Disease*, 7th ed, Vol. I, Scriver, C. R., Beaudet, A. L., Sly, W. S., and Valle, D., Eds., McGraw-Hill, New York, 1995, pp. 1461–1477.

Meister, A. and Tate, S. S., *Annu. Rev. Biochem.*, **45**, 559–604 (1976).

Meister, A., Tate, S. S., and Griffith, O. W., *Methods Enzymol.*, **77**, 237–253 (1981).

Nash, B. and Tate, S. S., *J. Biol. Chem.*, **257**, 585–588 (1982).

Nash, B. and Tate, S. S., *J. Biol. Chem.*, **259**, 678–685 (1984).

Nishikawa, A., Fujii, S., Sugiyama, T., and Taniguchi, N., *Anal. Biochem.*, **170**, 349–354 (1988).

Nishikawa, A., Gu, J., Fujii, S., and Taniguchi, N., *Biochim. Biophys. Acta.*, **1035**, 313–318 (1990).

Nishikawa, A., Ihara, Y., Hatakeyama, M., Kangawa, K., and Taniguchi, N., *J. Biol. Chem.*, **267**, 18199–18204 (1992).

Nishimura, M., Stein, H., Berges, W., and Teschke, R., *Biochem. Biophys. Res. Commun.*, **99**, 142–148 (1981).

Okamoto, T., Darbouy, M., Brouillet, A., Lahuna, O., Chobert, M. N., and Laperche, Y., *Biochemistry*, **33**, 11536–11543 (1994).

Palladino, M. A., Laperche, Y., and Hinton, B. T., *Biol. Reprod.*, **50**, 320–328 (1994).

Papandrikopoulou, A., Frey, A., and Gassen, H. G., *Eur. J. Biochem.*, **183**, 693–698 (1989).

Pawlak, A., Cohen, E. H., Octave, J. N., Schweickhardt, R., Wu, S. J., Bulle, F., Chikhi, N., Baik, J. H., Siegrist, S., and Guellaen, G., *J. Biol. Chem.*, **265**, 3256–3262 (1990).

Pawlak, A., Lahuna, O., Bulle, F., Suzuki, A., Ferry, N., Siegrist, S., Chikhi, N., Chobert, M. N., Guellaen, G., and Laperche, Y., *J. Biol. Chem.*, **263**, 9913–9916 (1988).

Pawlak, A., Wu, S. J., Bulle, F., Suzuki, A., Chikhi, N., Ferry, N., Baik, J. H., Siegrist, S., and Guellaen, G., *Biochem. Biophys. Res. Commun.*, **164**, 912–918 (1989).

PetitClerc, C., Shiele, F., Bagrel, D., Mahassen, A., and Siest, G., *Clin. Chem.*, **26**, 1688–1693 (1980).

Rajagopalan, S., Park, J. H., Patel, P. D., Lebovitz, R. M., and Lieberman, M. W., *J. Biol. Chem.*, **265**, 11721–11725 (1990).

Rajagopalan, S., Wan, D. F., Habib, G. M., Sepulveda, A. R., McLeod, M. R., Lebovitz, R. M., and Lieberman, M. W., *Proc. Natl. Acad. Sci. USA*, **90**, 6179–6183 (1993).

Rajpert-De-Meyts, E., Heisterkamp, N., and Groffen, J., *Proc. Natl. Acad. Sci. USA*, **85**, 8840–8844 (1988).

Razin, A. and Riggs, A. D., *Science*, **210**, 604–610 (1980).

Robertus, J. D., Kraut, J., Alden, R. A., and Birktoft, J. J., *Biochemistry*, **11**, 4293–4303 (1972).

Sakai, H., Sakabe, N., Sasaki, K., Hashimoto, W., Suzuki, H., Tachi, H., Kumagai, H., and Sakabe, K., *J. Biochem.*, **120**, 26–28 (1996).

Sakamuro, D., Yamazoe, M., Matsuda, Y., Kangawa, K., Taniguchi, N., Matsuo, H., Yoshikawa, H., and Ogasawara, N., *Gene*, **73**, 1–9 (1988).

Sawabu, N., Nakagen, M., Yoneda, M., Makino, H., Kameda, S., Kobayashi, K., Hattori, N., and Ishii, M., *Gann MonographCancer Res.*, **69**, 601–605 (1978).

Schasteen, C. S., Curthoys, N. P., and Reed, D. J., *Biochem. Biophys. Res. Commun.*, **112**, 564–570 (1983).

Sepulveda, A. R., Carter, B. Z., Habib, G. M., Lebovitz, R. M., and Lieberman, M. W., *J. Biol. Chem.*, **269**, 10699–10705 (1994).

Shi, Z. Z., Habib, G. M., Lebovitz, R. M., and Lieberman, M. W., *Gene*, **167**, 233–237 (1995).

Smith, T. K., Ikeda, Y., Fujii, J., Taniguchi, N., and Meister, A., *Proc. Natl. Acad. Sci. USA*, **92**, 2360–2364 (1995b).

Smith, T. K. and Meister, A., *FASEB J.*, **8**, 661–664 (1994).

Smith, T. K. and Meister, A., *J. Biol. Chem.*, **270**, 12476–12480 (1995a).

Stole, E. and Meister, A., *J. Biol. Chem.*, **266**, 17850–17857 (1991).

Stole, E., Seddon, A. P., Wellner, D., and Meister, A., *Proc. Natl. Acad. Sci. USA*, **87**, 1706–1709 (1990).

Stole, E., Smith, T. K., Manning, J. M., and Meister, A., *J. Biol. Chem.*, **269**, 21435–21439 (1994).

Sultan, A., Miyoshi, E., Ihara, Y., Nishikawa, A., Tukada, Y., and Taniguchi, N., *J. Biol. Chem.*, **272**, 2866–2872 (1996).

Suzuki, H., Kumagai, H., Echigo, T., and Tochikura, T., *J. Bacteriol.*, **171**, 5169–5172 (1989).

Szewczuk, A. and Connell, G. E., *Biochim. Biophys. Acta*, **106**, 352–367 (1965).

Taniguchi, N., *J. Biochem.*, **75**, 473–480 (1974).

Taniguchi, N., Iizuka, S., Zhe, Z. N., House, S., Yokosawa, N., Ono, N., Kinoshita, K., Makita, A., and Sekiya, C., *Cancer Res.*, **45**, 5835–5839 (1985).

Taniguchi, N., Saito, K., and Takakuwa, E., *Biochim. Biophys. Acta*, **391**, 265–271 (1975).

Taniguchi, N., Yokosawa, N., Iizuka, S., Sako, F., Tsukada, Y., Satoh, M., and Dempo, K., *Ann. N.Y. Acad. Sci.*, **417**, 203–212 (1983a).

Taniguchi, N., Yokosawa, N., Iizuka, S., Tsukada, Y., Sako, F., and Miyazawa, N., *Gann Monograph Cancer Res.*, **29**, 263–272 (1983b).

Tate, S. S., *FEBS Lett.*, **194**, 33–38 (1986).

Tate, S. S. and Galbraith, R. A., *J. Biol. Chem.*, **262**, 11403–11406 (1987b).

Tate, S. S. and Galbraith, R. A., *Biochem. Biophys. Res. Commun.*, **154**, 1167–1173 (1988).

Tate, S. S., *FEBS Lett.*, **54**, 319–322 (1975).

Tate, S. S. and Meister, A., *J. Biol. Chem.*, **249**, 7593–7602 (1974a).

Tate, S. S. and Meister, A., *Proc. Natl. Acad. Sci. USA*, **71**, 3329–3333 (1974b).

Tate, S. S. and Meister, A., *Proc. Natl. Acad. Sci. USA*, **74**, 931–935 (1977a).

Tate, S. S. and Meister, A., *Proc. Natl. Acad. Sci. USA*, **75**, 4806–4809 (1978).

Tate, S. S. and Meister, A., *Mol. Cell. Biochem.*, **39**, 357–368 (1981).

Tate, S. S. and Meister, A., *Methods Enzymol.*, **113**, 400–419 (1985).

Tate, S. S. and Nash, B., *FEBS Lett.*, **211**, 133–136 (1987a).

Tate, S. S. and Ross, M. E., *J. Biol. Chem.*, **252**, 6042–6045 (1977b).

Tateishi, N., Higashi, T., Nomura, T., Naruse, A., Nakamura, K., Shiozaki, H., and Sakamoto, Y., *Gann Monograph Cancer Res.*, **67**, 215–222 (1976).

Thompson, G. A. and Meister, A., *Proc. Natl. Acad. Sci. USA*, **72**, 1985–1988 (1975).

Thompson, G. A. and Meister, A., *Biochem. Biophys. Res. Commun.*, **71**, 32–36 (1976).

Thompson, G. A. and Meister, A., *J. Biol. Chem.*, **252**, 6792–6798 (1977).

Thompson, G. A. and Meister, A., *J. Biol. Chem.*, **254**, 2956–2960 (1979).

Tsuchida, S., Hoshino, K., Sato, T., Ito, N., and Sato, K., *Cancer Res.*, **39**, 4200–4205 (1979).

Tsutsumi, M., Sakamuro, D., Takada, A., Zang, S.-C., Furukawa, T., and Taniguchi, N., *Hepatology*, **23**, 1093–1097 (1996).

Visvikis, A., Thioudellet, C., Oster, T., Fournel-Gigleux, S., Wellman, M., and Siest, G., *Proc. Natl. Acad. Sci. USA*, **88**, 7361–7365 (1991).

Von Heijne, G., *J. Membr. Biol.*, **115**, 195–201 (1990).

Wetmore, L. A., Gerard, C., and Drazen, J. M., *Proc. Natl. Acad. Sci. USA*, **90**, 7461–7465 (1993).

Yamashita, K., Hitoi, A., Taniguchi, N., Yokosawa, N., Tsukada, Y., and Kobata, A., *Cancer Res.*, **43**, 5059–5063 (1983).

Yokosawa, N., Taniguchi, N., Tsukada, Y., and Makita, A., *Oncodev. Biol. Med.*, **4**, C71–C78 (1983).

ENZYMOLOGY OF BACTERIAL LYSINE BIOSYNTHESIS

By GIOVANNA SCAPIN
and JOHN S. BLANCHARD,

*Department of Biochemistry,
Albert Einstein College of Medicine,
Bronx, NY 10461*

CONTENTS

I. Introduction

Work's discovery in 1950 of the unusual amino acid *meso*-diaminopimelate in acid hydrolysates of *Corynebacterium diphtheriae* (Work, 1950), and subsequently in *Mycobacterium tuberculosis* (Work, 1951) led to the systematic study of its biosynthesis and the description of the bacterial lysine biosynthetic pathway (Gilvarg, 1960). Not only was diaminopimelate found to be the immediate

Advances in Enzymology and Related Areas of Molecular Biology, Volume 72:
Amino Acid Metabolism, Part A, Edited by Daniel L. Purich
ISBN 0-471-24643-3   © 1998 John Wiley & Sons, Inc.

precursor of L-lysine, but it was subsequently demonstrated to be a component of the bacterial cell wall (Cummins and Harris, 1956): specifically, the peptidoglycan portion. In *Escherichia coli*, and other bacteria, not only L-lysine, but also L-threonine, L-methionine, and L-isoleucine are derived from L-aspartate (Fig. 1). The common portions of the biosynthetic pathway include the enzymes aspartokinase and aspartate semialdehyde dehydrogenase, which generate aspartate semialdehyde. The reduction of aspartate semialdehyde by homoserine dehydrogenase yields homoserine, the precursor to threonine, methionine and isoleucine.

Bacteria have evolved three slightly different strategies for the biosynthesis of diaminopimelate, and thus lysine, from aspartate. After generation of aspartate semialdehyde, it is condensed with pyruvate by the action of the *dapA*-encoded synthase to generate dihydrodipicolinate. This compound is reduced by the *dapB*-encoded reductase to yield tetrahydrodipicolinate. The acyclic form of tetrahydrodipicolinate, L-2-amino-6-keto-pimelate, can be directly

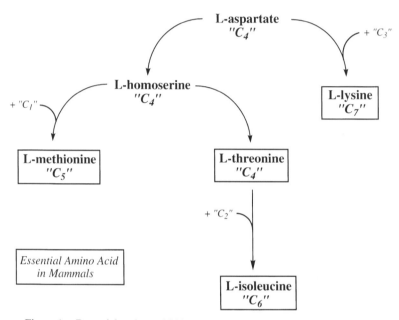

Figure 1.   Bacterial amino acid biosynthesis using aspartate as precursor.

converted to diaminopimelate by the reduced nicotinamide adenine denucleotide (NADPH)- and ammonia-dependent reductive amination of the 6-keto group, a reaction catalyzed by diaminopimelate dehydrogenase. DAP dehydrogenase is the only enzyme known to generate a D stereocenter during this chemical transformation, and is discussed in detail (Section X). Diaminopimelate dehydrogenase has very limited occurrence, having only been identified in *Corynebacterium glutamicum, Bacillus sphaericus*, and a *Pseudomonas* and *Brevibacterium* species. As an alternative to the direct reductive amination reaction, two alternative pathways exist in bacteria, which involve the *N*-acylation of L-2-amino-6-keto-pimelate to prevent cyclization. By far, the most prevalent pathway involves *N*-succinylation of the α-amino group by succinyl-CoA, catalyzed by the tetrahydrodipicolinate/succinylCoA *N*-succinyltransferase (Fig. 2). Once the acyclic form is stabilized by succinylation, a pyridoxal phosphate (PLP) dependent transaminase generates a second amino acid center of the L configuration, using L-glutamate as the amino group donor. The product of the transaminase, *N*-succinyl-L,L-diaminopimelate is desuccinylated by the *dapE*-encoded desuccinylase to generate succinate and L,L-diaminopimelate. The epimerization of this intermediate to form D,L-diaminopimelate is catalyzed by the *dapF*-encoded diaminopimelate epimerase, one of the rare examples of a non-PLP-dependent amino acid epimerase. The rationale for this unusual stereochemical inversion likely involves the requirement of the final enzyme in the pathway to specifically decarboxylate only one of the two amino acid centers. The reaction catalyzed by the *lysA*-encoded diaminopimelate decarboxylase is the only known reaction in which a PLP-dependent amino acid decarboxylase acts on a D-amino acid center. While many amino acid biosynthetic enzymes are clustered into multigenic operons, including the enzymes of isoleucine biosynthesis, the genes encoding the lysine biosynthetic enzymes are not clustered, and in fact are scattered over both the *E. coli* and *Haemophilus influenzae* circular chromosomes.

The alternate pathway involving *N*-acetylated intermediates instead of *N*-succinylated intermediates proceeds through an identical series of steps, and is, like the dehydrogenase pathway, extremely limited in its bacterial distribution. The deacylation of *N*-acetyldiaminopimelate has only been described in *Bacillus* species, with the acetylase pathway of *B. megaterium* being the best characterized system (Sundharadas and Gilvarg, 1967).

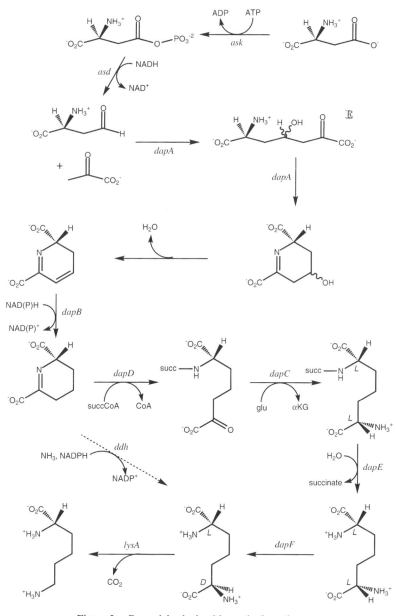

Figure 2.   Bacterial L-lysine biosynthetic pathway.

In contrast to bacterial lysine biosynthesis, yeast synthesize L-lysine via the α-ketoadipate pathway, in which acetyl CoA and α-ketoglutarate are condensed to form homocitric acid. Isomerization generates homoisocitric acid, which is oxidized to oxaloglutaric acid and decarboxylated to yield α-ketoadipate. α-Ketoadipate is transaminated to α-aminoadipate, reduced to α-aminoadipate semialdehyde, and condensed with L-glutamate to generate saccharopine, named for the yeast *Saccharomyces cerevisiae*, from which the compound was first identified. Oxidation and hydrolysis of the imine yields L-lysine and α-ketoglutarate (Fig. 3).

II. Aspartokinase and Aspartate Semialdehyde Dehydrogenase

In all species of bacteria, aspartokinase and aspartate semialdehyde dehydrogenase (EC 2.7.2.4 and 1.2.1.11) act sequentially to generate the early, common intermediates in L-threonine, L-isoleucine, L-methionine, and L-lysine biosynthesis. Aspartokinase has been the subject of intense genetic and biochemical interest due to the presence of three isozymes in *E. coli*, which are end-product inhibited by the three amino acid products: L-threonine, L-methionine, and L-lysine (Patte, 1996). The most abundant isozyme is the *thrA*-encoded aspartokinase I, a bifunctional aspartokinase-homoserine dehydrogenase. This 820 amino acid *E. coli* protein is composed of an amino terminal aspartokinase domain and a carboxy terminal homoserine dehydrogenase domain (Truffa-Bachi et al., 1968). Threonine exhibits competitive inhibition versus aspartate, and noncompetitive inhibition versus homoserine, but the inhibition curves are both sigmoidal suggesting cooperative binding to the homotetramer (see Cohen, 1985). Threonine and isoleucine also regulate the expression of the *thrA*-encoded aspartokinase. The *E. coli* *metL*-encoded aspartokinase II is also a bifunctional aspartokinase-homoserine dehydrogenase, which in contrast to aspartokinase I, is not inhibited by methionine. The enzyme exists as a homodimer of 809 amino acid monomers, and expression of the gene is repressed by methionine. Finally, the *lysC*-encoded aspartokinase III is a monofunctional aspartokinase, which is noncompetitively and cooperatively inhibited by lysine and other amino acids. The *E. coli* aspartokinase III is a homodimer of 449 amino acid monomers that are homologous to the amino terminal domains of the aspartokinase

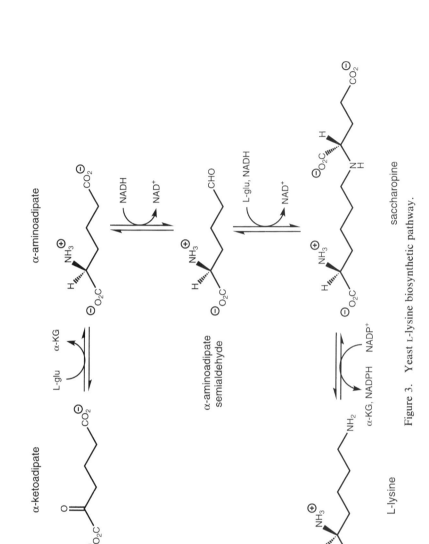

Figure 3.   Yeast L-lysine biosynthetic pathway.

284

I and II, and binds 2 mol of lysine per monomer cooperatively. The kinetic mechanism of the MgATP-dependent phosphorylation of the γ-carboxyl group of aspartate involves the random addition of substrates, followed by the ordered release of MgADP and aspartyl phosphate (Cohen, 1985). Little additional mechanistic information has been reported.

The *asd*-encoded aspartate semialdehyde dehydrogenase (EC 1.2.1.11) catalyzes the reversible NADPH-dependent reduction of aspartyl phosphate to generate aspartate semialdehyde. The *E. coli* gene has been sequenced (Haziza et al., 1982), and encodes a 367 amino acid monomer that exists as a homodimer in solution. The kinetic mechanism is random in the physiologically important direction of aspartyl phosphate reduction, but is ordered in the reverse direction, with $NADP^+$, inorganic phosphate (P_i), and aspartate semialdehyde binding in that order (Karsten and Viola, 1991). In the analogous reaction, catalyzed by glyceraldehyde-3-phosphate dehydrogenase, a cysteine residue has been implicated in catalysis, and the site-directed mutagenesis of Cys-135 to either an alanine or a serine residue (C135A or C135S) resulted in either a complete loss of activity, or a 300-fold reduction in activity compared to the wild-type enzyme (Karsten and Viola, 1992). These and other mutagenesis studies, and the analysis of the pH dependence of the kinetic parameters, support a chemical mechanism for the reaction catalyzed by aspartate semialdehyde dehydrogenase shown in Figure 4.

III.   Dihydrodipicolinate Synthase

The *dapA*-encoded dihydrodipicolinate synthase (DHDPS) (E.C.4.2.1.52) catalyzes the first unique step in lysine biosynthesis, the aldol condensation between pyruvate and aspartate semialdehyde. The activity was first described in extracts of *E. coli* in 1965 (Yugari and Gilvarg, 1965), and the enzyme was purified 5000-fold to apparent homogeneity from *E. coli* in 1970 (Shedlarsky and Gilvarg, 1970). Since it catalyzes the committed step in L-lysine biosynthesis, it is perhaps not surprising that all synthases studied to date are feedback inhibited by L-lysine. However, the extent of inhibition has been used to group the synthases into three classes: gram-positive bacterial synthases are only modestly inhibited by high concentrations of lysine; plant synthases are strongly inhibited at lysine con-

286

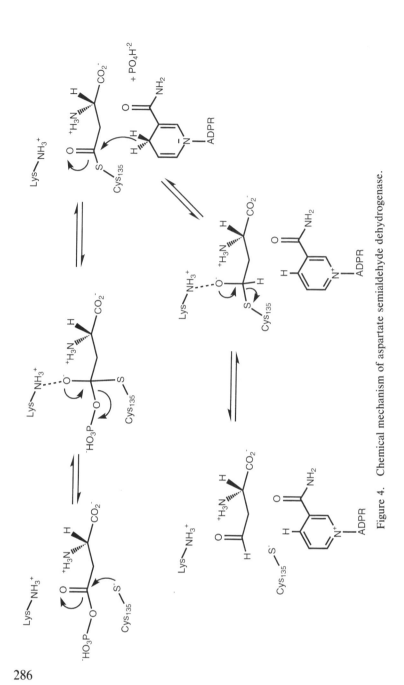

Figure 4. Chemical mechanism of aspartate semialdehyde dehydrogenase.

centrations less than 50 μM; while the *E. coli* and *B. sphaericus* enzymes are weakly inhibited by 0.2–1.0-mM concentrations of L-lysine. Early kinetic studies suggested that L-lysine was a competitive inhibitor versus aspartate semialdehyde, but more recent studies with the wheat enzyme suggest that lysine is a bona fide allosteric inhibitor (Shaver et al., 1996). These kinetic studies have recently been confirmed by structural studies discussed below.

The native *E. coli* enzyme was reported to exhibit a molecular weight of 134,000 Da, now known to be a homotetramer of 31,372-Da monomers (Richaud et al., 1986). These early studies identified a lysine residue as being involved in the initial binding of pyruvate. Stoichiometric reduction of the presumptive Schiff base by NaBH$_4$ resulted in the loss of enzyme activity, incorporation of radiolabel from pyruvate into an enzyme-bound form, and the generation of N-(1-carboxyethyl)-lysine after acid hydrolysis of the inactivated enzyme. The demonstration of tritiated pyruvate exchange in the absence of aspartate semialdehyde suggested that the Schiff base could catalyze pyruvate enolization, generating a carbanion equivalent for nucleophilic attack on the aldehydic carbon of aspartate semialdehyde. Cyclization and dehydration yields the product, dihydrodipicolinate. Numerous steady-state kinetic studies revealed parallel initial velocity patterns when pyruvate and aspartate semialdehyde (ASA), were varied, suggesting a ping–pong type of kinetic mechanism. Binding of pyruvate to the active site lysine residue and enolization of the Schiff base yields a proton as the first "product." If proton release were irreversible, then the parallel initial velocity pattern would be expected, as has been recently documented (Karsten, 1997). Binding of ASA to the eneamine form of the enzyme, followed by carbon–carbon bond formation yields the bound imine form of 2-keto-4-hydroxy-6-aminopimelic acid. Cyclization via transimination yields the cyclic 4-hydroxy-tetrahydrodipicolinate, the compound recently identified as the enzyme product (Blickling et al., 1997). Dehydration occurs nonenzymatically, with the formation of the *cis*-α,β-unsaturation enforced by the cyclic imine. These studies suggest a chemical mechanism for the dihydrodipicolinate reaction shown in Figure 5.

The three-dimensional structure of the *E. coli* dihydrodipicolinate synthase has been solved using X-ray crystallography at 2.5-Å resolution (Mirwaldt et al., 1995). The asymmetric unit is composed of

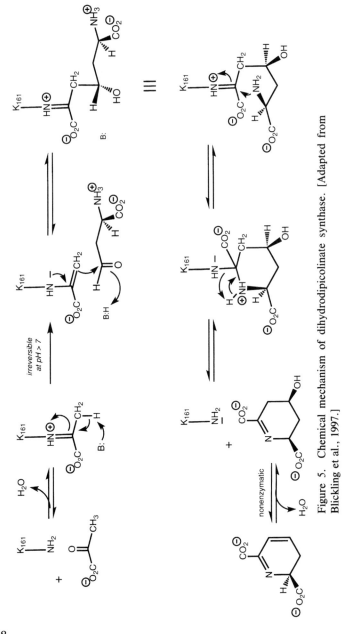

Figure 5. Chemical mechanism of dihydrodipicolinate synthase. [Adapted from Blickling et al., 1997.]

288

two monomers, and the tetramer is generated using crystallographic symmetry operations. The tetramer has a large solvent-filled cavity at its center (Fig. 6). Each monomer is composed of two domains: The amino terminal domain assumes an eight-stranded parallel α/β barrel ranging from residues 1–224, while the carboxy-terminal domain is composed predominantly of α-helices ranging from residues 224–292. The catalytic site, as defined by the position of Lys-161, is at the carboxy-terminal end of the barrel, with the ε-amino group of Lys-161 almost exactly centered in the barrel. This site is the position of all catalytic sites in such parallel α/β barrel enzymes (Farber and Petsko, 1990), and is commonly used to stabilize developing negative charges on intermediates, or in the transition state, for enzyme catalyzed reactions. In a continuation of these studies, Farber & Petsko (1990) successfully soaked a number of substrates and substrate analogues into the apoenzyme, and determined the structures of many of these complexes (Blickling et al., 1997). The pyruvyl complex of the enzyme showed the expected formation of the covalent Schiff base with Lys-161, and additionally suggested that Tyr-133 may act as a general acid in the dehydration reaction,

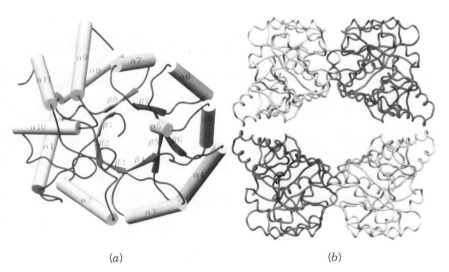

(a)   (b)

Figure 6.   Three-dimensional structure of *E. coli* dihydrodipicolinate synthase monomer (*a*) and tetramer (*b*). (Coordinate filename 1DHP.pdb.)

based on the distance between the phenolic oxygen and the C2 of bound pyruvate (3.4 Å). The C1 carboxyl group makes an interaction with the Thr-44 side-chain hydroxyl group. Proton removal accompanying eneamine formation generates the nucleophile, which attacks the aldehydic carbon, and it is likely that the protonated base may assist in protonating the (4S)-alkoxide. By using the substrate analogue, succinate semialdehyde, and pyruvate, the crystalline enzyme generates a product analogue, 2-keto-(4S)-hydroxy-pimelate, bound as the Schiff base at the active site. The 7-carboxyl group is held in place by interactions with Arg-138 and Asn-248, and the 4-hydroxyl group is within hydrogen-bonding distance of the Lys-161 imine nitrogen and a backbone oxygen of Gly-186. The structure of an enzyme-2,6-pyridine dicarboxylate complex revealed that the two carboxylates were interacting with Arg-138 and Thr-45, the result of a conformational change in which interactions between the C1-carboxyl and Thr-44 are replaced by interactions with Thr-45.

Of substantial interest was the determination of the structure of the L-lysine complex of DHDPS. Two molecules of L-lysine were shown to bind at the monomer–monomer interface, with interactions between each lysine molecule and residues of both monomers, potentially explaining the observed cooperative binding of L-lysine to the tetramer. The α-carboxyl group of bound L-lysine interacts predominantly with the phenolic oxygen of Tyr-106, and the α-amino group makes extensive interactions with Asn-80 and Glu-84. The ε-amino group makes interactions with His-53 and His-56, and all of these residues are observed to move from their position in the apoenzyme upon lysine binding. Sequence alignments (Mirwaldt et al., 1995) of the derived protein sequences from plants and bacteria show that while Asn-80 and Tyr-106 are conserved in all synthases, Glu-84 is only present in synthases that exhibit strong lysine inhibition, and is replaced by a threonine residue in gram-positive bacteria that are weakly inhibited by lysine. In support of the critical role of Glu-84 in L-lysine binding to the synthase, is the recent report that mutations of this residue in the maize synthase result in the synthesis of a synthase that has lost sensitivity to L-lysine inhibition (Shaver et al., 1996). This finding suggests that mutant synthases that are not feedback inhibited by lysine may be useful to generate transgenic plants with increased levels of free lysine for animal feedstock, allowing for this required, and limiting, amino acid to be more easily acquired.

IV. Dihydrodipicolinate Reductase

The *dapB*-encoded dihydrodipicolinate reductase (E.C.1.3.1.26) catalyzes the thermodynamically favorable NAD(P)H-dependent reduction of dihydrodipicolinic acid (DHDP) to tetrahydrodipicolinic acid (THDP), the second step in the lysine biosynthetic pathway. The reductase activity was initially identified in *E. coli* mutants (M-203) that required diaminopimelic acid and lysine in their growth medium, and were subsequently shown to lack the reductase (Farkas and Gilvarg, 1965). The enzyme was subsequently purified 1900-fold (Tamir and Gilvarg, 1974) to homogeneity. The enzyme exhibited a broad pH optimum around 7, and the enzyme's native molecular weight was found to be 110 kDa by gel filtration chromatography and sedimentation velocity measurements. The enzyme was shown to have a high affinity for its substrates, with reported K_m values of 9 and 10 μM for dihydrodipicolinate and NADPH, respectively. The reductase was inhibited by phosphate ions and by cyclic dicarboxylic compounds, including dipicolinic acid and isophtalic acid, which were shown to be competitive inhibitors versus dihydrodipicolinate. These data suggested that it was a cyclic form of the substrate that was bound by the enzyme. Isotopic-transfer studies suggested that the mechanism of reduction involved direct hydride transfer of the (4R) hydrogen to the substrate (Tamir and Gilvarg, 1974).

About the same time, the identification, purification, and characterization of a dihydrodipicolinate reductase from sporulating *Bacillus subtilis* was reported (Kimura, 1975; Kimura and Goto, 1975). The *B. subtilis* reductase was reported to be a homotetramer of about 77kDa molecular weight by gel filtration chromatography and sodium decyl sulfate (SDS)-gel electrophoresis. The purified enzyme showed a visible absorbance spectrum typical of a flavoprotein, and the prosthetic group was identified by thin-layer and paper chromatography to be flavin mononucleotide (FMN). The flavin content was calculated to be 2 mol of FMN per enzyme tetramer. The kinetic mechanism was shown to be ping–pong, and spectroscopic evidence supported the role of FMN in transferring reducing equivalents from

NADPH to dihydrodipicolinate (Kimura and Goto, 1975). The K_m values for DHDP and NADPH were 770 and 72 μM, respectively. These differences between the reductases in the nonsporulating *E. coli* and *B. subtilis* presumably reflect differences in their metabolic roles, possibly in spore formation (Kimura, 1975).

The first *dapB* gene sequence was reported in 1984 for the *E. coli* enzyme (Bouvier et al., 1984). The DNA sequence predicted a 273 residue polypeptide, with a subunit mass of 28,798 Da. To date, eight additional sequences have been reported: *B. subtilis* (Henner et al., 1990), *Brevibacterium lactofermentum* (Pisabarro et al., 1992), *Corynebacterium glutamicum* (Eikmanns, 1992), *Mycobacterium leprae* (Staffen and Cole, 1994), *Haemophilus influenzae* (Fleischmann et al., 1995), *Pseudomonas syringae* (Liu and Shaw, 1997), *Mycobacterium tuberculosis* (Pavelka et al., 1997), and *Synechocystis sp.* (Tabata, 1997). They all encode reductases of between 246 and 286 residues. All share several regions of homology, in particular the motif ([8]V/I)(A/G)(V/I)XGXXGXXG[18], located at the extreme N-terminus of the protein, and the motif [157]E(L/A)HHXXKXDAPSGTA[171] (numbers are for the *E. coli* sequence). The first motif is found in a variety of dinucleotide-dependent dehydrogenases, and in the three-dimensional structure of the *E. coli* enzyme (Scapin et al., 1995) is part of the nucleotide-binding domain. The second region has been suggested as a potential substrate-binding region on the basis of the conserved positively charged residues (Pavelka et al., 1997) and its location in the three-dimensional structure of the *E. coli* enzyme.

The most thoroughly characterized reductase is from *E. coli*, and this recombinant enzyme (rDHPR) has been expressed and purified to homogeneity in large quantities (Reddy et al., 1995). Initial velocity, product, and dead-end inhibition studies support a sequential kinetic mechanism. Steady-state K_m values for NADPH and DHDP were determined to be 8 and 50 μM, respectively. The stereochemistry of hydride transfer was reinvestigated using both [4S-] and [4R-4-^{3}H]-NADPH and determining the radioactivity transferred to the chromatographically separate products. The results show that *E. coli* DHPR catalyzes the transfer of the *pro-R* hydrogen to the substrate, confirming the previously reported stereochemistry of transfer (Tamir and Gilvarg, 1974). The mechanism of C—C double bond reduction, in particular the position of the hydride transfer from NADPH, was studied using [4R-4-^{2}H]-NADPH, and converting the unstable enzymatic product, tetrahydrodipicolinate, to the stable *meso*-diaminopimelate (DAP) by the action of diaminopimelate dehydrogenase. These results were used to support the chemical mechanism shown in Figure 7.

The reaction is initiated by hydride ion transfer from the C4R position of NADPH to the C4 position of DHDP. There is no primary deuterium kinetic isotope effect exhibited by [4R-4-^{2}H]-NADPH on the maximum velocity, suggesting that hydride transfer is rapid relative to other steps. The intermediate eneamine may be stabilized by interaction with appropriately positioned groups, and tautomerization and protonation at C3 yields the product THDP. A large solvent kinetic isotope effect on the maximum velocity suggests that protonation of the eneamine may be rate limiting, and supports a stepwise mechanism for the reduction.

The three-dimensional structure of the recombinant *E. coli* enzyme in complex with NADPH was initially solved and refined to 2.2-Å resolution (Scapin et al., 1995). In solution, *E. coli* DHPR has been reported to be tetramer of identical subunits, and although DHPR crystals contain one molecule per asymmetric unit, the tetramer can be generated by crystallographic 222 symmetry. Figure 8 shows a ribbon diagram of the *E. coli* DHPR monomer (*a*) and tetramer (*b*). The *E. coli* DHPR monomer is composed of two domains: an N-terminal dinucleotide binding domain and a C-terminal substrate binding domain. The dinucleotide binding domain is composed of four α-helices ([A1–A3 and A6 in Fig. 8(*a*)] and 7 β-strands (B1–B6 and B11) making up a Rossman-fold.

The dinucleotide is bound at the C-terminal end of the β-strands, in an extended conformation. Atoms of the nicotinamide mononucleotide are part of an extensive hydrogen-bond network that define the orientation of the nicotinamide ring with respect to the substrate-binding region and confirm the determined stereochemistry of hydride transfer (Reddy et al., 1995). The adenine ring makes few interactions with the enzyme. The *E. coli* enzyme exhibits quite unusual pyridine nucleotide substrate specificity, showing only modest selectivity for phosphorylated and nonphosphorylated dinucleotide substrates (Tamir and Gilvarg, 1974). The DHPR uses NADH slightly better than NADPH (the relative V/K value for NADH is approximately two-fold higher). The interaction of *E. coli* rDHPR with different pyridine nucleotide cofactors was further investigated

294

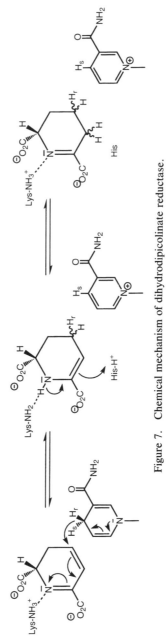

Figure 7. Chemical mechanism of dihydrodipicolinate reductase.

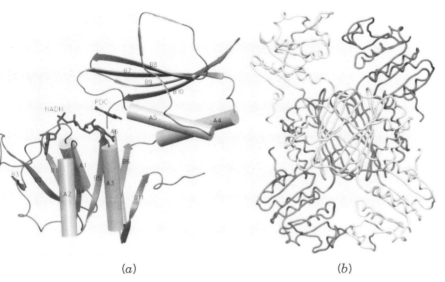

(a)   (b)

Figure 8.   Three-dimensional structures of the *E. coli* dihydrodipicolinate reductase monomer (*a*) and tetramer (*b*) (coordinate filename 1DIH.pdb.)

using isothermal titration calorimetry to determine the thermodynamic parameters of binding. In addition, a structural analysis of the binary enzyme–nucleotide complexes was performed to determine the molecular basis for the different affinities (Reddy et al., 1996b). In the adenosyl ribose region, both a basic residue (Arg-39) capable of interacting with the negatively charged phosphate of NADPH, and an acidic residue (Glu-38), capable of interacting with the 2′- and 3′-hydroxyls of NADH are observed (Reddy et al., 1996b). This region of the enzyme, representing the loop connecting B2 and B3, and the length of the peptide connecting the two halves of the Rossman-fold varies among the *dap*B sequences of various organisms. The Arg-39 residue is present in the aligned sequences of the *E. coli, H. influenzae*, and *P. syringae* reductases, while it is absent from the reductases of *M. tuberculosis, M. leprae, B. lactofermentum*, and *C. glutamicum* (Pavelka et al., 1997). All enzymes have the conserved acidic residue corresponding to the *E. coli* Glu-38 position. This finding suggests that the presence of both the acidic and basic residues determines the dual specificity, and predicts that the

H. influenzae and *P. syringae* reductases will exhibit dual specificity, while those enzymes having only the Glu-38 residue will be NADH specific.

The C-terminal substrate-binding domain contains an open, mixed β-sandwich, formed by two α-helices and four β-strands (A4 and A5, and B7–B10 in Fig. 8a). Residues of the substrate-binding domain are additionally involved in the formation of the tetramer: The 16 β-strands form a flattened central β-barrel, surrounded by eight α-helices and anchored by four loops, each one belonging to one monomer and wrapping around the mixed β-sheet of the neighboring monomer. The four dinucleotide-binding domains extend from the core region of the tetramer, and from their considerably higher temperature factors, compared to the core region, they appear to be rather flexible. In all the crystal structures of the binary complexes, the distance between the nicotinamide C4 and the putative DHDP binding domain (8 Å) is too long for hydride transfer. It was proposed (Scapin et al., 1995) that a hinge movement around the two short loops connecting the two domains could be responsible for a domain reorientation that would allow for catalysis. This proposal has recently received support from hydrogen-exchange studies of the reductase in the absence and presence of substrates and inhibitors. Analysis of hydrogen exchange by electrospray ionization mass spectrometry has identified certain regions of the enzyme whose exchange kinetics are slowed in the presence of NADH and the inhibitor, 2,6-pyridine dicarboxylate (2,6-PDC; Wang et al., 1997). In addition, regions identified in the three-dimensional structure as representing the hinges between the two domains were shown to exhibit reduced exchange in the presence of both NADH and 2,6-PDC, suggesting that the domains close following substrate binding.

Recently, the three-dimensional structure of the ternary complex of *E. coli* DHPR with NADH and the inhibitor 2,6-PDC has been solved to 2.6-Å resolution (Scapin et al., 1997). This crystal form contains one tetramer per asymmetric unit. Although the structure of the N- and C-terminal domains are very similar to that of the binary complex, their relative position are different in the binary and ternary complex. Overlay of either the N- or the C-terminal regions shows a rigid body movement of the two domains around two hinge regions (Phe-129 and Ser-239). The sequential binding of the nucleotide and inhibitor causes a rotation of the dinucleotide-binding domain of about 16° with respect to the C-terminal domain, bringing the nicotinamide ring and the bound 2,6-PDC to within 3.5 Å. The inhibitor makes a number of hydrogen-bonding interactions with main- and side-chain atoms of residues of the DAP binding motif (His-160, Lys-163, Ser-169, Thr-170). The side-chain Nε of His-159 is located about 4 Å away from carbon C3 of the inhibitor, and it may represent the general acid in the reaction. Alternatively, a water molecule that has been located in the binding site 3.0 Å from the His-159 Nε and 4.0 Å from the C3 atom of PDC may act as general acid, upon "activation" by His-159. Site-directed mutants of His-159 and Lys-163 have been prepared and support the roles described (Scapin et al., 1997) (Fig. 9).

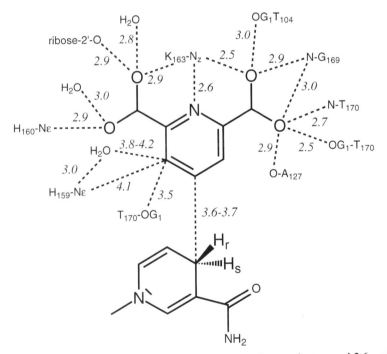

Figure 9.   Interactions between *E. coli* dihydrodipicolinate reductase and 2,6-pyridine dicarboxylate. Distances are in angstroms.

V. Tetrahydrodipicolinate *N*-Succinyltransferase

The activity of the *dapD*-encoded tetrahydrodipicolinate *N*-succinyltransferase (EC 2.3.1.117) commits intermediate flux into the widely distributed succinylase pathway of L-lysine biosynthesis. The earliest suggestion that the cyclic product of the THDP reaction might be converted into a stable, acyclic intermediate, came in 1959 from the isolation and identification of *N*-succinyl-diaminopimelate (Gilvarg, 1959) from a mutant *E. coli* strain that accumulated this intermediate. This finding suggested that succinylation of the α-amino group of the acyclic form of THDP, L-2-amino-6-keto-pimelate, would provide the free keto group for subsequent transamination, a reaction demonstrated in 1961 (see below). The *dapD* gene was thought to be present at approximately 4 min on the *E. coli* chromosome, and in 1984, the gene was cloned and sequenced (Richaud et al., 1984). The gene encoded an open reading frame of 274 amino acids (monomer molecular weight 30,040), and expression of the gene resulted in increased succinyl transferase activity.

The native *E. coli* enzyme has been purified and extensively characterized (Simms et al., 1984). The purification to homogeneity required a 1900-fold purification from wild-type *E. coli*, and the subunit molecular weight by SDS–PAGE analysis was about 31,000 Da. The purified enzyme was shown to catalyze the reversible transfer of the succinyl group from succinyl-CoA to tetrahydrodipicolinate, although the forward, physiologically important reaction was 380 times faster than the reverse reaction at the pH optimum of 8.2. The enzyme appeared to be sensitive to thiol oxidation, and the succinyltransferase reaction could be inhibited by cobalt and copper ions, as well as the mercurial, pCMBS. The reported K_m values for THDP and succinyl-CoA were 24 and 15 μM, respectively. The enzyme was proposed to exist as a homodimer, based on the observed sedimentation coefficient in sucrose density gradients.

The specificity of the enzyme for substrates and inhibitors was subsequently investigated (Berges et al., 1986a). L-2-Aminopimelate was identified as an alternate substrate which, lacking the 6-keto group, existed exclusively in an acyclic form (Fig. 10). Similarly, the 6-hydroxyl derivative was shown to be as good a substrate as 2-aminopimelate, but longer or shorter chain homologs had essentially no activity. The cyclic analogue, 3,4-dihydro-2*H*-1,4-thiazine-3,5-dicarboxylate, was shown to be a good substrate for the succiny-

tetrahydrodipicolinate

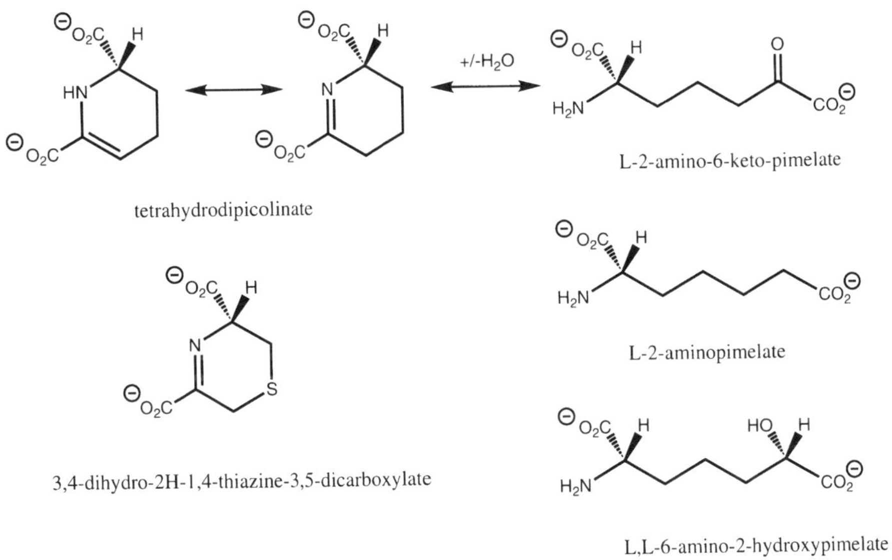

L-2-amino-6-keto-pimelate

3,4-dihydro-2H-1,4-thiazine-3,5-dicarboxylate

L-2-aminopimelate

L,L-6-amino-2-hydroxypimelate

Figure 10.   Substrates for tetrahydrodipicolinate/succinyl-CoA N-succinyltransferase.

lase, and like the natural substrate could exist in a number of cyclic and acyclic tautomeric forms (imine, eneamine, keto, hydrated keto). This data, plus the measured slow rate of ring opening of THDP by reaction with trinitrobenzenesulfonate, suggested that the enzyme must catalyze the ring-opening reaction of cyclic imine substrates.

Support for this proposal came from the finding that 2-hydroxytetrahydropyran-2,6-dicarboxylate (2-HTHP) was a potent competitive inhibitor versus THDP, exhibiting a K_i value of 58 nM, 400 times lower than the K_m for THDP. These authors (Berges, et al., 1986a) proposed that 2-HTHP was acting as a transition state analogue of the hydration reaction, and proposed a reaction mechanism that incorporated this feature (Fig. 11). Binding of the cyclic imine, followed by stereospecific hydration, yields the carbinolamine shown below. The obvious similarity of this intermediate to 2-HTHP rationalizes the strong inhibition exerted by this compound. Succinyl transfer from succinyl-CoA to the secondary amine generates the

300

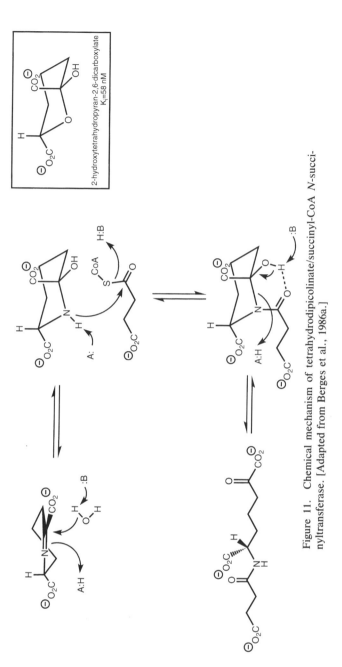

Figure 11. Chemical mechanism of tetrahydrodipicolinate/succinyl-CoA *N*-succinyltransferase. [Adapted from Berges et al., 1986a.]

N-succinylated intermediate, which decomposes to generate the *N*-succinylated, α-ketoacid product.

The three-dimensional structure of the *N*-succinyltransferase has recently been reported at 2.2 Å (Beaman et al., 1997). Primary sequence studies had earlier noted that the transferase was homologous to a family of enzymes that bore a signature motif, termed the "hexapeptide repeat," or leucine patch (Vaara, 1992). The principal feature of this motif is the presence of multiply repeated sequences, [LIV]-[GAED]-X-X-[STAV]-X, which has been structurally characterized in two previous cases, the *E. coli* UDP-*N*-acetylglucosamine acyltransferase (Raetz and Roderick, 1995) and the *Methanosarcina thermophila* carbonic anhydrase (Kisker et al., 1996). The homologous hexapeptide repeat sequences of both of these enzymes form an unusual left-handed β-helix. Both enzymes exist as trimers, as does the *dapD*-encoded tetrahydrodipicolinate-*N*-succinyl transferase. The main chain trace of one monomer of the *N*-succinyltransferase and the active trimer are shown below (Fig. 12). The active site was proposed to be at each monomer–monomer interface, since this is where both pCMBS and cobalt(II), inhibitors of the reaction, were

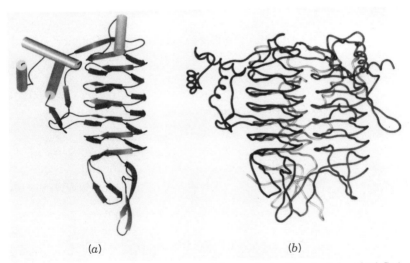

(a)   (b)

Figure 12.   Three-dimensional structure of the tetrahydrodipicolinate/succinyl-CoA *N*-succinyltransferase monomer (*a*) and trimer (*b*). (Coordinate filename 1TDT.pdb.)

bound. More recent studies of the enzyme crystallized in the presence of 2-aminopimelate and CoA have confirmed the presence of these two molecules at the monomer–monomer interface. The CoA molecule assumes a "J-shaped" conformation, with multiple interactions between the 3'-phosphate and pyrophosphoryl groups of CoA and a number of lysine and arginine residues from both monomers. The pantothenyl moiety is extended in a completely "trans" conformation, and lies parallel to the long axis of the β-helix and perpendicular to the parallel β-strands. Similarly, L-2-aminopimelate is bound in an extended conformation, with the α-amino group a reasonable distance from the thiol of CoA. There is no evidence for the binding of the D isomer, in spite of the use of the racemate in crystallization, nor is there any obvious way to accommodate cyclic inhibitors, such as 2-HTHP (Fig. 12).

Two significant conformational changes accompany substrate binding, the movement of a small loop that interacts with the amino acid substrate, and the massive ordering of the carboxy terminal 18 amino residues. The carboxyl terminus is barely visible in the structure of the unliganded enzyme, but appears as clear electron density in the ternary enzyme–CoA-2-aminopimelate complex. With the exception of the above noted carbonic anhydrase, the vast majority of enzymes that exhibit the "hexapeptide repeat" motif catalyze acyl group transfer to either amines or alcohols, and the conserved left-handed β-helix structural motif may represent a common pantothenyl-binding structure. Additional structural studies will confirm this hypothesis.

VI. *N*-Succinyl-L,L-Diaminopimelate Aminotransferase

Almost nothing is known about the aminotransferase that catalyzes the transfer of the amino group from L-glutamate to *N*-succinyl-L-2-amino-6-ketopimelate to generate ketoglutarate and *N*-succinyl-L,L-diaminopimelate (EC 2.6.1.17). Furthermore, no sequence of any *dapC* gene has been reported from any source, although several reports have incorrectly identified the *dapD* gene as the *dapC* gene. The enzyme was first partially purified from *E. coli* (Peterkofsky and Gilvarg, 1961), and could be purified approximately 110-fold (Peterkofsky, 1962). More recently, the enzyme has been purified 1500-fold to homogeneity (Cox et al., 1996). The enzyme has a subunit mass of 39,896, and exists as a homodimer. The enzyme is absolutely specific for L-glutamate as the amino donor, but will use a number of N-acyl ketopimelates. Reported K_m and k_{cat} values for the natural substrate, N-succinyl-L-2-amino-6-ketopimelate are 180 μM and 86^{-1} (Cox et al., 1996). A reasonable chemical mechanism for the reaction is shown in Figure 13.

Hydrazino analogues of N-acyl-diaminopimelates are extremely potent inhibitors of the transaminase. Both the N-succinyl-α-hydrazino- and N-Cbz-α-hydrazino analogues were slow, tight-binding inhibitors, exhibiting K_i^* values of 22 and 54 nM, respectively. Although the mechanism of action was not investigated, a possible mechanism involves the formation of the stable PLP-inhibitor hydrazone.

VII. N-Succinyl-L,L-Diaminopimelate Desuccinylase

Along with the preceding enzyme in the lysine biosynthetic pathway, the dapE-encoded N-succinyl-L,L-diaminopimelate desuccinylase (E.C.3.5.1.18) is poorly characterized. The substrate of the enzyme was first identified in 1959 (Gilvarg, 1959), but the enzyme was not purified until quite recently (Gelb et al., 1990). Enzyme levels do not appear to be regulated by lysine levels in the media, and the enzyme was purified to homogeneity from E. coli K12 grown in rich media. The enzyme was purified 7100-fold, and appeared as a mixture of dimeric and tetrameric 40,000-Da monomers. The enzyme requires a divalent metal ion for activity, with cobalt being more effective than zinc on the basis of a 2.2-fold higher maximum velocity in the presence of cobalt. Steady-state kinetic studies were performed to determine the K_m value for N-succinyl-L,L-DAP of 400 μM, and a maximum velocity of 16,000 min^{-1}. The substrate specificity of the enzyme was evaluated by preparing a number of N-succinylated diaminopimelate isomers and analogues, as shown in Figure 14.

Analogues in which the N-succinylated stereocenter was of the D-configuration were not substrates. Of the four compounds shown in Figure 14, only N-succinyl-L,L- and N-succinyl-D,L-diaminopimelate were reasonable substrates, with the N-succinyl-L-2-aminopimelate and N-succinyl-L-lysine being very poor substrates. A reasonable chemical mechanism is shown below. In this proposed mechanism,

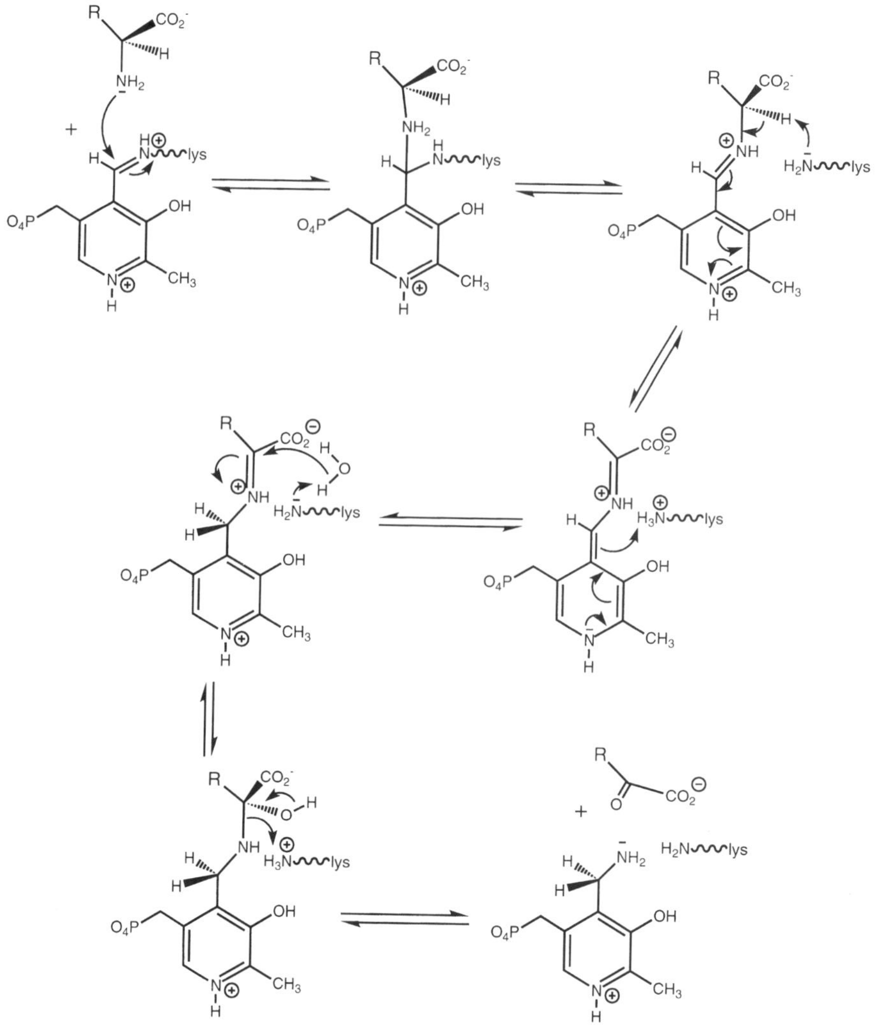

$R = {}^-O_2C\text{-}(CH_2)_2\text{-}$ (L-glutamate)

$= {}^-O_2C\text{-}(CH_2)_2\text{-}CO\text{-}NH\text{-}CH(CO_2{}^-)\text{-}(CH_2)_3\text{-}$ (N-succinyl-L,L-diaminopimelate)

Figure 13.   Chemical mechanism of L-glutamate/N-succinyl-2-amino-6-keto-pimelic acid aminotransferase.

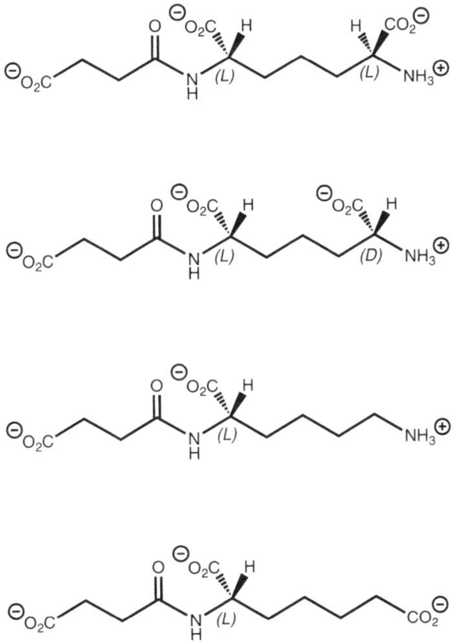

	rel V_{max}	K_m (mM)
	1.00	0.4
	0.3	
	0.00016	
	0.00036	16

Figure 14.   Substrates for N-succinyl-L,L-diaminopimelate desuccinylase.

the metal ion functions to lower the pK value of the attacking water molecule, and the base-catalyzed deprotonation of the metal-bound water generates the active nucleophile. Hydroxide ion attack at the amide carbonyl generates the sp^3-hybridized intermediate, although it is unclear how this tetrahedral intermediate may be stabilized. Productive decomposition of this intermediate is likely to require general acid assistance (Fig. 15) to generate the observed products and regenerate the appropriately deprotonated catalytic residue.

The amino acid sequences of five bacterial desuccinylases have been reported, and alignment of the sequences from *E. coli* (375 amino acids), *H. influenzae* (377 amino acids), *C. glutamicum* (369 amino acids), *M. jannaschii* (410 amino acids), and *M. leprae* (337 amino acids) reveals a highly homologous consensus sequence in the first 100 residues. This consensus pattern has also been identified in several other metal-dependent enzymes that catalyze the hydroly-

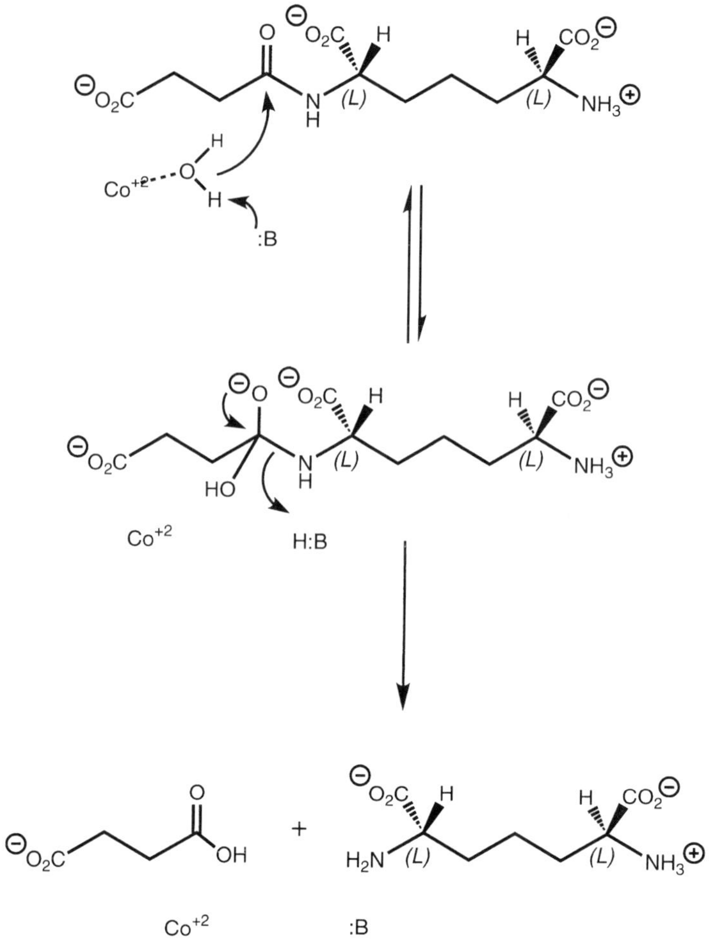

Figure 15. Chemical mechanism of N-succinyl-L,L-diaminopimelate desuccinylase.

sis of amide linkages (Meinnel et al., 1992). These include the *E. coli argE*-encoded acetylornithinase, the *Pseudomonas cpg2*-encoded carboxypeptidase G2, the *Lactobacillus pepV*-encoded carnosinase, the bacterial *pepT*-encoded tripeptidase, and both the yeast *yscS*-encoded carboxypeptidase and the mammalian aminoacylase-1. This finding suggests that the conserved amino acid sequence may be structurally important for metal ligation and catalysis. The three-dimensional structure of the *Pseudomonas cpg2*-encoded carboxypeptidase G2 has very recently been determined (Rowsell et al., 1997) to 2.5 Å. The structure revealed the presence of two zinc atoms at the presumptive active site that share a bridging water/hydroxide ligand, and also demonstrated that the conserved glutamate, aspartate and histidine residues were involved in metal ligation.

VIII. l,l-Diaminopimelate Epimerase

The *dapF*-encoded diaminopimelate epimerase (E.C.5.1.1.7) catalyzes the interconversion of l,l-DAP, the product of the *dapE*-encoded desuccinylase reaction, and d,l-DAP, the substrate for the *lysA*-encoded decarboxylase. The enzyme was initially identified in crude extracts of *E. coli* in 1957 by Work and co-workers using paper chromatographic identification of the formation of d,l-DAP from l,l-DAP (Antia et al., 1957) and only purified to homogeneity from *E. coli* in 1984 (Wiseman and Nichols, 1984). In this latter study, formation of d,l-DAP from l,l-DAP was determined spectrophotometrically by coupling the product to diaminopimelate dehydrogenase, which is specific for the d,l-isomer (see Section X). The homogeneous enzyme was purified 6000-fold in 3% yield and shown to behave as a monomer on gel permeation chromatography. The subunit molecular weight determined by SDS–PAGE was 34,000, as subsequently confirmed by the cloning and sequencing of the gene from *E. coli* (Richaud and Printz, 1988). Although earlier studies had yielded ambiguous data on the presence of pyridoxal phosphate as a cofactor in the epimerization reaction, the homogeneous enzyme exhibited no visible absorbance indicative of the presence of PLP, neither was the reaction stimulated by exogenously added PLP, nor was the reaction inhibited by hydrazine or hydroxylamine, reagents known to inhibit PLP-dependent enzymes. This result suggested that the reaction might be analogous to the previously described reaction catalyzed by proline racemase (Cardinale and Abeles, 1968), in which PLP was shown not to be required. In fact, as was described for proline racemase, a reactive thiol was shown to be essential for the activity of DAP epimerase. Enzyme activity was rapidly lost at pH 8 when the enzyme was stored in the absence of dithiothreitol (DTT), and the enzyme could be rapidly and stoichiometrically

(1.2 labeled iodoacetamide molecules per 34,000 monomer) inactivated by iodoacetamide. The alkylation of the thiol could be protected against by the addition of substrate, suggesting the presence of a cysteine residue at or near the active site. Subsequently, reaction of the purified, recombinant *E. coli* epimerase with the active site-directed aziridine derivative, 2-(4-amino-4-carboxybutyl)-2-aziridine-carboxylic acid, demonstrated the irreversible and covalent inactivation of the epimerase. Tryptic digestion followed by peptide separation and sequencing demonstrated that Cys-73 was alkylated by this compound (Higgins et al., 1989). This residue is conserved in the aligned sequences of four bacterial *dapF*-encoded epimerases (Scapin et al., unpublished).

The specific activity of the *E. coli* enzyme has been reported to be between 108 and 192 units/mg, the latter obtained from the overexpressed enzyme, and likely to be a more reliable value. The detailed kinetic evaluation of the reaction has been reported for the wild-type enzyme. In the D,L to L,L-DAP direction, the k_{cat} was 67 s^{-1} and the K_m for D,L-DAP is 360 μM. In the L,L- to D,L-DAP direction, the k_{cat} was 84 s^{-1} and the K_m for L,L-DAP is 160 μM. The Haldane equation requires that the ratio of the k_{cat}/K_m values equal the equilibrium constant for the reaction. The ratio calculated from these kinetic constants (2.8) is close to the observed equilibrium constant of 2, reflective of the unique statistical distribution of isomers. Monotritiated D,L-DAP, specifically labeled at the D position, was prepared using DAP dehydrogenase and [^{3}H]NADPH, and used to measure the washout of tritium from the D position during epimerization. In addition to demonstrating a tritium kinetic isotope effect of about 5 on the washout of tritium into solvent, intermolecular transfer of tritium into the product, L,L-DAP could not be demonstrated (Wiseman and Nichols, 1984). While not unambiguous, these data were the first data to argue against a one-base mechanism. The mechanistic distinction between one- and two-base isomerization, racemization, and epimerization mechanisms were first noted by Rose (1966), and the different patterns of hydrogen exchange were used to characterize the two-base mechanism of proline racemase (Cardinale and Abeles, 1968; Rudnick and Abeles, 1975). In the case of DAP epimerase, tritium from solvent is incorporated into product four times more rapidly than into substrate, regardless of whether one begins with L,L- or D,L-DAP. Quantitative analysis of the kinetics of tritium incorporation allowed the authors to calculate primary tritium kinetic isotope effects for the cleavage of the α-C—H bond from either the D- or L-amino acid center. They calculated $^{T}V/K$ values for cleavage of the D isomer equal to 5.9 and for cleavage of the corresponding L isomer of 3.3, and concluded that the difference in these values reflected different transition state fractionation factors for the proton abstraction step. These authors (Wiseman and Nichols, 1984) reasoned that the two bases responsible for either L- or D- α-proton abstraction were chemically different, and suggested that the uniquely low fractionation factor of the catalytic thiol, relative to other potential bases (e.g., carboxylates or imidazole), could account for this difference. A reasonable chemical mechanism can be formulated from these results as depicted in Figure 16.

Because of the carbanionic nature of the intermediate or transition state for the epimerization reaction, the synthesis and evaluation of 3-fluorodiaminopimelate isomers as inhibitors of diaminopimelate epimerase was performed (Gelb et al., 1990). Of the four isomers prepared (Fig. 17), the two that contain the (3S)-fluoro substituent were good substrates and were rapidly epimerized, but eliminated HF only very slowly. On the other hand, the two isomers that contained the (3R)-fluoro substituent were not epimerized, but rapidly eliminated HF. The mechanistic consequence of HF elimination from either the L,L- or D,L-(3R)-3-fluoro-DAPs were the generation of the achiral eneamine, which after tautomerization to the imine, and hydrolysis, generated the L-2-amino-6-ketopimelate (also referred to as THDP in its cyclic form; see above). The demonstration of the formation of this compound relied on the use of diaminopimelate dehydrogenase in the presence of ammonia and NADPH to reductively aminate the 6-keto group to generate D,L-DAP. These experiments add an additional layer of stereochemical complexity to an already stereochemically stringent enzyme reaction.

Few effective inhibitors have been designed and tested against the purified epimerase. It is not clear that the *dapF*-encoded epimerase is an essential gene in *E. coli*, since its disruption has no obvious lethal consequence to the organism (Richaud et al., 1987). While such mutants accumulate L,L-DAP intracellularly, it has been proposed that other enzymes may act to epimerize the L,L-DAP to generate suffi-

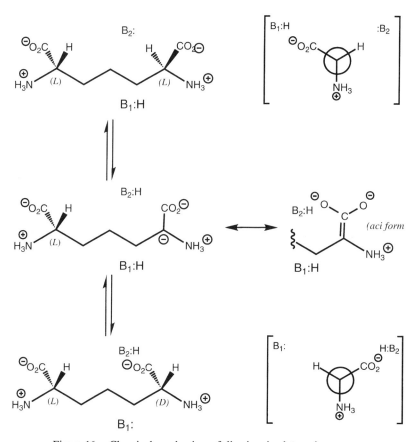

Figure 16.   Chemical mechanism of diaminopimelate epimerase.

cient D,L-DAP for incorporation into the peptidoglycan component of the cell wall. As described above, the aziridine derivative is a potent, active-site directed irreversible inhibitor of the epimerase, although its antibacterial properties have not been explored.

IX.   D,L-**Diaminopimelate Decarboxylase**

The activity of the *lysA*-encoded D,L-diaminopimelate decarboxylase (E.C.4.1.1.20) was first demonstrated in crude extracts of *E. coli* in 1952 (Dewey and Work, 1952). The enzyme has been purified from a number of bacteria and plants, and catalyzes the only PLP-

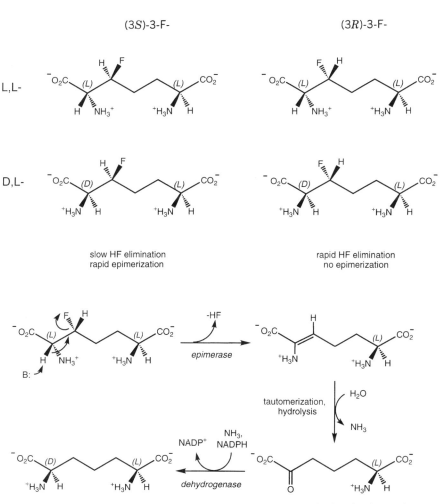

Figure 17. Reaction of fluorinated diaminopimelates with diaminopimelate epimerase. [Adapted from Gelb et al., 1990.]

dependent decarboxylation of an α-amino acid of D configuration. In addition to this unique substrate stereospecificity, it has been shown that both the *B. sphaericus* and wheat germ decarboxylases catalyze the decarboxylation reaction with inversion of configuration (Asada et al., 1981, Kelland et al., 1985). This is in contrast to all other stereochemical outcomes for PLP-dependent decarboxylases.

Essentially no bona fide mechanistic studies have appeared for dia-minopimelate decarboxylase, in spite of renewed interest in the potential antibacterial properties of inhibitors of this enzyme (see below).

In the last dozen years, a number of sequences of bacterial *lysA* genes have appeared, including those from *E. coli* (Stragier et al., 1983), *C. glutamicum* (Yeh et al., 1988b), *P. aeruginosa* (Martin et al., 1988), *B. subtilis* (Yamamoto et al., 1989), and *M. tuberculosis* (Andersen and Hansen, 1993). In addition, the *lysA* genes of *H. influenzae* and *M. jannaschii* have been identified from genome se-quencing projects by homology to those sequences reported above (Fleischmann et al., 1995; Bult et al., 1996). Alignment of these bac-terial decarboxylase sequences reveals three conserved lysine resi-dues, one of which is likely to be the lysine that binds the PLP cofactor in an aldimine linkage. The bacterial diaminopimelate decar-boxylase sequences are also homologous to other basic amino acid decarboxylases, including both bacterial and mammalian ornithine decarboxylases and arginine decarboxylase. An extensive alignment of PLP-dependent enzymes has been used to separate all such en-zymes into four types, based on primary sequence homology (Alex-ander et al., 1994). Diaminopimelate decarboxylase is a member of the third type, which includes eukaryotic ornithine decarboxylases, plant and bacterial arginine decarboxylases, and bacterial alanine racemases (Grishin et al., 1994). Both sequence alignment and sec-ondary structural predictions have been used to suggest that the type 3 PLP-containing enzymes assume an eight stranded α/β barrel. The recent determination of the high-resolution three-dimensional struc-ture of *B. stearothermophilus* alanine racemase (Shaw et al., 1997) provides support for this prediction. The racemase consists of an amino-terminal α/β barrel domain composed of residues 1–240, and a carboxyl-terminal composed of β strands. The lysine residue in-volved in Schiff base formation with the cofactor is located at the carboxyl terminus of the first of the eight central beta strands that make up the barrel.

The lysine involved in Schiff base formation in mouse ornithine decarboxylase is Lys-69, equivalent to Lys-58 in the *E. coli* diami-nopimelate decarboxylase. A reasonable chemical mechanism for diaminopimelate decarboxylase is shown in Figure 18. By precedent with all PLP-dependent enzymes, the first step is the transimination

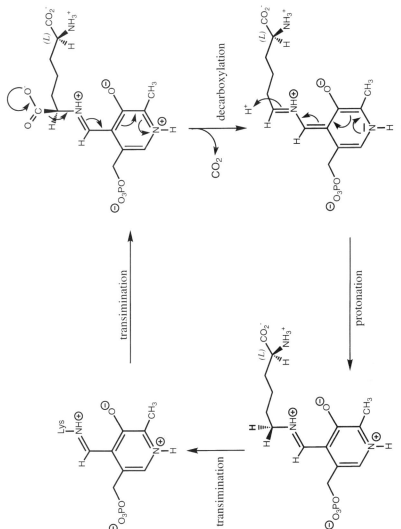

Figure 18. Chemical mechanism of diaminopimelate decarboxylase.

313

sequence leading to the replacement of the internal aldimine linkage between the cofactor and Lys-58 with the corresponding aldimine linkage between the amino acid substrate and PLP. Decarboxylation proceeds from a complex in which the carboxyl group is oriented such that the carboxyl-α-carbon bond is perpendicular to the plane of the conjugated π system. The carbanion generated as a result of decarboxylation is stabilized by resonance delocalization of the electron pair into the pyridinium ring nitrogen, forming the quininoid tautomer. This charge neutralization is assisted by a strong hydrogen bond between the pyridinium nitrogen and a conserved aspartate residue in PLP-dependent enzymes. There are six conserved acidic residues among the diaminopimelate and ornithine decarboxylases, and recent mutagenesis studies of eukaryotic ornithine decarboxylases suggests that the glutamate residue equivalent to E268 in the *E. coli* sequence interacts with the pyridinium nitrogen of the cofactor (Osterman et al., 1995). The quininoid tautomer is protonated with overall stereochemical inversion, leading to the formation of the L-lysine-PLP imine, which is transiminated prior to product release.

The chemistry catalyzed by PLP-dependent enzymes has made them the subject of numerous mechanism-based inhibitor studies. Based on similar studies of ornithine decarboxylase, fluoromethyl-diaminopimelates have been synthesized and shown to be inhibitors of the epimerase (see above). The β,γ-unsaturated *meso*-diaminopimelate has been synthesized and shown to be a modest inhibitor of the decarboxylase (Giroden et al., 1986). More recently, inhibitors in which the carboxylic moiety undergoing decarboxylation has been replaced by the corresponding phosphonate have been prepared and shown to inhibit the decarboxylase, as well as the dehydrogenase and epimerase (Song et al., 1994). Unfortunately, these compounds are both poor inhibitors of the enzyme and exhibit negligible antibacterial activity against *E. coli* or *Salmonella*. However, tripeptides containing the phosphonate analogue exhibit some antibacterial activity, reminiscent of successful antibacterial studies in which L-2-aminopimelate was incorporated into tripeptides, presumably to enhance transport of these molecules across the cell wall and membrane (Berges et al., 1986b). The determination of the three-dimensional structure of the decarboxylase may allow some of these studies to be resuscitated.

X. D,L-Diaminopimelate Dehydrogenase

In the dehydrogenase variant of the lysine biosynthetic pathway, the intermediate tetrahydrodipicolinate, common to all three pathways, is converted in a single step to *meso*-diaminopimelate by *meso*-DAP dehydrogenase (DapDH; EC 1.4.1.16). The DapDH belongs to the family of amino acid dehydrogenases, a group of $NAD(P)^+$-dependent enzymes that catalyze the oxidative deamination of an amino acid to its ketoacid and ammonia (reviewed in Brunhuber and Blanchard, 1994). The DapDH represents the most unusual of these dehydrogenases, being uniquely specific for a D amino acid stereocenter. This enzyme activity was first reported from *B. sphaericus* (Misono et al., 1976), suggesting that this variant of the lysine biosynthetic pathway was present in *B. sphaericus* (Misono et al., 1979), a report subsequently confirmed (White, 1983) and shown to be the principal route for lysine synthesis in *B. sphaericus*. In other bacteria, including *C. glutamicum*, the dehydrogenase pathway exists, as does the more common succinylase pathway (Schrumpf, 1991). In *C. glutamicum*, 30% of lysine is synthesized via the dehydrogenase pathway, while the remainder is derived from succinylated intermediates (Sonntag, 1993). The dehydrogenase pathway is not widely present in bacteria, and DapDH activity has only been found in some *Brevibacteria, Pseudomonas, Bacilli*, and *Corynebacteria* species.

The DapDH has been isolated, purified, and enzymatically characterized from *B. sphaericus, Brevibacterium* sp., and *C. glutamicum* (Misono and Soda, 1980; Misono et al., 1986a; Misono et al., 1986b). These bacterial dehydrogenases have been shown to be homodimers with a native molecular weight of about 70,000 Da, and are highly specific for *meso*-DAP and NADPH as substrates. L,L- and D,D-diaminopimelate are competitive inhibitors versus *meso*-DAP for the *Bacillus* and *Corynebacterium* enzymes (Misono and Soda, 1980; Misono et al., 1986b) with K_i values in the millimolar range. The *Brevibacterium* and the *Corynebacterium* enzymes can use NAD^+ as a slow substrate (3% of the activity observed with $NADP^+$), while the *Bacillus* enzyme exhibits no activity with NAD^+ (Misono and Soda, 1980).

The majority of kinetic and mechanistic study work has been performed with the *Bacillus* enzyme (Misono and Soda, 1980), but some characterization of the dehydrogenases from *Brevibacterium* and *Corynebacterium* has been reported (Misono et al., 1986a,b). All enzymes exhibit similar Michaelis constants for substrates and products, and similar pH optima for both the oxidative deamination and reductive amination reactions (10.5, 9.8, 10.5 and 7.5, 7.6, and 8.5 for the *Bacillus, Corynebacterium*, and *Brevibacterium* enzymes, respectively; Misono and Soda 1980b; Misono et al., 1986a,b). For the *Bacillus* enzymes, the equilibrium constant was determined to be $4.5 \times 10^{-14} M^2$ at pH 7.29 and 25°C, making diaminopimelate synthesis energetically favorable. All three enzymes are inactivated by bulky mercurials and iodoacetic acid, suggesting the presence of a reactive cysteine at or near the active site. However, modification of the sulfhydryl group of the *Bacillus* enzyme with small and uncharged thiol alkylating reagents had no affect on the activity of the enzyme, suggesting that the sulfhydryl group does not play a role in catalysis (Misono et al., 1981). For both the *Bacillus* and the *Corynebacterium* dehydrogenases, steady-state kinetic studies support an ordered kinetic mechanism, with $NADP^+$ binding followed by *meso*-DAP binding, and ammonia, L-2-amino-6-ketopimelate and NADPH released in this order. No detailed mechanistic studies have been reported, and it has been assumed that the reaction mechanism is similar to those previously reported for other amino acid dehydrogenases. For the *Bacillus* enzyme, the stereospecificity of hydride transfer has been investigated using stereospecifically labeled [4S-4-^{3}H]-NADPH. Based on the magnitude of the tritium kinetic isotope effect, the authors suggested that the enzyme catalyzed the transfer of the *pro-S* hydrogen atom, although the transfer of the [^{3}H] label to the product, diaminopimelate, was not demonstrated (Misono and Soda, 1980) (Fig. 19).

Inhibitors have been synthesized and tested with the *B. sphaericus* diaminopimelate dehydrogenase, using compounds that were either substrate analogues or potential transition state analogues for the enzyme (Lam et al., 1988). These studies have confirmed previous reports of the strict specificity for *meso* isomers and the undesirability of substitutions at C3—C5 (Misono and Soda, 1980). A recent report (Abbot et al., 1994) describing the synthesis and testing of heterocyclic diaminopimelate derivatives, designed as conformationally restricted transition state analogues, showed these com-

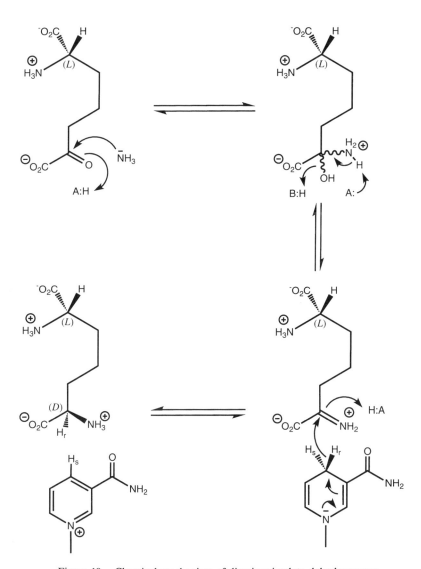

Figure 19.   Chemical mechanism of diaminopimelate dehydrogenase.

pounds to be potent inhibitors of the dehydrogenase. One isoxazoline isomer is a potent inhibitor of the dehydrogenase reaction in
both the forward and reverse direction at pH 7.8, exhibiting a K_i
value in the micromolar range, compared to millimolar values for
the K_m of diaminopimelate. This isoxazoline is a noncompetitive
inhibitor versus diaminopimelate and a competitive inhibitor versus
THDP. At higher pH values (10.5), the isoxazoline K_i value increases
dramatically.

The only primary sequence of a diaminopimelate dehydrogenase
has been determined for the *C. glutamicum* enzyme, whose *ddh* gene
has been cloned and sequenced (Ishino et al., 1987). The *ddh* gene
encodes the 320 amino acid dehydrogenase (molecular weight =
35,200 Da), which has been overexpressed in *E. coli*, purified to
homogeneity, and crystallized (Reddy et al., 1996). The three-dimensional structure of the binary enzyme-NADP$^+$ complex has been
reported (Scapin et al., 1996), and the structure of the enzyme–*meso*-
DAP complex has also been solved (Scapin et al., 1998). The crystalline enzyme is a homodimer with each monomer composed of three
domains; an amino-terminal dinucleotide binding domain, a carboxyl-terminal substrate binding domain, and a third domain that
has been shown to be involved in dimer formation (Fig. 20). The
relative positions of the substrate-binding and dinucleotide binding
domains are different in the NADP$^+$ and *meso*-DAP complexes.

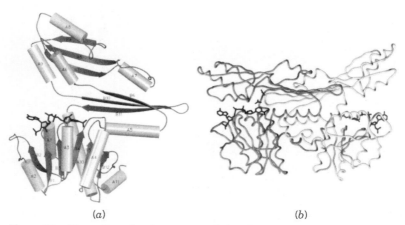

(a)   (b)

Figure 20. Three-dimensional structure of the diaminopimelate dehydrogenase
monomer (a) and dimer (b). (Coordinate filename 1DAP.pdb.)

In the enzyme–NADP$^+$ complex, the protein assumes an "open" conformation that has been described as a "binding" conformation. In the enzyme–*meso*-DAP complex, the protein assumes a "closed" conformation due to the relative movement of the nucleotide and substrate binding domains toward each other, and has been ascribed to an "active" conformation of the enzyme (Scapin et al., 1996). NADP$^+$ is bound at the C-terminal end of the dinucleotide binding domain, and the 2'-phosphate group makes seven hydrogen-bonding interactions with protein atoms. This large number of interactions provides an explanation for the high specificity of *C. glutamicum* diaminopimelate dehydrogenase for NADPH, as has been described for other NADPH-specific enzymes (Karplus and Schulz, 1989). The nicotinamide carboxamido group is rigidly bound within a pocket, and the C4R hydrogen faces the substrate binding domain. This result suggest that the *C. glutamicum* diaminopimelate dehydrogenase catalyzes the transfer of the (4R) hydrogen to substrate, in contrast with the *pro-S* stereospecificity reported for the *B. sphaericus* enzyme (Misono and Soda, 1980).

The bound *meso*-diaminopimelate is found within a cavity formed by main c- and side-chain atoms of residues that belong to all three protein domains (Scapin et al., 1998). In the closed conformation, the D stereocenter of diaminopimelate is located about 2.8 Å from the C4R of the dinucleotide. The hydrogen-bond donor–acceptor composition of the binding site only allows one possible orientation for the D stereocenter of *meso*-diaminopimelate. The L-amino acid stereocenter binding pocket shows a similar arrangement of hydrogen-bond donors and acceptors, which permits the discrimination between L and D enantiomers at this site.

Diaminopimelate dehydrogenase shares a similar fold to the functionally related bacterial amino acid dehydrogenase, glutamate dehydrogenase (Baker et al., 1992) and leucine dehydrogenase (Baker et al., 1995). All three enzymes are composed of two domains, one responsible for nucleotide binding and the other for amino acid substrate binding, separated by a cleft identified as the active site. However, while in both glutamate and leucine dehydrogenases the carboxyl terminal domain is the nucleotide binding domain, in the case of DapDH, the amino terminal domain represents the nucleotide binding domain. In fact, DapDH shows the highest degree of structural organization and similarity to the *E. coli* dihydrodipicolinate reductase (DHPR), the enzyme immediately preceding DapDH in the diaminopimelate/lysine biosynthetic pathway (Scapin et al., 1996). This structural homology is true despite the fact that the two enzymes share only 17% sequence identity. The reactions carried out by dihydrodipicolinate reductase and diaminopimelate reductase are similar, catalyzing the pyridine nucleotide-dependent reduction of the cyclic imine, tetrahydrodipicolinate, or the imine generated from ammonia and 2-amino-6-keto-pimelate, respectively. Both enzymes position their reducible substrates with respect to the bound nucleotide to allow for transfer of the *pro-R* hydrogen atom as a hydride ion. These structural, mechanistic, and stereochemical similarities between the two enzymes suggest that they may be evolutionarily related.

XI. Summary

The widespread occurrence of the bacterial biosynthetic pathway described in the preceding sections has suggested to many authors that inhibitors of the enzymes in this pathway may have antibacterial properties (Girodeau et al., 1986). The absence of the enzymes in this pathway in mammals and the dual biosynthetic importance of the pathway, in both protein and cell wall biosynthesis, bolster these arguments. Recent insertional inactivation of the *ask* gene has demonstrated that the biosynthesis of diaminopimelate is an essential function in *Mycobacterium smegmatis*, and that organisms lacking the *ask* gene product suffer "DAP-less" death (Pavelka and Jacobs, 1996). The demonstrated clinical effectiveness of inhibitors of enzymes involved in cell wall biosynthesis, particularly peptidoglycan synthesis, suggests that a more thorough enzymological and structural characterization of these unique enzymes may yield new classes of inhibitors with therapeutic efficacy. The recent appearance of multiply antibiotic-resistant bacterial pathogens increases the urgency to find new and effective antibacterial compounds, with broad spectrum antibacterial activity.

Acknowledgments

We would like to acknowledge the National Institutes of Health for continuing support of this work (AI33696). We would also like to thank Dr. S. Blickling, Dr. R. Huber, Dr. W. Karsten, and Dr. R. Viola for providing us with unpublished information.

References

Abbot, S. D., Lane-Bell, P., Sidhu, K. P. S., and Vederas, J. C., *J. Am. Chem. Soc.*, **116**, 6513 (1994).

Alexander, F. W., Sandmeier, E., Mehta, P. K., and Christen, P., *Eur. J. Biochem.*, **219**, 953–960 (1994).

Andersen, A. B. and Hansen, E. B., *Gene*, **124**, 105–109 (1993).

Antia, M., Hoare, D. S., and Work, E. *Biochem. J.* **65**, 448 (1957).

Asada, Y., Tanizawa, K., Sawada, S., Suzuki, T., Misono, H., and Soda, K., *Biochemistry*, **20**, 6881–6886 (1981).

Baker, P. J., Britton, K. L., Engel, P. C., Farrants, G. W., Lilley, K. S., Rice, D. W., and Stillman, T. J., *Proteins: Structure, Function Genet.*, **12**, 75 (1992).

Baker, P. J., Turnbull, A. P., Sedelnikova, S. E., Stillman, T. J., and Rice, D. W., *Structure*, **3**, 693 (1995).

Beaman, T. W., Binder, D. A., Blanchard, J. S., and Roderick, S. L. *Biochemistry* **36**, 489 (1997).

Berges, D. A., DeWolf, W. E., Dunn, G. L., Newman, D. J., Schmidt, S. J., Taggart, J. J., and Gilvarg, C., *J. Biol. Chem.*, **261**, 6160 (1986a).

Berges, D. A., DeWolf, W. E., Dunn, G. L., Grappel, S. F., Newmann, S. J., Taggart, J. J., and Gilvarg, C., *J. Med. Chem.*, **29**, 89 (1986b).

Blickling, S., Renner, C., Laber, B., Pohlenz, H.-D., Holak, T. A., and Huber, R., *Biochemistry*, **36**, 24–33 (1997).

Bouvier, J., Richaud, C., Richaud, F., Patte, J.-C., and Stragier, P., *J. Biol. Chem.*, **259**, 14829 (1984).

Brunhuber, N., M., W., and Blanchard, J. S., *Crit. Rev. Biochem. Mol. Biol.*, **29**, 415–467 (1994).

Bult et al., *Science*, **273**, 1058–1073 (1996).

Cardinale, G. J. and Abeles, R. H., *Biochemistry*, **7**, 3970 (1968).

Chrystal, E. J. T., Couper, L., and Robins, D. J., *Tetrahedron*, **51**, 10241 (1995).

Cirrilo, J. D., Weisbrod, T. R., Banerjee, A., Bloom, B. R., and Jacobs, W. R., Jr., *J. Bacteriol.*, **176**, 4424 (1994).

Cohen, G. N., *Methods Enzymol.*, **113**, 596–599 (1985).

Cox, R. J., Lam, L. K. P., Sherwin, W. A., and Vederas, J. C., *J. Am. Chem. Soc.*, **118**, 7449 (1996).

Cremer, J., Eggeling, L., and Sahm, H., *Mol. Gen. Genet.*, **220**, 478 (1991).

Cremer, J., Treptow, C., Eggeling, L., and Sahm, H., *J. Gen. Microbiol.*, **134**, 3221–3229 (1988).

Cummins, C. S. and Harris, H., *J. Gen. Microbiol.*, **141**, 583–600 (1956).

Dewey, D. L. and Work, E., *Nature (London)*, **169**, 533–534 (1952).

Eikmanns, B. J., Direct GenBank submission, accession number X67737 (1992).

Farber, G. and Petsko, G., *Trends Biochem. Sci.*, **15**, 228–234 (1990).

Farkas, W. and Gilvarg, C., *J. Biol. Chem.*, **240**, 4717–4722 (1965).

Fleischmann, R. D. et al., *Science*, **269**, 496–512 (1995).

Gelb, M. H., Lin, Y., Pickard, M. A., Song, Y., and Vederas, J. C., *J. Am. Chem. Soc.*, **112**, 4932 (1990).

Gerhart, F., Higgens, W., Tardif, C., and Ducep, J.-B., *J. Med. Chem.*, **33**, 2157 (1990).

Gilvarg, C., *J. Biol. Chem.* **234**, 2955 (1959).

Gilvarg, C., *Fed. Proc.*, **19**, 948–952 (1960).

Gilvarg, C. and Weinberger, S., *J. Bacteriol.*, **101**, 323 (1970).

Girodeau, J.-M., Agouridas, C., Masson, M., Pineau, R., and Le Goffic, F., *J. Med. Chem.*, **29**, 1023–1030 (1986).

Grishin, N. V., Phillips, M. A., and Goldsmith, E. J., *Proten Sci.*, **4**, 1291–1304 (1994).

Haziza, C., Stragier, P., and Patte, J. C., *EMBO J.*, **1**, 379–384 (1982).

Henner, D., Gollnick, P., and Moir, A., *Proc. Sixth Int. Symp. Genet. Ind. Microorg.*, **6**, 657–665 (1990).

Higgins, W., Tardif, C., Richaud, C., Krivanek, M. A., and Cardin, A., *Eur. J. Biochem.*, **186**, 137 (1989).

Ishino, S., Mizukami, K., Yamaguchi, K., Katsumata, R., and Araki, K., *Nucleic Acids Res.* **15**, 3917 (1987).

Karplus, P. A. and Schulz, G. E., *J. Mol. Biol.*, **210**, 163–180 (1989).

Karsten, W. E., *Biochemistry*, **36**, 1730–1739 (1997).

Karsten, W. E. and Viola, R. E., *Biochim. Biophys. Acta*, **1077**, 209–219 (1991).

Karsten, W. E. and Viola, R. E., *Biochim. Biophys. Acta*, **1121**, 234–238 (1992).

Kelland, J. G., Palcic, M. M., Pickard, M. A., and Vederas, J. C., *Biochemistry*, **24**, 3263–3267 (1985).

Kimura, K., *J. Biochem.*, **77**, 405–413 (1975).

Kimura, K. and Goto, T., *J. Biochem.*, **77**, 415–420 (1975).

Kisker, C., Schindelin, H., Alber, B. E., Ferry, J. G., and Rees, D., *EMBO J.*, **15**, 2323–2330 (1996).

Lam, L. K. P., Arnold, L. D., Kalantar, T. H., Kelland, J. G., Lanebell, P. M., Palcic, P. M., Pickard, M. A., and Vederas, J. C., *J. Biol. Chem.*, **263**, 11814 (1988).

Lin, Y., Myhrman, R., Schrag, M. L., and Gelb, M. H., *J. Biol. Chem.*, **263**, 1622 (1988).

Lippert, B., Metcalf, B. W., Jung, M. J., and Casara, P., *Eur. J. Biochem.*, **74**, 441 (1977).

Liu, L. X. and Shaw, P. D., *J. Bacteriol.*, **179**, 507–513 (1997).

Martin, C., Cami, B., Yeh, P., Stragier, P., Parsot, C., and Patte, J.-C., *Mol. Biol. Evol.*, **5**, 549–559 (1988).

Meinnel, T., Schmitt, E., Mechulam, Y., and Blanquet, S., *J. Bacteriol.*, **174**, 2323–2331 (1992).

Metcalf, B. W., Bey, P., Danzin, C., Jung, M. J., Casara, P., and Vevert, J. P., *J. Am. Chem. Soc.*, **100**, 2551 (1978).

Mirwaldt, C., Korndorfer, I., and Huber, R., *J. Mol. Biol.*, **246**, 227–239 (1995).

Misono, H., Nagasaki, S., and Soda, K., *Agric. Biol. Chem.*, **45**, 1455–1460 (1981).

Misono, H., Ogasawara, M., and Nagasaki, S., *Biol. Chem.*, **50**, 1329–1330 (1986a).

Misono, H., Ogasawara, M., and Nagasaki, S., *Agric. Biol. Chem.*, **50**, 2729–2734 (1986b).

Misono, H. and Soda, K., *J. Biol. Chem.*, **255**, 10599 (1980).

Misono, H., Togawa, H., Yamamoto, T., and Soda, K., *Biochem. Biophys. Res. Commun.*, **72**, 89–93 (1976).

Misono, H., Togawa, H., Yamamoto, T., and Soda, K., *J. Bacteriol.*, **137**, 22–27 (1979).

Osterman, A. L., Kinch, L. N., Grishin, N. V., and Phillips, M. A., *J. Biol. Chem.*, **270**, 11797–11802 (1995).

Patte, J.-C., Neidhart, F. C., Ed. In *Escherichia coli and Salmonella typhimurium: Cellular and Molecular Biology*, Vol 1, American Society of Microbiology, Washington, DC 1996 pp. 528–541.

Pavelka, M. S., Jr., and Jacobs, W. R., Jr., *J. Bacteriol.*, **178**, 6496–6507 (1996).

Pavelka, M. S. Jr., Weisbrod, T. R., and Jacobs, W. R. Jr., *J. Bacteriol.*, **179**, 2777–2782 (1997).

Peterkofsky, B., *Methods Enzymol.*, **5**, 853–858 (1962).

Peterkofsky, B. and Gilvarg, C., *J. Biol. Chem.*, **236**, 1432 (1961).

Pissabarro, A., Malumbres, M., Mateos, L. M., Oguiza, J. A., and Martin, J. F., *J. Bacteriol.*, **175**, 2743 (1992).

Raetz, C. R. H. and Roderick, S. L., *Science*, **270**, 997 (1995).

Reddy, S. G., Sacchettini, J. C., and Blanchard, J. S., *Biochemistry*, **34**, 3492 (1995).

Reddy, S. G., Scapin, G., and Blanchard, J. S., *Proteins: Structure Function Genet.*, **25**, 514–516 (1996a).

Reddy, S. G., Scapin, G., and Blanchard, J. S., *Biochemistry*, **35**, 13294 (1996b).

Richaud, C., Higgins, W., Mengin-Lecreuix, D., and Stragier, P., *J. Bacteriol.*, **169**,, 1454–1459 (1987).

Richaud, C., Richaud, F., Martin, C., Haziza, C., and Patte, J. C., *J. Biol. Chem.*, **259**, 14824 (1984).

Richaud, F., Richaud, C., Ratet, P., and Patte, J. C., *J. Bacteriol.*, **166**, 297–300 (1986).

Richaud, C. and Printz, C., *Nucleic Acids Res.*, **16**, 10367 (1988).

Rose, I. A., *Annu. Rev. Biochem.*, **35**, 23–56 (1966).

Rowsell, S., Pauptit, R. A., Tucker, A. D., Melton, R. G., Blow, D. M., and Brick, P., *Structure*, **5**, 337–347 (1997).

Rudnick, G. and Abeles, R. H., *Biochemistry*, **14**, 4515–4522 (1975).

Scapin, G., Blanchard, J. S., and Sacchettini, J. C., *Biochemistry*, **34**, 3502 (1995).

Scapin, G., Reddy, S. G., and Blanchard, J. S., *Biochemistry*, **35**, 13540 (1996).

Scapin, G., Reddy, S. G., Zheng, R., and Blanchard, J. S., *Biochemistry* **36**, (1997).

Scapin, G., Cirilli, M., Reddy, S. G., Gao, Y., Vederas, J. C., and Blanchard, J. S. (1998), submitted.

Schrumpf, B., Schwarzer, Kalinowski, J., Puhler, A., Eggeling, L., and Sahm, H., J. Bacteriol., **173**, 4510 (1991).

Shaver, J. M., Bittel, D. C., Sellner, J. M., Frisch, D. A., Somers, D. A., and Gengenbach, B., *Proc. Natl. Acad. Sci. USA*, **93**, 1962–1966 (1996).

Shaw, J. P., Petsko, G. A., and Ringe, D., *Biochemistry*, **36**, 1329–1342 (1997).

Shedlarski, J. G. and Gilvarg, C., *J. Biol. Chem.*, **245**, 1362 (1970).

Simms, S. A., Voige, W. H., and Gilvarg, C., *J. Biol. Chem.*, **259**, 2734 (1984).

Song, Y., Niederer, D., Lane-Bell, P. M., Lam, L. K. P., Crawley, S., Palcic, M. M., Pickard, M. A., Preuss, D. L., and Vederas, J. C., *J. Org. Chem.*, **59**, 5784–5793 (1994).

Sonntag, K., Eggeling, L., De Graaf, A. A., and Sahm, H., *Eur. J. Biochem.*, **213**, 1325–1331 (1993).

Staffen, B. and Cole, S. T., *Mol. Microbiol.*, **5**, 517–534 (1994).

Stragier, P. Danos, O., and Patte, J.-C., *J. Mol. Biol.*, **168**, 321–331 (1983).

Sundharadas, G. and Gilvarg, C., *J. Biol. Chem.*, **242**, 3983–3988 (1967).

Tabata, S., Direct GenBank submission, Accession Number D90899 (1997).

Tamir, H. and Gilvarg, C., *J. Biol. Chem.*, **249**, 3034–3040 (1974).

Truffa-Bachi, P. van Rapenbusch, R., Janin, J., Gros, C., and Cohen, G. N., *Eur. J. Biochem.*, **5**, 73–80 (1968).

Vaara, M., *FEMS Microbiol. Letts.*, **97**, 249 (1992).

Wang, F. Blanchard, J. S., and Tang, X.-J., *Biochemistry*, **36**, 3755–3759 (1997).

Weinberger, S. and Gilvarg, C., *J. Bacteriol.*, **101**, 323 (1970).

White, P. J., *J. Gen. Microbiol.*, **129**, 739 (1983).

White, P. J. and Kelly, B., *Biochem. J.*, **96**, 75 (1965).

Wiseman, J. S. and Nichols, J. S., *J. Biol. Chem.*, **259**, 8907–8914 (1984).

Work, E., *Nature (London)*, **165**, 74–75 (1950).

Work, E., *Biochem. J.*, **49**, 17–23 (1951).

Yamamoto, J., Shimizu, M., and Yamane, K., *Nucleic. Acids Res.*, **17**, 10105 (1989).

Yeh, P., Sicard, A. M., and Sinskey, A. J., *Mol. Gen. Genet.*, **212**, 105–111 (1988).

Yeh, P., Sicard, A. M., and Sinskey, A. J., *Mol. Gen. Genet.*, **212**, 112–119 (1988).

Yugari, Y. and Gilvarg, C., *J. Biol. Chem.* **240**, 4710 (1965).

COLLAGEN HYDROXYLASES AND THE PROTEIN DISULFIDE ISOMERASE SUBUNIT OF PROLYL 4-HYDROXYLASES

By KARI I. KIVIRIKKO
and TAINA PIHLAJANIEMI, *Collagen Research Unit, Biocenter and Department of Medical Biochemistry, University of Oulu, Finland*

CONTENTS

Advances in Enzymology and Related Areas of Molecular Biology, Volume 72: Amino Acid Metabolism, Part A, Edited by Daniel L. Purich
ISBN 0-471-24643-3   © 1998 John Wiley & Sons, Inc.

I. Introduction

Prolyl 4-hydroxylase (EC 1.14.11.2, procollagen-proline, 2-oxo-glutarate, 4-dioxygenase), prolyl 3-hydroxylase (EC 1.14.11.7, pro-collagen-proline, 2-oxoglutarate, 3-dioxygenase), and lysyl hydrox-ylase (EC 1.14.11.4, procollagen-lysine, 2-oxoglutarate, 5-dioxygenase) catalyze the formation of 4-hydroxyproline, 3-hy-droxyproline, and hydroxylysine in collagens and certain other pro-teins with collagen-like amino acid sequences. It has been known since the early isotopic studies of Stetten and Shoenheimer (1944) and Stetten (1949) that free 4-hydroxyproline is not incorporated into proteins but that their 4-hydroxyproline is derived from proline, which is hydroxylated in a peptide-bound form. The 3-hydroxypro-line and hydroxylysine in proteins are correspondingly not derived from free 3-hydroxyproline and hydroxylysine but from proline and lysine that are likewise hydroxylated only in peptide-bound form. Prolyl 4-hydroxylase is widely distributed in nature and has been identified in a number of vertebrate, invertebrate, and plant sources. The other two enzymes are also found in many animal sources, but they have not been identified in plants.

Prolyl 4-hydroxylase, prolyl 3-hydroxylase, and lysyl hydroxylase belong to the group of 2-oxoglutarate dioxygenases, of which prolyl 4-hydroxylase was one of the earliest to be discovered, is the best investigated, and has often been used as a model for studying the other enzymes in this group. All three collagen hydroxylases require Fe^{2+}, 2-oxoglutarate, O_2, and ascorbate (Fig. 1). The 2-oxoglutarate is stoichiometrically decarboxylated during hydroxylation, with one atom of the O_2 molecule being incorporated into succinate while the other is incorporated into the hydroxy group formed on the proline or lysine residue.

The aim of this chapter is to review current information on the collagen hydroxylases. Because of the extensive literature on these enzymes, this chapter will concentrate primarily on recent advances. Much of the early work on prolyl 4-hydroxylase was performed in the laboratories of S. Udenfriend and D. J. Prockop, and for detailed references to the early work the reader is referred especially to the reviews written by these two scientists and their collaborators (Car-dinale and Udenfriend, 1974; Prockop et al., 1976). Additional refer-ences to the early work and work published subsequent to those two articles can be found in other reviews (Kuttan and Radhakrishnan,

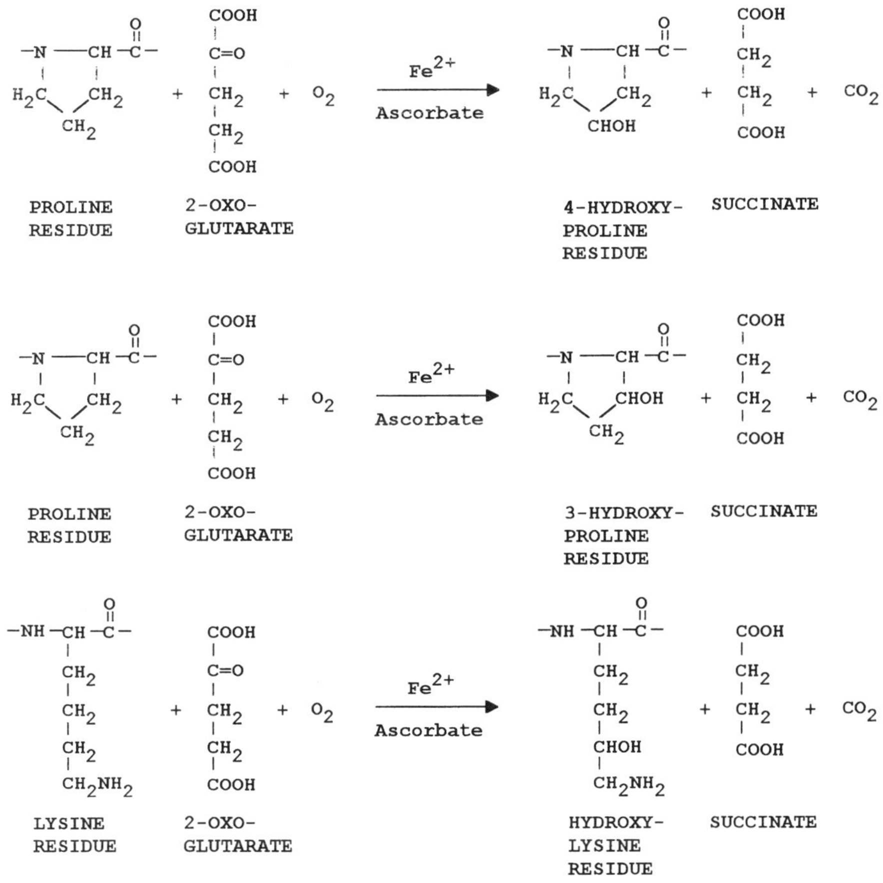

Figure 1. The reactions catalyzed by prolyl 4-hydroxylase, prolyl 3-hydroxylase, and lysyl hydroxylase.

1973; Adams and Frank, 1980; Kivirikko and Myllylä 1980, 1985; Kivirikko et al., 1989, 1990, 1992). Reviews are also available on the occurrence of 4-hydroxyproline in plants and the corresponding plant enzyme (Kuttan and Radhakrishnan, 1973; Lamport, 1977; Adams and Frank, 1980; Fincher et al., 1983; Showalter et al., 1989; Kivirikko et al., 1992; Prescott, 1993) and on methodological aspects of the collagen hydroxylases (Kivirikko and Myllylä 1982, 1987), which will not be discussed here.

II.   Occurrence and Functions of 4-Hydroxyproline,
3-Hydroxyproline, and Hydroxylysine in Animal Proteins

The great majority of the 4-hydroxyproline, 3-hydroxyproline, and hydroxylysine in animal proteins is found in various collagens, in addition to which 4-hydroxyproline and hydroxylysine are found in the collagen-like domains of several other proteins containing collagenous sequences but not defined as collagens (Table 1). 4-Hydroxyproline is also found in elastin, whereas 3-hydroxyproline has not been identified in any other vertebrate protein except collagens. A single 4-hydroxyproline residue is also present in hydroxyproline-lysyl-bradykinin and hydroxyproline luteinizing hormone-releasing hormone, while a single hydroxylysine residue is found in anglerfish somatostatin-28 (Table 1).

The collagens are a family of closely related but distinct extracellular matrix proteins. All collagen molecules consist of three polypeptide chains, called α chains. Each of these chains is coiled into a left-handed helix, and the three helical chains are then wrapped around each other into a right-handed triple helix so that the final structure is a rope-like rod. In some collagen types all three α chains of the molecule are identical, while in other types the molecule contains two or even three different α chains. All collagen α chains are characterized by the presence of repeating -Gly-X-Y- sequences, but they differ in the precise amino acids present and their length. Most collagens also contain noncollagenous domains in addition to the actual collagenous domains. For detailed information on collagen structure, types, and biosynthesis, the reader is referred to a number of reviews (van der Rest and Garrone, 1991; Burgeson and Nimni, 1992; Hulmes, 1992; Kielty et al., 1993; Kivirikko, 1993, 1995; van der Rest and Bruckner, 1993; Mayne and Brewton, 1993; Pihlajaniemi and Rehn, 1995; Prockop and Kivirikko, 1995).

At least 19 proteins containing altogether more than 30 distinct α chains are now defined as collagens, and more than 10 additional animal proteins have collagen-like domains (Table 1). All collagen molecules form supramolecular aggregates that are stabilized in part by interactions between the triple-helical domains. The most abundant collagens form extracellular fibrils and are hence known as fibril-forming collagens. Others form network-like structures, bind to the surface of collagen fibrils, form beaded filaments, act as anchor-

TABLE 1

Vertebrate Proteins Containing 4-Hydroxyproline, 3-Hydroxyproline, or Hydroxylysine[a]

Protein	4-Hyp	3-Hyp	Hyl
Collagens (types I–XIX)	+	+	+
C1q of complement	+	–	+
Acetylcholinesterase	+	–	+
Pulmonary surfactant proteins SP-A and SP-D	+	–	+
Mannan-binding protein	+	–	+
Conglutinin	+	–	+
Collectin-43	+	–	+
Ficolins	+	–	+
Macrophage scavenger receptors (types I and II)	+	–	+
Macrophage receptor MARCO	+	–	+
Adipose specific collagen-like factor apM1	+	–	+
Src-homologous-and-collagen (SHC) protein	+	–	+
Elastin	+	–	–
[Hydroxyproline]-lysyl-bradykinin	+	–	–
[Hydroxyproline^9]-luteinizing hormone-releasing hormone	+	–	–
Somatostatin 28 (anglerfish)	–	–	+

[a] For references, see text. The presence of the 4-Hyp and Hyl residues in some of these proteins is so far only based on predictions from cDNA data.

329

ing fibrils for the skin, or are transmembrane proteins (van der Rest and Garrone, 1991; Burgeson and Nimni, 1992; Hulmes, 1992; Kielty et al., 1993; Kivirikko, 1993, 1995; van der Rest and Bruckner, 1993; Mayne and Brewton, 1993; Pihlajaniemi and Rehn, 1995; Prockop and Kivirikko, 1995).

Figures for 4-hydroxyproline, 3-hydroxyproline, and hydroxylysine content have been reported for the collagenous (i.e., triple helical) domains of most collagen polypeptide chains (for a summary of the detailed values, see Kivirikko et al., 1992). The $(Gly-X-Y)_n$ sequences in the polypeptide chains of type I collagen, the most abundant collagen, contain about 1000 amino acid residues, with about 100 of the X position amino acids being proline and about 100 of the Y position amino acids 4-hydroxyproline. The 4-hydroxyproline content shows only small, though distinct, differences between the various collagen types, whereas the 3-hydroxyproline and hydroxylysine contents vary markedly, 3-hydroxyproline ranging from 0 to more than 10 residues per 1000 amino acids and hydroxylysine from about 5–70. Further variations in 3-hydroxyproline and hydroxylysine content are found within the same collagen type between tissues and even in the same tissue in different physiological and pathological states (Kivirikko and Myllylä, 1980; Kivirikko et al., 1992).

The group of proteins containing collagenous sequences but not defined as collagens (Table 1) includes the subcomponent C1q of complement (Reid, 1979), the tail structure of acetylcholinesterase (Rosenberry et al., 1982), the pulmonary surfactant proteins SP-A and SP-D (Floros et al., 1986; Lu et al., 1992), mannan binding protein (Drickamer et al., 1986; Colley and Baenziger, 1987; Reid, 1993), conglutinin (Davis and Lachmann, 1984, Jensenius et al., 1994), collectin-43 (Jensenius et al., 1994), ficolins (Ichijo et al., 1993; Matsushita et al., 1996), type I and type II macrophage scavenger receptors (Kodama et al., 1990; Rohrer et al., 1990), an additional macrophage receptor termed MARCO (Elomaa et al., 1995), an adipose specific collagen-like factor apM1 (Maeda et al., 1996), and a src-homologous-and-collagen (SHC) protein (Pelicci et al., 1992; Thomas et al., 1995). The presence of 4-hydroxyproline and hydroxylysine in most of these proteins has been established by amino acid analysis, but in some cases their presence is so far based on predictions from cDNA data. None of these proteins has been reported to contain 3-hydroxyproline (Kivirikko and Myllylä, 1980; Kivirikko et al., 1992).

Elastin differs from the proteins described above in that it has no collagen-like triple-helical domain, but its polypeptide chain does feature repeating -Gly-X-Y- sequences, which contain 4-hydroxyproline but no hydroxylysine (Kivirikko and Myllylä, 1980; Kivirikko et al., 1992; Rosenbloom, 1993). The 4-hydroxyproline content of elastin varies greatly, usually being about 10–25 residues per 1000 amino acids (Rosenbloom, 1993), but ranging up to about 50 in special situations (Kivirikko et al., 1992).

4-Hydroxyproline has been identified in vertebrate proteins almost exclusively in the Y positions of the repeating -X-Y-Gly- sequences (the repeating -Gly-X-Y- sequences are written here as -X-Y-Gly- due to the properties of the three collagen hydroxylases, see Section V). The only known exceptions are single triplets of -X-4Hyp-Ala- in two of the polypeptide chains of the subcomponent Clq of human complement and in some collagen α chains (Kivirikko et al., 1992). Hydroxylysine is correspondingly found almost exclusively in the Y positions of the repeating -X-Y-Gly- sequences, but in some fibril-forming collagens the short nontriple-helical regions at the ends of the α chain contain one sequence of -X-Hyl-Ala- or -X-Hyl-Ser- (Kivirikko and Myllylä, 1980; Kivirikko et al., 1992). Although 4-hydroxyproline and hydroxylysine are found almost exclusively in proteins containing repeating -X-Y-Gly- sequences, the kinin mixture in human plasma, urine, and ascitic fluid contains both lysyl bradykinin and small amounts of hydroxyproline-lysyl-bradykinin in which a single 4-hydroxyproline residue is present in the sequence -Pro-4Hyp-Gly- (Maeda et al., 1988; Sasaguri et al., 1988). The hydroxyproline luteinizing hormone-releasing hormone contains a single 4-hydroxyproline residue in the sequence -Arg-4Hyp-Gly- (Gautron et al., 1991). Correspondingly, anglerfish somatostatin-28 contains a single hydroxylysine residue in the sequence -Trp-Hyl-Gly- (Andrews et al., 1984; Spiess and Noe, 1985). 3-Hydroxyproline has been identified in collagens only in the sequence -Gly-3Hyp-4Hyp-Gly-.

The hydroxylation of many of the lysine residues in the Y positions of the -X-Y-Gly- triplets is incomplete, and a number of lysine residues in the Y positions of some collagens are not hydroxylated at all (Kivirikko and Myllylä, 1980; Kivirikko et al., 1992). Such variation in the degree of hydroxylation of lysine residues explains the differences in hydroxylysine content to be found within the same collagen type from various sources. This variation also in part explains the differences in hydroxylysine content found between collagen types, but additional factors involved are differences in the total numbers of lysine residues incorporated per 1000 amino acids and differences in the distribution of the incorporated lysine residues between the X and Y positions. The differences in 3-hydroxyproline content within a given collagen type and between collagen types are likewise due to variation in the extent of 3-hydroxylation of proline residues in the X positions of the -Pro-4Hyp-Gly- sequences.

Earthworm cuticle collagen differs from all the other collagens studied in that more than 90% of the incorporated proline residues are found as 4-hydroxyproline. Earthworm cuticle collagen also differs from the others in that most, if not all, of the 4-hydroxyproline residues are confined to the X positions of the repeating -X-Y-Gly-sequences (Kivirikko and Myllylä, 1980; Kivirikko et al., 1992).

3-Hydroxyproline has recently been identified in the sequences -Val-3Hyp-Asp-, -Val-3Hyp-Glu-, and -Tyr-3Hyp-Tyr- of the secreted cathepsin L-like proteinases of the trematode *Fasciola hepatica* (Wijffels et al., 1994). As these sequences show no similarity to those around the 3-hydroxyproline residues in collagens, it seems likely that the prolyl 3-hydroxylase present in this parasite is quite different from the collagen prolyl 3-hydroxylase (Wijffels et al., 1994).

The function of the hydroxy group of the 4-hydroxyproline residues is to stabilize the collagen triple helix under physiological conditions. Nonhydroxylated collagen α chains can fold into a triple helix at low temperatures, but the midpoint of thermal transition from helix to coil (T_m) in the case of type I procollagen molecules consisting of nonhydroxylated polypeptide chains is only 24°C (Berg and Prockop, 1973a; Jimenez et al., 1973), about 15°C lower than that for molecules consisting of fully hydroxylated polypeptide chains (i.e., chains in which all the Y position prolines have been converted to 4-hydroxyproline). Thus nonhydroxylated collagen polypeptide chains cannot form triple-helical molecules *in vivo*, and an almost complete 4-hydroxylation of proline residues in the Y positions of the -X-Y-Gly- triplets is required for the formation of a molecule that is stable at 37°C. It is probable that 4-hydroxyproline has a similar function in the other proteins with collagen-like triple helices, whereas its role in elastin is unknown (Kivirikko and Myllylä, 1980; Kivirikko et al., 1992).

The means by which 4-hydroxyproline stabilizes the triple helix are now known in detail. Its hydroxy group forms two kinds of bond via a water molecule, either between two polypeptide chains or within the same chain in the triple helix (Brodsky and Ramshaw, 1997).

The hydroxylysine residues of collagens have two important functions: their hydroxy groups act as attachment sites for carbohydrate units, and they are essential for the stability of the intermolecular collagen cross-links. The carbohydrate is present in part as the monosaccharide galactose and in part as the disaccharide glucosylgalactose, the structure of the disaccharide unit with its peptide attachment being 2-O-α-D-glucopyranosyl-O-β-D-galactopyranosylhydroxylysine (Kivirikko and Myllylä, 1979). The extent of glycosylation of hydroxylysine residues and the ratio of galactosylhydroxylysine to glucosylgalactosylhydroxylysine vary markedly between collagen types, and even within the same collagen type from various sources (Kivirikko and Myllylä, 1979). The hydroxylysine residues of most if not all of the other proteins containing collagen-like domains are also glycosylated, mostly by the disaccharide glucosylgalactose, but the possible presence of hydroxylysine-linked carbohydrate units has not yet been studied in the case of some proteins.

The intermolecular cross-links of collagens are formed by condensation between an aldehyde derived from a lysine or hydroxylysine residue and the ε-amino group of a second unmodified lysine, hydroxylysine or glycosylated hydroxylysine residue (Hulmes, 1992; Kielty et al., 1993; Kivirikko, 1995). The cross-links formed from a hydroxylysine-derived aldehyde are much more stable than those from a lysine-derived aldehyde (Kivirikko et al., 1992; Hulmes, 1992; Kielty et al., 1993; Kivirikko, 1995). With time, and depending on whether the initially oxidized residue was lysine or hydroxylysine, these difunctional cross-links undergo further intramolecular and intermolecular reactions to yield several kinds of mature trifunctional and tetrafunctional cross-links. Two such well-characterized trifunctional compounds are hydroxylysyl-pyridinoline, which is formed from two hydroxylysine-derived aldehydes and one hydroxylysine residue, and lysyl-pyridinoline, in which one of the three residues is derived from lysine instead of hydroxylysine (Hulmes, 1992; Kielty et al., 1993; Kivirikko, 1995). The significance of lysine hydroxylation for the stabilization of collagen cross-links is clearly demonstrated by the marked changes in the mechanical properties of certain tissues seen in patients with mutations in the gene for lysyl hydroxylase that lead to a deficiency in lysyl hydroxylase activity (Section XI). The cross-links derived from hydroxylysine appear to be unique to collagens.

The function of 3-hydroxyproline is unknown at present.

III. General Molecular Properties of Collagen Hydroxylases

A. PROLYL 4-HYDROXYLASES

Prolyl 4-hydroxylase was first purified to near homogeneity by conventional procedures (Cardinale and Udenfriend, 1974; Prockop et al., 1976; Kivirikko and Myllylä, 1980; Kivirikko et al., 1992). Two affinity chromatography procedures were subsequently developed that made it possible to isolate large quantities of pure enzyme (Berg and Prockop, 1973b; Tuderman et al., 1975), and the hydroxylase has been purified to homogeneity by these procedures or their modifications (Kivirikko and Myllylä, 1982, 1987) from many animal sources (Kivirikko et al., 1992), including human tissues (Kuutti et al., 1975).

Prolyl 4-hydroxylase from all the vertebrate sources studied is an $\alpha_2\beta_2$ tetramer with a molecular weight of about 240,000 and consisting of two types of monomer with molecular weights of about 63,000 (α subunit) and 58,000 (β subunit). The tetramer has no interchain disulfide bonds, but intrachain disulfide bonds appear to be essential for the α subunits to maintain the native structure necessary for their association (Kivirikko and Myllylä, 1980; Kivirikko et al., 1992).

The α subunit of prolyl 4-hydroxylase has been cloned from humans (Helaakoski et al., 1989), rat (Hopkinson et al., 1994), mouse (Helaakoski et al., 1995), chicken (Bassuk et al., 1989), and the nematode *Caenorhabditis elegans* (Veijola et al., 1994). Prolyl 4-hydroxylase has long been assumed to be only one type of enzyme, with no isoenzymes, but an isoform of the α subunit has recently been cloned from humans (Annunen et al., 1997) and mouse (Helaakoski et al., 1995) and has been termed the $\alpha(II)$ subunit. Correspondingly, the previously known α subunit is now called the $\alpha(I)$ subunit, and the

$[\alpha(I)]_2\beta_2$ and $[\alpha(II)]_2\beta_2$ tetramers are known as types I and II enzymes, respectively (Helaakoski et al., 1995). Recent data on expression in insect cells (see below) argue against the presence of a mixed $\alpha(I)\alpha(II)\beta_2$ tetramer (Annunen et al., 1997). The mRNA for the new $\alpha(II)$ subunit, and also the corresponding polypeptide, are expressed in a variety of tissues, the type I enzyme typically representing about 70–90% and the type II enzyme 10–30% of the total prolyl 4-hydroxylase activity (Helaakoski et al., 1995; Annunen et al., 1997). Surprisingly, however, the type II enzyme was found to be the main prolyl 4-hydroxylase form in chondrocytes and capillary endothelial cells (Annunen et al., 1998).

Additional heterogeneity in the forms of the prolyl 4-hydroxylase tetramer is caused by alternative splicing of RNA transcripts for the $\alpha(I)$ subunit (Helaakoski et al., 1989, 1995). Two types of mRNA for this α subunit are found in human (Helaakoski et al., 1989) and mouse (Helaakoski et al., 1995) tissues at least differing with respect to the splicing of sequences corresponding to two consecutive, homologous 71 bp exons in the respective gene (Helaakoski et al., 1994). The sequences corresponding to these two exons are spliced in a mutually exclusive manner, and thus the $\alpha(I)$ subunit always has sequences corresponding to one of these two exons but never to both. Still further heterogeneity in the forms of the prolyl 4-hydroxylase tetramer is caused by variable extents of glycosylation of the $\alpha(I)$ and $\alpha(II)$ subunits. Both types of subunit contain two potential attachment sites for asparagine-linked oligosaccharides (Helaakoski et al., 1989, 1995), and the extent of glycosylation of these two sites is highly variable (Kedersha et al., 1985a,b).

Molecular cloning and nucleotide sequencing of the β subunit of prolyl 4-hydroxylase indicated, surprisingly, that this polypeptide is identical to the enzyme protein disulfide isomerase (PDI) (Pihlajaniemi et al., 1987; Parkkonen et al., 1988). Moreover, the β subunit was found to have PDI activity even when present in the native prolyl 4-hydroxylase tetramer (Koivu et al., 1987). The PDI (EC 5.3.4.1) catalyzes thiol–disulfide interchange *in vitro*, leading to net protein disulfide formation, reduction or isomerization, depending on the reaction conditions, and is the *in vivo* catalyst of disulfide bond formation in the biosynthesis of various secretory and cell surface proteins (Noiva and Lennarz, 1992; Freedman et al., 1994; Freedman, 1995). Subsequent work has demonstrated that the PDI/

β polypeptide has several additional functions, as will be described in Section IX.B. This polypeptide has now been cloned from many sources (Edman et al., 1985; Pihlajaniemi et al., 1987), (Section IX.A).

Although the $\alpha_2\beta_2$ structure of the vertebrate prolyl 4-hydroxylases has been known for more than 25 years, all attempts to associate an enzyme tetramer from its subunits *in vitro* have been unsuccessful (Kivirikko et al., 1992). It has nevertheless been possible to produce a fully active recombinant prolyl 4-hydroxylase tetramer in insect cells by means of baculovirus vectors (Vuori et al., 1992b). The vertebrate prolyl 4-hydroxylase tetramers produced in insect cells are indistinguishable from those isolated from vertebrate tissues in terms of their molecular and catalytic properties (Vuori et al., 1992b; Helaakoski et al., 1995). The prolyl 4-hydroxylase tetramer has also been produced in a cell-free translation/translocation system of dog pancreas microsomes (John et al., 1993), but the amount of tetramer formed is so small that its enzyme activity cannot be determined. When an α subunit is expressed in insect cells alone, without the PDI/β subunit, it forms insoluble aggregates with no prolyl 4-hydroxylase activity (Vuori et al., 1992b; Helaakoski et al., 1995). Thus one function of the PDI/β subunit in the prolyl 4-hydroxylase tetramer appears to be that of keeping the α subunits in solution (for further discussion of this aspect see Section IX.B).

Although the α subunit of prolyl 4-hydroxylase from *C. elegans* resembles the vertebrate α subunits in its amino acid sequence (Section VIII), its expression in insect cells together with either the human or *C. elegans* PDI/β polypeptide led to formation of a prolyl 4-hydroxylase αβ dimer (Veijola et al., 1994, 1996a). The catalytic properties of this dimer are very similar to those of the vertebrate prolyl 4-hydroxylase tetramers. It is currently unknown whether any of the other nonvertebrate prolyl 4-hydroxylases are likewise dimers rather than tetramers. It may be noted that not all forms of prolyl 4-hydroxylase have the PDI/β subunit, as the enzyme from unicellular and multicellular green algae is an α monomer antigenically related to the α subunit of the vertebrate enzymes (Kaska et al., 1987, 1988).

B. LYSYL HYDROXYLASE

Lysyl hydroxylase has been purified to homogeneity from chick embryos and human placental tissues by means of two affinity column procedures (Turpeenniemi-Hujanen et al., 1980, 1981). The active enzyme is an α_2 dimer with a molecular weight of about 190,000 in gel filtration and consisting of only one type of monomer with a molecular weight of about 85,000 (Turpeenniemi-Hujanen et al., 1980, 1981; Myllylä et al., 1988).

Molecular cloning and nucleotide sequencing have been reported for the chick (Myllylä et al., 1991), human (Hautala et al., 1992a; Yeowell et al., 1992), and rat (Armstrong and Last, 1995) lysyl hydroxylase subunit. A surprising finding was that no significant homology was detected between the overall primary structures of lysyl hydroxylase and the two types of subunit of prolyl 4-hydroxylase, in spite of the marked similarities between the catalytic properties of these two enzymes. Lysyl hydroxylase also differs from the animal prolyl 4-hydroxylases in that it does not require any PDI/β subunit, and the purified lysyl hydroxylase protein has no disulfide isomerase activity (Myllylä et al., 1989). The chick, rat, and human lysyl hydroxylase polypeptides all contain four potential attachment sites for asparagine-linked oligosaccharide units (Myllylä et al., 1991; Hautala et al., 1992a; Yeowell et al., 1992; Armstrong and Last, 1995), and a considerable heterogeneity is found in the extent of the glycosylation of the enzyme polypeptide (Myllylä et al., 1988; Pirskanen et al., 1996). Some of the asparagine-linked carbohydrate units of lysyl hydroxylase, unlike those of prolyl 4-hydroxylase, are essential for maximal enzyme activity (Myllylä et al., 1988; Pirskanen et al., 1996). The catalytic properties of a recombinant enzyme expressed in insect cells are virtually identical to those of the enzyme isolated from vertebrate tissues (Armstrong and Last, 1995; Pirskanen et al., 1996; Krol et al., 1996).

Many findings suggest that lysyl hydroxylase may have tissue-specific isoenzymes. As indicated in Section II, large differences in the extent of lysine hydroxylation are found between the collagen types, and even within the same collagen type from different tissues. The collagen hydroxylysine deficiency observed in patients with mutations in the gene for this enzyme (Section XI) also shows a wide variation in degree from tissue to tissue. Recently molecular cloning was reported for one isoenzyme (lysyl hydroxylase 2) that is highly expressed in pancreas and muscle (Valtavaara et al., 1997). The tissue-specific isoenzymes may show differences with respect to hydroxylation of various collagen types, but some data also argue against the existence of collagen type-specific isoenzymes (Kivirikko et al., 1992). Indirect evidence has been put forward for the presence of a separate lysyl hydroxylase (Royce and Barnes, 1985) acting on the lysines found in the sequences -X-Lys-Ala- and -X-Lys-Ser- in the short noncollagenous regions at the ends of the α chains in some collagens (Section II), but no direct data are available on the existence of this presumed enzyme (Kivirikko and Myllylä, 1980; Kivirikko et al., 1992).

C. PROLYL 3-HYDROXYLASE

Prolyl 3-hydroxylase and prolyl 4-hydroxylase activities have been shown to be derived from separate proteins (Risteli et al., 1977; Tryggvason et al., 1977). Prolyl 3-hydroxylase has been purified to about 5000-fold from an ammonium sulfate fraction of chick embryo extract (Tryggvason et al., 1979), but it has not been isolated as a homogeneous protein. Its molecular weight is about 160,000 in gel filtration (Tryggvason et al., 1979), but its subunit structure is unknown, and it has not yet been cloned.

IV. Intracellular Sites of Hydroxylation

Prolyl 4-hydroxylase, the free PDI/β polypeptide and lysyl hydroxylase have all been shown by a variety of techniques to reside within the lumen of the endoplasmic reticulum (Kivirikko et al., 1992; Kellokumpu et al., 1994), and prolyl 3-hydroxylase is likewise located within this cell compartment (Kivirikko et al., 1992). Prolyl 4-hydroxylase and the PDI/β polypeptide are soluble endoplasmic reticulum luminal proteins (Kivirikko et al., 1992), whereas lysyl hydroxylase appears to be a luminally oriented peripheral membrane protein that associates with the membrane via weak electrostatic interactions (Kivirikko et al., 1992; Kellokumpu et al., 1994). The location of the prolyl 4-hydroxylase tetramer and the PDI/β polypeptide agrees with the presence of a -Lys-Asp-Glu-Leu sequence as the C-terminus of the PDI/β polypeptide (Section IX.A), a sequence that has been shown to be both necessary and sufficient for the retention of a number of soluble proteins within the lumen of the endoplasmic reticulum (Pelham, 1990). It has also been demonstrated that deletion of the C-terminal -Lys-Asp-Glu-Leu sequence from the PDI/β polypeptide leads to the secretion of considerable amounts of both the free polypeptide and an active prolyl 4-hydroxyl-

ase tetramer from insect cells expressing a recombinant human prolyl 4-hydroxylase (Vuori et al., 1992c).

The procollagen polypeptide chains are synthesized on membrane-bound ribosomes and pass through the membranes into the
lumen of the endoplasmic reticulum in the course of being assembled
(Fig. 2). Most of the hydroxylation of proline and lysine residues
occurs in the form of cotranslational modifications while the polypeptide chains are being synthesized, but all these reactions are continued as posttranslational modifications after the release of the complete polypeptide chains from the ribosomes (Cardinale and
Udenfriend, 1974; Prockop et al., 1976; Kivirikko and Myllylä, 1980;
Kivirikko et al., 1992), until formation of the triple helix (Fig. 2)
prevents any further hydroxylation (Section V). The biological substrates for the collagen hydroxylases are thus proline or lysine residues in appropriate sequences of a growing or newly synthesized
polypeptide chain. An active system appears to exist for the transport of 2-oxoglutarate (Tschank et al., 1988) and ascorbate (Peterkofsky et al., 1987) into the lumen of the endoplasmic reticulum through
microsomal membranes.

V. Peptide Substrates

The collagen hydroxylases hydroxylate proline or lysine residues
only in peptide linkages, and none of them acts on the corresponding
free amino acid. Prolyl 4-hydroxylase does not hydroxylate
tripeptides with the structure Gly-X-Pro or Pro-Gly-X, whereas
tripeptides with the structure X-Pro-Gly do serve as substrates (Cardinale and Udenfriend, 1974; Prockop et al., 1976; Kivirikko and
Myllylä, 1980; Kivirikko et al., 1992). Polytripeptides with the structure $(Pro-Pro-Gly)_n$ are good substrates, with only the prolines preceding glycine being hydroxylated (Kivirikko and Prockop, 1967a;
Cardinale and Udenfriend, 1974; Prockop et al., 1976; Kivirikko and
Myllylä, 1980, Kivirikko et al., 1992). In agreement with this position
specificity, polytripeptides with the structure $(Ala-Pro-Gly)_n$ are
good substrates, whereas those with the structure $(Pro-Ala-Gly)_n$ are
either very poor substrates or completely inactive (Kivirikko et al.,
1969; Rao and Adams, 1978). These findings and additional data
(Cardinale and Udenfriend, 1974; Prockop et al., 1976; Kivirikko
and Myllylä, 1980; Kivirikko et al., 1992) agree with the presence
of 4-hydroxyproline in collagens and other proteins containing this

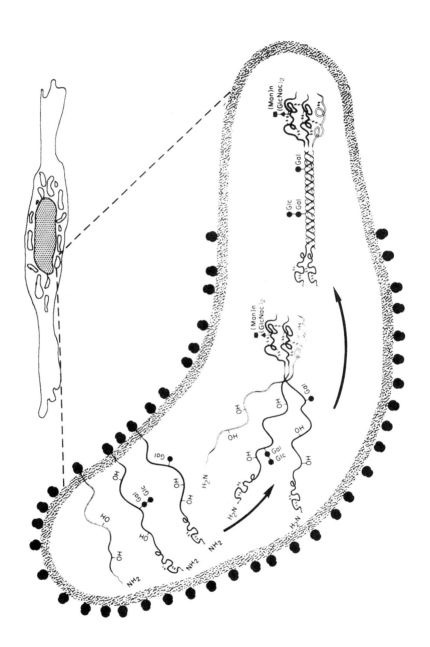

amino acid only in the Y positions of the -X-Y-Gly- triplets (Section II), and demonstrate that the minimum sequence requirement for a vertebrate prolyl 4-hydroxylase is fulfilled by an -X-Pro-Gly- triplet. However, glycine can well be replaced by β-alanine, an amino acid that is larger but does not contain a side chain (Kivirikko and Myllylä, 1980; Kivirikko et al., 1992). The tetrapeptides -Pro-Pro-Ala-Pro- and -Pro-Pro-Gln-Pro- also serve as substrates, although they are hydroxylated only at a very low rate (Atreya and Ananthanarayanan, 1991). The few cases of single -X-4Hyp-Ala- sequences found in the subcomponent Clq of complement and in some collagen polypeptide chains (Section II) agree with these data and indicate that the glycine in the -X-Pro-Gly- sequence can in some exceptional situations be replaced by other amino acids.

Prolyl 4-hydroxylase from the subcuticular epithelium of earthworms differs in its sequence requirements from all other animal prolyl 4-hydroxylases studied in that $(Pro-Ala-Gly)_n$ is a far better substrate for it than $(Ala-Pro-Gly)_n$ (Kivirikko and Myllylä, 1980; Kivirikko et al., 1992). This specificity agrees with the unique occurrence of 4-hydroxyproline in the X positions of the -X-Y-Gly- sequences in earthworm cuticle collagen.

The interaction of peptide substrates with prolyl 4-hydroxylase is further affected by the amino acid in the X position of the triplet to be hydroxylated and by other nearby amino acids, the peptide chain length, and the peptide conformation. Current information concerning the effects of the amino acid in the X position and amino acids in other parts of the peptide has been summarized elsewhere (Kivirikko et al., 1992). No detailed conclusions can be reached, however, as only a relatively small number of peptides have been studied and as other properties of the peptides, such as their lengths, have also

←——————————————————————————————————————

Figure 2. Schematic representation of the intracellular events of collagen biosynthesis within the lumen of the endoplasmic reticulum. The scheme shows that most of the hydroxylation of proline and lysine residues and glycosylation of hydroxylysine residues occur while the polypeptide chains are being synthesized, but all these reactions are continued after the release of the complete polypeptide chains from the ribosomes until formation of the triple helix prevents any further hydroxylation or hydroxylysine-linked glycosylation. [Modified from Kivirikko et al., 1989 with permission.]

varied. Proline in the X position appears to give a particularly high maximal velocity (V), while alanine, leucine, arginine, valine, and glutamate give lower values for V in this order (Kivirikko et al., 1992). The presence of glycine or sarcosine in the X position of (X-Pro-Gly)$_n$ polytripeptides prevents hydroxylation, but the data obtained with polytripeptides do not necessarily hold good for a single -Gly-Pro-Gly- sequence, as some collagens do contain a few -Gly-4Hyp-Gly- sequences (Kivirikko et al., 1992). The effects of amino acids outside the triplet to be hydroxylated have been studied especially with analogues of bradykinin that contain a single -Pro-Pro-Gly- sequence (Cardinale and Udenfriend, 1974), but it is not known to what extent the effects found with this short peptide can be related to those occurring with long peptides having glycine in every third position.

The chain length of the peptide has a major effect on the K_m, which decreases with increasing chain length (Cardinale and Udenfriend, 1974; Prockop et al., 1976; Kivirikko and Myllylä, 1980; Kivirikko et al., 1992). The K_m of type I prolyl 4-hydroxylase for the tripeptide Pro-Pro-Gly, for example, is about 20 mM, whereas the K_m for (Pro-Pro-Gly)$_5$ is about 2 mM and that for (Pro-Pro-Gly)$_{20}$ is 50 μM. All these values are expressed as the concentration of the -X-Pro-Gly- units (Kivirikko and Myllylä, 1980; Kivirikko et al., 1992). In the case of peptides with the structure (Pro-Pro-Gly)$_n$, the increase in chain length does not appear to influence the V of the reaction (Kivirikko and Myllylä, 1980; Kivirikko et al., 1992). Type I protocollagen, a biologically prepared protein consisting of nonhydroxylated collagenous polypeptide chains with molecular weights of about 150,000, has a particularly low K_m, about 0.2 μM when expressed as above, but the V for this protein is not significantly different from that for the short (Pro-Pro-Gly)$_n$ polytripeptides (Kivirikko and Myllylä, 1980; Kivirikko et al., 1992). The K_m values of the type II prolyl 4-hydroxylase for all the peptide substrates tested are about three-to-six times those of the type I enzyme, the K_m of the former for type I protocollagen being about 1 μM (Annunen et al., 1997). These K_m values of prolyl 4-hydroxylases for protocollagen are very similar to those of the two other collagen hydroxylases (see below).

The conformation of the peptide has a marked effect in that the triple-helical conformation of the collagenous peptides completely prevents hydroxylation (Kivirikko and Myllylä, 1980; Kivirikko et al., 1992). Studies on the solution structures of X-Pro-Gly-Y tetrapeptides have indicated that these contain a poly(L-proline) II conformation in the X-Pro segment, while the Pro-Gly segment forms a β turn (Chopra and Ananthanarayanan, 1982; Lee et al., 1984; Atreya and Ananthanarayanan, 1991). Based on these and additional data, it has been proposed (Atreya and Ananthanarayanan, 1991) that the poly(L-proline) conformation in the X-Pro segment of nascent collagenous peptides and their models may be necessary at the binding site on prolyl 4-hydroxylases, while the β turn in the Pro-Gly segment may be essential for hydroxylation at the catalytic site. The peptide binding site may extend from the catalytic site located in the interior of the enzyme outward to the surface (Atreya and Ananthanarayanan, 1991). Productive binding of the substrate, and thus effective concentration of the β turn at the catalytic site, would increase with an increase in the chain length of the peptide, as for a given concentration, shorter peptides would be distributed between the inner and outer binding sites while a contiguous longer peptide would simultaneously bind at both these sites (Atreya and Ananthanarayanan, 1991). The model would explain, at least in part, the effect of the peptide chain length on hydroxylation, and also the finding that triple-helical peptides do not act as substrates (Atreya and Ananthanarayanan, 1991). Denatured collagenous polypeptides would contain both the poly(L-proline) structure in the X-Pro segment and the β turn in the Pro-Gly segment, and would thus be hydroxylated (Atreya and Ananthanarayanan, 1991). In elastin and related peptides such as Gly-Val-Pro-Gly-Val, a β-strand structure may be interchangeable with the poly(L-proline) II structure (Atreya and Ananthanarayanan, 1991). It should be noted, however, that little direct data are currently available to support this model.

Lysyl hydroxylase resembles prolyl 4-hydroxylase in its substrate requirements. It does not act on the tripeptide Lys-Gly-Pro, whereas the tripeptide Ile-Lys-Gly is hydroxylated (Kivirikko et al., 1972). Many other peptides containing an -X-Lys-Gly- sequence also serve as substrates indicating that the minimum sequence requirement is fulfilled by an -X-Lys-Gly- triplet (Kivirikko and Myllylä, 1980; Kivirikko et al., 1992). However, lysyl hydroxylase also acts on arginine-rich histone, which does not contain any -X-Lys-Gly- sequence, but does contain sequences such as -X-Lys-Ala- and -X-Lys-Ser- (Kivirikko and Myllylä, 1980; Kivirikko et al., 1992). This agrees with the occurrence of hydroxylysine in the sequences -X-Hyl-Ala- and -X-Hyl-Ser- in the short noncollagenous regions at the ends of some collagen α chains (Section II).

Interaction with lysyl hydroxylase, like that with prolyl 4-hydroxylase, is further affected by the amino acid sequence around the lysine residue to be hydroxylated, the peptide chain length, and the peptide conformation (Kivirikko and Myllylä, 1980; Kivirikko et al., 1992). As in the case of prolyl 4-hydroxylase, the chain length of the peptide appears to influence primarily the K_m, whereas the V of the reaction seems to be unaffected (Kivirikko et al., 1992). The K_m values for (Ile-Lys-Gly)$_2$, (Ile-Lys-Gly)$_3$-Phe, and (Ile-Lys-Gly)$_5$-Phe, for example, are 4, 1, and 0.14 mM, respectively, all expressed as concentrations of the -X-Lys-Gly- units (Kivirikko et al., 1972; Kikuchi and Tamiya, 1982). The K_m for type I protocollagen is particularly low, about 0.2 μM, when expressed as above (Ryhänen and Kivirikko, 1974).

The conformation of the peptide substrate has a crucial effect, in that the triple-helical conformation of the collagenous peptides completely prevents lysine hydroxylation (Kivirikko et al., 1973; Ryhänen et al., 1974; Kivirikko and Myllylä, 1980; Kivirikko et al., 1992). As in the case of prolyl 4-hydroxylase, an extended poly(L-proline) II conformation in the peptide substrate may interact at the binding site on lysyl hydroxylase, while a β turn in the Lys-Gly segment may be necessary for hydroxylation at the catalytic site (Jiang and Ananthanarayanan, 1991; Ananthanarayanan et al., 1992).

Prolyl 3-hydroxylase does not hydroxylate proline residues in type I protocollagen, a protein consisting of nonhydroxylated collagenous polypeptide chains (Tryggvason et al., 1977). However, when all the Y position prolines of this protein are first hydroxylated to 4-hydroxyproline, it will subsequently serve as a good substrate for prolyl 3-hydroxylase (Tryggvason et al., 1977). Thus the main or only substrate sequence for prolyl 3-hydroxylase is probably -Pro-4Hyp-Gly-, whereas a -Pro-Pro-Gly- sequence is probably not hydroxylated (Tryggvason et al., 1977). This agrees with the presence of 3-hydroxyproline in collagens only in the sequence -Gly-3Hyp-4Hyp-Gly- (Section II).

The 3-hydroxylation of various -Pro-4Hyp-Gly- triplets is probably affected by the amino acids in the adjacent triplets, as in the

case of the other collagen hydroxylases. Longer peptides are likewise better substrates for prolyl 3-hydroxylase than shorter ones,
and the triple-helical conformation prevents hydroxylation (Risteli
et al., 1977). The K_m for fully 4-hydroxylated type I procollagen as
a substrate is about 0.4 μM when expressed as the concentration
of -Pro-4Hyp-Gly- units (Kivirikko et al., 1992).

VI.   Cosubstrates and Reaction Mechanisms

All three collagen hydroxylases require Fe^{2+}, 2-oxoglutarate, O_2
and ascorbate (Fig. 1). The 2-oxoglutarate is stoichiometrically decarboxylated during hydroxylation, whereas ascorbate is not consumed stoichiometrically and the enzymes can catalyze their reactions for a number of cycles in its absence. As described below, the
reaction requiring ascorbate is an uncoupled decarboxylation of 2-
oxoglutarate, that is, decarboxylation without subsequent hydroxylation (Fig. 3).

The apparent K_m values of the various collagen hydroxylases for
their cosubstrates are very similar (Table 2). The values shown for
the vertebrate types I and II prolyl 4-hydroxylases were determined
for the human and mouse enzymes, respectively (Helaakoski et al.,
1995), except that the K_m for O_2 was measured only for the chick
enzyme (Myllylä et al., 1977). The K_m values for the other cosubstrates of the chick enzyme are virtually identical to those shown in
Table 2 (Myllylä et al., 1977). The K_m values of lysyl hydroxylase
shown in Table 2 were determined for the human enzyme (Turpeenniemi-Hujanen et al., 1981; Pirskanen et al., 1996), and the values
for the chick enzyme are again essentially identical (Puistola et al.,
1980a). The values obtained with synthetic peptide substrates are
usually slightly higher than with biologically prepared protocollagen
substrates, as shown in Table 2 for chick prolyl 4-hydroxylase (Kivirikko and Prockop, 1967b). Thus the values for prolyl 3-hydroxylase
are probably not significantly different from those for the prolyl 4-
hydroxylases.

The prolyl 4-hydroxylase $\alpha_2\beta_2$ tetramers have two catalytic sites.
This assumption is based on the findings that two Fe^{2+} atoms are
present in one enzyme tetramer (de Jong and Kemp, 1982) and that
each of the two α subunits contains a 2-oxoglutarate binding site (de
Waal et al., 1988) and a peptide binding site (de Waal et al., 1985;

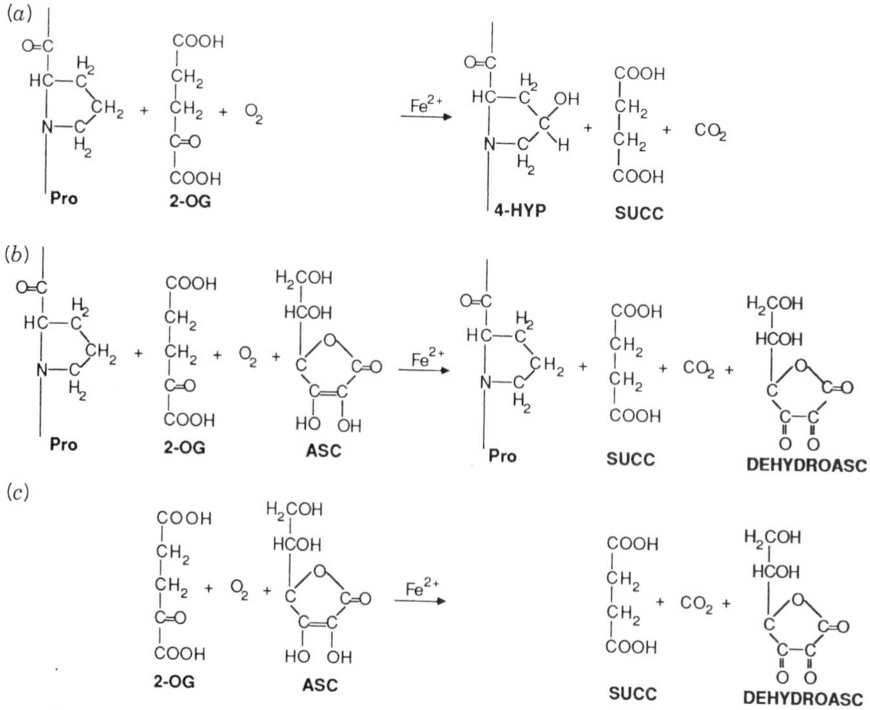

Figure 3. Schematic representation of the reactions catalyzed by prolyl 4-hydroxylase. The 2-oxoglutarate is stoichiometrically decarboxylated during the hydroxylation reaction, which does not need ascorbate (a). The enzyme also catalyzes an uncoupled decarboxylation of 2-oxoglutarate without subsequent hydroxylation of the peptide substrate. Ascorbate is stoichiometrically consumed in the uncoupled reaction cycles, which may occur either in the presence (b) or absence (c) of the peptide substrate. Pro = proline; 4-Hyp = 4-hydroxyproline; 2-Og = 2-oxoglutarate; Succ = succinate; Asc = ascorbate; Dehydroasc = dehydroascorbate. [Reproduced from Kivirikko et al., 1989 with permission.]

Helaakoski, Myllylä, Günzler, Tripier, Henke, and Kivirikko, unpublished observations). Most, if not all, of the sequences contributing to the catalytic sites are located in the α subunits. Data supporting this location have been obtained in photoaffinity labeling experiments with analogues of 2-oxoglutarate (de Waal et al., 1987, 1988) and the peptide substrate (de Waal et al., 1985) and in binding studies

TABLE 2

Apparent K_m Values for the Cosubstrates of Vertebrate Collagen Hydroxylases[a]

		K_m for Cosubstrate		
Enzyme and Substrate	Fe^{2+} (μM)	2-Oxoglutarate (μM)	O_2 (μM)	Ascorbate (μM)
Type I prolyl 4-hydroxylase				
Synthetic peptide substrate	4	22	43	330
Biological substrate	2	5	ND[b]	100
Type II prolyl 4-hydroxylase				
Synthetic peptide substrate	4	12	ND[b]	340
Lysyl hydroxylase				
Synthetic peptide substrate	2	100	45	350
Prolyl 3-hydroxylase				
Biological substrate	2	3	30	120

[a] The values shown for types I and II prolyl 4-hydroxylases with the synthetic peptide substrate were determined for the recombinant human and mouse enzymes, respectively (Helaakoski et al., 1995), except that the K_m for O_2 was determined only for the chick enzyme (Myllylä et al., 1977). The values shown for lysyl hydroxylase were likewise determined for the recombinant human enzyme (Pirskanen et al., 1996), except that the K_m for O_2 is for the nonrecombinant human enzyme (Turpeenniemi-Hujanen et al., 1981). The values for type I prolyl 4-hydroxylase with the biologically prepared protocollagen substrate (Kivirikko and Prockop, 1967b) and for prolyl 3-hydroxylase (Tryggvason et al., 1979) were obtained for the chick enzymes by varying the concentration of only the cosubstrate to be studied, and are therefore probably slightly too low.

[b] ND = not determined.

with two suicide inactivators that act as analogues of 2-oxoglutarate (Günzler et al., 1987) and the peptide substrate (Günzler et al., 1988a). The only data suggesting that some of the sequences contributing to the catalytic sites or found close to these sites may be located in the PDI/β subunits come from experiments with a suicide inactivator that acts as an ascorbate analogue (Günzler et al., 1988b). Most of the binding of this inactivator occurred in the α subunit, but some was also found in the PDI/β subunit. Most of the inhibitory monoclonal antibodies likewise react with the α subunit, but some do react with the PDI/β subunit (Höyhtyä et al., 1984; Bai et al., 1986). It may be noted that the prolyl 4-hydroxylase monomer from unicellular and multicellular green algae (Kaska et al., 1987, 1988), which does not have any PDI/β subunit (Section III.A), must have all the sequences contributing to the catalytic site in its α subunit.

   Kinetic studies on the prolyl 4-hydroxylase and lysyl hydroxylase reactions have demonstrated that both involve an ordered binding of Fe^{2+}, 2-oxoglutarate, O_2, and the peptide substrate to the enzyme and an ordered release of the reaction products, in which Fe^{2+} is not released between catalytic cycles (Myllylä et al., 1977, 1978; Tuderman et al., 1977; Puistola et al., 1980a,b; Günzler et al., 1986). Based on these and other data, a detailed stereochemical mechanism has been proposed for prolyl 4-hydroxylase in which the catalytic cycle can be divided into two half-reactions: initial generation of the reactive hydroxylating species and its subsequent utilization for 4-hydroxyproline formation (Hanauske-Abel and Günzler, 1982; Hanauske-Abel, 1991).

   The stereochemical mechanism assumes that the Fe^{2+} is located in a pocket (Hanauske-Abel and Günzler, 1982; Hanauske-Abel, 1991) coordinated with the enzyme by three side chains (L_1-L_3 in Fig. 4). The pocket may be highly hydrophobic, as has recently been shown for a related enzyme, isopenicillin N synthase (Roach et al., 1995). Early data suggested that some of the side chains L_1-L_3 might be cysteine residues (Kivirikko et al., 1992), but recent site-directed mutagenesis studies have shown that four out of the five cysteine residues present in the α(I) subunit of human prolyl 4-hydroxylase are probably involved in two intrachain disulfide bonds that are essential for tetramer assembly or stability (John and Bulleid, 1994; Lamberg et al., 1995), and mutation of the only remaining cysteine

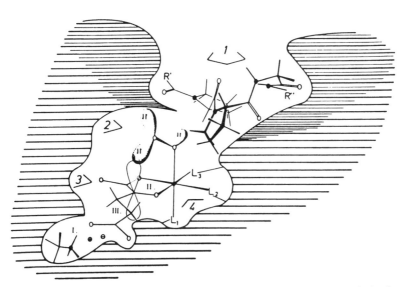

Figure 4. Binding of the cosubstrates and the peptide substrate at a catalytic site of prolyl 4-hydroxylase. The iron (4) is assumed to be located in a pocket coordinated with the enzyme by three side chains (L_1-L_3). 2-Oxoglutarate (3) is bound to subsites I–III in its energetically most stable, staggered conformation. Molecular oxygen (2) is thought to be bound end-on in an axial position, leading to a superoxide-like structure. The peptide substrate (1) is sterically oriented to participate in the hydroxylation reaction stereospecifically. [Reproduced from Majamaa et al., 1984 with permission.]

residue to serine has only a minor effect on the catalytic activity (Lamberg et al., 1995).

Experiments on the inactivation of collagen hydroxylases by diethyl pyrocarbonate and the prevention of this inactivation by cosubstrates of the reaction have suggested that histidine residues are functional at the catalytic sites, probably Fe^{2+} binding sites (Myllylä et al., 1992). A search for conserved amino acids within the sequences of several 2-oxoglutarate dioxygenases and a related enzyme, isopenicillin N synthase, (Carr et al., 1986; Samson et al., 1987; Holdsworth et al., 1987; Diekman and Fischer, 1988; Helaakoski et al., 1989; Kovacevic et al., 1989; McGarvey et al., 1990; Kovacevic and Miller, 1991; Matsuda et al., 1991; Myllylä et al., 1991) demonstrated a weak homology within two histidine-containing motifs, His-1 and His-2 (Fig. 5), located about 50–70 amino acids apart (Myllylä et al., 1992), the histidines concerned being residues 412 and 483 in the α(I) subunit of human prolyl 4-hydroxylase and residues 638 and 690 in human lysyl hydroxylase (this numbering differs from that used originally, which began with the first residue of the signal peptide). Subsequent site-directed mutagenesis studies of the histidine residues present in the α (I) subunit of human prolyl 4-hydroxylase demonstrated that mutation of either His-412 or 483 to serine inactivated the enzyme completely but had no effect on tetramer assembly or binding of the tetramer to poly(L-proline), suggesting that neither of these mutations caused any major overall changes in enzyme structure (Lamberg et al., 1995). Mutation of His-501 inactivated the enzyme by about 95%, whereas mutation of many other histidines had either a more minor effect or no effect at all (Lamberg et al., 1995). As analyses of isopenicillin N synthase by a variety of techniques had suggested that all three Fe^{2+} binding ligands in that enzyme may be histidines (Jiang et al., 1991; Ming et al., 1991; Randall et al., 1993), it was initially considered possible that histidine 501 might provide the third Fe^{2+} binding ligand in prolyl 4-hydroxylase, even though its mutation did not inactivate the enzyme completely (Lamberg et al., 1995). Subsequent determination of the crystal structure of isopenicillin N synthase (Roach et al., 1995) indicated, however, that although two of the iron binding ligands are histidines corresponding to those in the His-1 and His-2 motifs, the third ligand is a conserved aspartic acid residue located in position $+2$ with respect to the histidine present in the His-1 motif (Fig. 5). A corresponding aspartic acid residue is found both in the α subunits of all prolyl 4-hydroxylases studied and in lysyl hydroxylase (Fig. 5), and it was demonstrated recently that mutation of Asp-414 to alanine in the human α(I) subunit completely inactivates type I prolyl 4-hydroxylase (Myllyharju and Kivirikko, 1997). Mutation of this aspartate to glutamate increased the K_m for Fe^{2+} about 15-fold with no change in the V of the reaction (Myllyharju and Kivirikko, 1997). Furthermore, it was demonstrated that mutation of His-638 or 690 to serine or Asp-640 to alanine completely inactivates human lysyl hydroxylase (Pirskanen et al., 1996). It thus seems very likely that the three Fe^{2+} binding ligands in the α(I) subunit of human prolyl 4-hydroxylase are His-412, Asp-414, and His-483 (Fig. 6), while the

His-1 Motif

		Motif	
1	DAOC/DAC synthase	PHYDLSTIT	-59 aa -
2	DAOC synthase	PHYDLSMVT	-59 aa -
3	DAC synthase	PHYDLSIIT	-59 aa -
4	IPN synthase	WHEDVSLIT	-54 aa -
5	LH	PHHDASTFT	-52 aa -
6	4-PH, α-subunit	PHFDFARKD	-71 aa -
7	Plant E3	AHTDAGGLI	-56 aa -
8	Plant TOM13	AHTDAGGLI	-56 aa -
9	Plant E8	QHTDIGFVT	-56 aa -
10	H6H	GHYDGNLIT	-49 aa -

His-2 Motif

1	HRVKSPGRDQVGSSRTSSVFLRP
2	HHVAAPRRDQAGSSRTSSVFLRP
3	HHVRSPGAGMEGSDRTSSVFLRP
4	HRVKW.....VNEERQSLPFVNL
5	HEGL.PT...TKGTRYIAVSIDP
6	HAAC.PV...LVGNKWVSNKLHE
7	HRVVAQ....TDGNRMSLASYNP
8	HRAISN....NVGSRMSITCFGE
9	HRVIAQ....TDGTRMSLASIDP
10	HRVV......TDPTRDRVSITLI

Figure 5. Comparison of amino acid sequences showing the proposed His-1 and His-2 motifs in a number of enzymes. The sequences shown are for deacetoxycephalosporin C/deacetylcephalosporin C (DAOC/DAC) synthase from *Cephalosporium acremomium* (Samson et al., 1987), DAOC synthase from *Streptomyces clavuligerus* (Kovacevic et al., 1989), DAC synthase from *Streptomyces clavuligerus* (Kovacevic and Miller, 1991), isopenicillin N (IPN) synthase from *Penicillum chrysogenum* (Carr et al., 1986), lysyl hydroxylase (LH) from chick (Myllylä et al., 1991), the type I α subunit of prolyl 4-hydroxylase (4-PH) from human (Helaakoski et al., 1989), plant E3 from avocado (McGarvey et al., 1990), plant TOM13 (Holdsworth et al., 1987) and E8 (Diekman and Fischer, 1988) from tomato, and hyoscyamine 6β hydroxylase (H6H) from *Hyoscyamus niger* (Matsuda, 1991). White letters on a black background indicate identity and black letters on a gray background indicate similarity. The invariant histidine residues in each motif and the aspartic acid residue located in position +2 with respect to the histidine present in the His-1 motif are indicated by asterisks. The numbers are the numbers of amino acids (aa) between the two motifs. Similar amino acids: Gly = Ala = Ser; Ala = Val; Val = Ile = Leu = Met; Ile = Leu = Met = Phe = Tyr = Trp; Lys = Arg = His; Asp = Glu = Gln = Asn; Ser = Thr = Glu = Asn. [Slightly modified from Myllylä et al., 1992 with permission.]

corresponding ligands in human lysyl hydroxylase are His-638, Asp-640, and His-690.

The Fe^{2+} in isopenicillin N synthase further possesses between catalytic cycles two water molecules occupying coordination sites directed into the hydrophobic cavity within the enzyme and a fourth protein ligand, a glutamine residue, which is not conserved through other members of the family of related enzymes (Roach et al., 1995). It therefore seems likely that the Fe^{2+} of the collagen hydroxylases also contains water molecules and an additional protein ligand occupying coordination sites involved in the subsequent binding of the 2-oxoglutarate and O_2 molecules.

2-Oxoglutarate is a highly specific requirement (Rhoads and Udenfriend, 1968; Cardinale and Udenfriend, 1974; Prockop et al., 1976; Kivirikko and Myllylä, 1980; Kivirikko et al., 1992). Nevertheless, it can be replaced by 2-oxoadipinate, although the K_m for the latter is markedly higher (Majamaa et al., 1984, 1985). The K_m of 2-oxoglutarate for lysyl hydroxylase is significantly higher than for the other collagen hydroxylases (Table 2), suggesting that there may be some differences in the structures of the 2-oxoglutarate binding sites between collagen hydroxylases. The 2-oxoglutarate binding site can be divided into three distinct subsites (Hanauske-Abel and Günzler, 1982; Majamaa et al., 1984; Hanauske-Abel, 1991). Subsite I is a positively charged side chain of the enzyme that ionically binds the C5 carboxyl group of 2-oxoglutarate, subsite II consists of two cis-positioned coordination sites of the enzyme-bound Fe^{2+} and is chelated by the C1-C2 moiety, while subsite III involves a hydrophobic binding site in the C3-C4 region of the cosubstrate (Fig. 4). Site-directed mutagenesis studies (Myllyharju and Kivirikko, 1997) suggest that the residue forming subsite I in the $\alpha(I)$ subunit of human prolyl 4-hydroxylase is Lys-493 (Fig. 6). Mutation of this residue to alanine or histidine inactivated the enzyme completely, whereas mutation to arginine inactivated it by 85% under standard assay conditions and increased the K_m for 2-oxoglutarate by about 15-fold (Myllyharju and Kivirikko, 1997). The K_i values of the Lys-493 Arg enzyme for competitive inhibitors (Section VII) that become bound at all subsites of the 2-oxoglutarate binding site were likewise increased, whereas those for compounds that cannot bind to subsite I were unaltered (Myllyharju and Kivirikko, 1997).

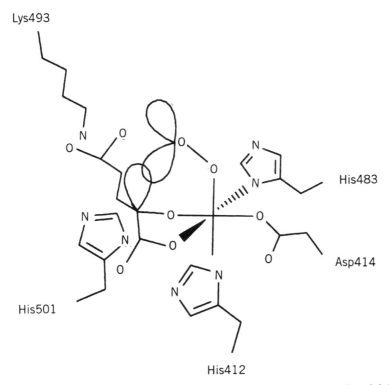

Figure 6. Critical residues at a cosubstrate binding site of human type I prolyl 4-hydroxylase. The Fe^{2+} is bound to the $\alpha(I)$ subunit by three residues, His-412, Asp-414, and His-483. The 2-oxoglutarate binding site can be divided into two main sub-sites: subsite I consists of Lys-493, which ionically binds the C5 carboxyl group of 2-oxoglutarate, while subsite II consists of two cis-positioned equatorial coordination sites of the enzyme-bound Fe^{2+} and is chelated by the C1 carboxyl and C2 oxo functions. Molecular oxygen bound to the iron is also shown. His-501 is an additional important residue at the cosubstrate binding site, probably involved in both coordination of the C1 carboxyl group of 2-oxoglutarate to Fe^{2+} and the decarboxylation of this cosubstrate. [Reproduced from Myllyharju and Kivirikko, 1997 with permission.]

Molecular oxygen is assumed to be bound to the Fe^{2+} end-on in an axial position (Fig. 4), producing a species characterized by doubly occupied orbitals, the dioxygen unit (Hanauske-Abel and Günzler, 1982; Hanauske-Abel, 1991). One of the electron-rich orbitals of the dioxygen unit is sterically directed into the electron-depleted orbital at C2 of the 2-oxoglutarate bound to the iron atom [Fig. 7(a)], so that the C2 undergoes rehybridization from its sp^2 hybridized planar oxo structure to an sp^3 hybridized tetrahedral transition state, forming a covalent bond with the noncooridinated atom of the dioxygen [Fig. 7(b)]. This weakens both the C—C bond in the 2-oxoglutarate and the O—O bond in the dioxygen unit (Hanauske-Abel and Günzler, 1982; Hanauske-Abel, 1991). Decarboxylation will then occur simultaneously with cleavage of the O—O bond, and the original C2 of 2-oxoglutarate, which has become C1 of succinate, returns to the sp^2 hybridization. At the same time a highly reactive Fe—O complex, a ferryl ion, is formed [Fig. 7(c)], which hydroxylates the proline or lysine residue in the peptide substrate in the second half-reaction (Hanauske-Abel and Günzler, 1982; Hanauske-Abel, 1991). This event renews the Fe^{2+} at the catalytic site and concludes the catalytic cycle.

Histidine 501 appears to be an additional critical residue at the catalytic site, as its mutation to serine, asparagine, or glutamine decreases the V by about 95%, while mutation to lysine or arginine inactivates by 85–90% (Lamberg et al., 1995; Myllyharju and Kivirikko, 1997). The K_m values of all the His-501 mutant enzymes for 2-oxoglutarate were increased about two- to three-fold, whereas the K_m values for Fe^{2+}, ascorbate, and peptide substrates were the same as those of the wild-type enzyme (Lamberg et al., 1995; Myllyharju and Kivirikko, 1997). A surprising finding was that the rate of the uncoupled decarboxylation [Fig. 3(c)] was decreased far less than the rate of the complete reaction, and thus the percentage of the rate of the uncoupled decarboxylation was increased up to 12-fold with all the histidine 501 mutants but not with any other mutant studied so far (Myllyharju and Kivirikko, 1997). These and additional data suggest that histidine 501 plays two roles at the catalytic site (Fig. 6). It directs the orientation of the C1 carboxyl group of 2-oxoglutarate to the active iron center and accelerates the breakdown of the tetrahedral ferryl intermediate [Fig. 7(b)] to succinate, CO_2 and ferryl ion (Myllyharju and Kivirikko, 1997).

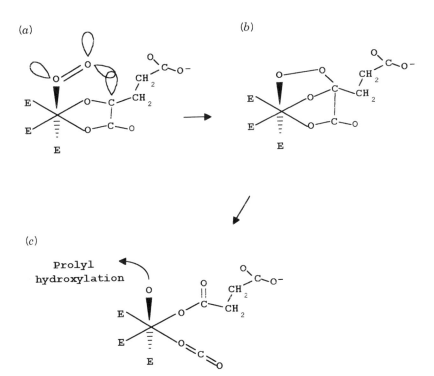

Figure 7.   Schematic representation of the first half-reaction of prolyl 4-hydroxylase that takes place within the coordination sphere of the catalytic site Fe^{2+}. (a) One of the electron-rich orbitals of the dioxygen unit overlaps with the electron-depleted orbital at the planar aliphatic C2 of the 2-oxoglutarate bound to the iron atom. The electronically unusual states of these orbitals are due to interaction between the ligands and the central ion, that is, the end-on coordination of O_2, the chelation of 2-oxoglutarate via its C1-C2 moiety, and specific orbitals of the Fe^{2+} that facilitate electron redistribution. Consequently, the nucleophilicity of the noncoordinated atom of dioxygen is enhanced, as is the susceptibility of C2 to nucleophilic attack. (b) Nucleophilic attack on C2 generates a tetrahedral intermediate, with loss of the double bond in the dioxygen unit and of double bond characteristics in the oxo-acid moiety. (c) Elimination of CO_2 coincides with the formation of succinate and an iron–oxo complex, which hydroxylates the proline residue in the peptide substrate in the second half-reaction E, enzyme. [Slightly modified from Hanauske-Abel, 1991 with permission.]

Ascorbate is not required in the reaction scheme discussed above, and the collagen hydroxylases can complete many catalytic cycles at an essentially maximal rate in its absence (Tuderman et al., 1977; Myllylä et al., 1978; Puistola et al., 1980a). Hydroxylation then ceases, however, and ascorbate is required to reactivate the enzyme (Myllylä et al., 1978; de Jong et al., 1982). The reaction requiring ascorbate is an uncoupled decarboxylation of 2-oxoglutarate [Fig. 3(c)], in which ascorbate is consumed stoichiometrically with the decarboxylation of 2-oxoglutarate (Myllylä et al., 1984; de Jong and Kemp, 1984). In these uncoupled reaction cycles the reactive iron–oxo complex is probably converted to $Fe^{3+} \cdot O^-$ making the enzyme unavailable for new catalytic cycles until reduced by ascorbate (Günzler et al., 1988a; de Jong and Kemp, 1984; Myllyä et al., 1984). The collagen hydroxylases catalyze uncoupled decarboxylation cycles even in the presence of a saturating concentration of their peptide substrates [Fig. 3(b)], and certain nonhydroxylatable peptide sequences increase the rate of uncoupled decarboxylation (Tuderman et al., 1977; Rao and Adams, 1978; Myllylä et al., 1984). The biological peptide substrates of collagen hydroxylases contain many nonhydroxylatable sequences in addition to the substrate sequences. An interaction between some of the former sequences and the catalytic site may cause an uncoupled decarboxylation event. It is thus probable that the main biological function of ascorbate in the collagen hydroxylase reactions, both *in vitro* and *in vivo*, is that of serving as an alternative oxygen acceptor in the uncoupled decarboxylation cycles (Myllylä et al., 1984).

Ascorbate is a highly specific requirement, but it can be replaced by cysteine or dithiothreitol to a minor extent (Myllylä et al., 1978; Puistola et al., 1980a). A low hydroxylation rate has also been found in some cultured cells in the absence of ascorbate (Kivirikko and Myllylä, 1980; Kivirikko et al., 1992), the reductant possibly being a microsomal protein that contains cysteinyl–cysteine (Chauhan et al., 1985). The ascorbate binding site of the collagen hydroxylases probably also contains the two cis-positioned equatorial coordination sites of the enzyme-bound iron discussed above, and is thus partially identical to the binding site of 2-oxoglutarate (Majamaa et al., 1986). In agreement with this suggestion, modifications of the ring atoms of ascorbate that abolish the capacity to bind iron destroy the cosubstrate activity, as in L-galactono γ-lactone (Majamaa et al.,

1986) and 3-methoxy-ʟ-ascorbate (Tschank et al., 1994). Ascorbate can be fully replaced by many derivatives differing only in their side chain such as ᴅ-isoascorbate, 5,6-O-isopropylidene ʟ-ascorbate, and 5-deoxy-ʟ-ascorbate, but the presence of a very large hydrophobic substituent or a positively charged group in the side chain increases the K_m (Majamaa et al; 1986; Tschank et al., 1994).

It is still not known how the collagen hydroxylases act on their biological substrates or other extended polypeptides with a number of hydroxylatable proline or lysine residues. A mechanism has been proposed in which two large peptide binding sites on the enzyme act processively and bind to the same polypeptide substrate (de Waal and de Jong, 1988; de Jong et al., 1991). This would prevent dissociation of the enzyme–substrate complex between successive interactions of a long peptide with multiple substrate sites (de Waal and de Jong, 1988; de Jong et al., 1991). Nevertheless, the kinetic constants of the C. elegans prolyl 4-hydroxylase $\alpha\beta$ dimer, which contains only one catalytic site, for long peptide substrates are identical to those of vertebrate prolyl 4-hydroxylase $\alpha_2\beta_2$ tetramers (Veijola et al., 1996a), a finding that is not consistent with the proposed processive mechanism.

VII. Inhibitors and Inactivators: Potentials for the Development of Antifibrotic Drugs

The formation of scars and fibrous tissue is part of the normal beneficial process of healing after injury, but in some situations the fibrous material accumulates in excessive amounts, leading to impairment of the normal functioning of the affected tissue. The central role of collagen in fibrosis of the liver, lungs, kidneys, and other organs has prompted attempts to develop drugs that inhibit collagen accumulation. The prolyl 4-hydroxylase reaction would seem a particularly suitable target for antifibrotic therapy, as inhibition of this reaction will prevent collagen triple helix formation and thus lead to a nonfunctional protein that is rapidly degraded. Many attempts have therefore been made to develop inhibitors of this reaction (for reviews, see Kivirikko and Majamaa, 1985; Kivirikko and Savolainen, 1988; Kivirikko et al., 1992; Franklin, 1995).

Inhibitors are now known for collagen hydroxylases with respect to their peptide substrates and all the cosubstrates. Many peptides are competitive inhibitors with respect to the peptide substrates (Kivirikko et al., 1992). Poly(L-proline) has been regarded as a well-recognized, effective competitive inhibitor of prolyl 4-hydroxylase, the K_i decreasing with increasing chain length so that poly(L-proline) with a molecular weight of 15,000 has a K_i of 0.02 μM (Prockop and Kivirikko, 1969). This inhibition has been assumed to be related to the requirement of a poly(L-proline) II conformation in the X-Pro segment of nascent collagen peptides and their models at the peptide binding site of prolyl 4-hydroxylases (Section V). Recent characterization of the vertebrate type II prolyl 4-hydroxylase nevertheless indicated, surprisingly, that this enzyme is inhibited by poly(L-proline) only at very high concentrations, the difference in K_i between the vertebrate type I and II enzymes being about 1000-fold (Helaakoski et al., 1995). The C. elegans prolyl 4-hydroxylase dimer (Veijola et al., 1994, 1996a) and prolyl 4-hydroxylase from Ascaris lumbricoides (Fujimoto and Prockop, 1969) resemble the vertebrate type II enzyme in this respect. These findings indicate that there must be some differences in the structure of the peptide binding site between various prolyl 4-hydroxylases. Nevertheless, the K_m values of the vertebrate type II enzyme for various polypeptide substrates are only about three-to-six times those of the type I enzyme (Helaakoski et al., 1995; Annunen et al., 1997), suggesting that various prolyl 4-hydroxylases have only minor differences in the requirements concerning the conformation of the polypeptide substrates at their binding and catalytic sites.

Many divalent cations inhibit collagen hydroxylases competitively with respect to Fe^{2+}, the most effective of these being Zn^{2+}, with a K_i of about 1 μM for prolyl 4-hydroxylase and lysyl hydroxylase (Kivirikko and Majamaa, 1985; Kivirikko and Savolainen, 1988; Kivirikko et al., 1992). Superoxide dismutase active copper chelates such as $Cu(acetylsalicylate)_2$ and $Cu(lysine)_2$ inhibit them competitively with respect to O_2, the K_i of prolyl 4-hydroxylase and lysyl hydroxylase for these two compounds being 30 μM (Myllylä et al., 1979). Inhibition probably takes place by dismutation of the activated form of oxygen at the catalytic site. Superoxide dismutase itself does not do this, however, probably because it cannot reach the activated form of oxygen in the catalytic site pocket (Kivirikko et al., 1992).

The most potent inhibitors with respect to 2-oxoglutarate (Fig. 8)

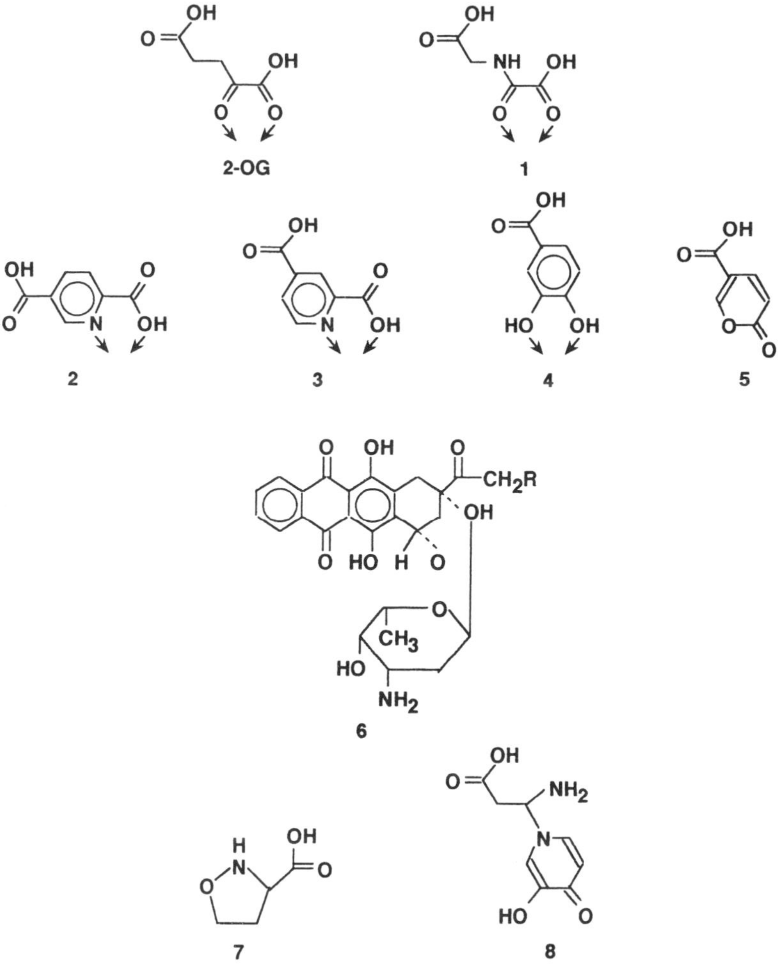

Figure 8.   Structures of certain compounds that inhibit or inactivate prolyl 4-hydroxylase. Oxalylglycine (1), pyridine 2,5-dicarboxylate (2), pyridine 2,4-dicarboxylate (3), and 3,4-dihydroxybenzoate (4) inhibit it competitively with respect to 2-oxoglutarate (2-OG), 3,4-dihydroxybenzoate also being a competitive inhibitor with respect to ascorbate. Coumalic acid (5), doxorubicin (6), and peptides containing 5-oxaproline (7) probably inactivate the enzyme by a suicide mechanism. Mimosine (8) is also shown.

all have functional groups that can bind to the Fe^{2+} at the catalytic site, and these compounds can also interact at the other subsites of the 2-oxoglutarate binding site (Majamaa et al., 1984, 1985, 1986; Cunliffe and Franklin, 1986). Particular interest has been focused on pyridine 2,4-dicarboxylate and its derivatives in recent years (Hanauske-Abel and Günzler, 1982; Majamaa et al., 1984; Hanauske-Abel, 1991). The K_i of pyridine 2,4-dicarboxylate for prolyl 4-hydroxylase is about 2 μM, while its K_i for prolyl 3-hydroxylase is 3 μM and that for lysyl hydroxylase 50 μM (Majamaa et al., 1984, 1985). A higher K_i for lysyl hydroxylase than for the prolyl hydroxylases is found with almost all the aliphatic and aromatic 2-oxoglutarate analogues (Majamaa et al., 1984, 1985), and may be related to the higher K_m of 2-oxoglutarate for lysyl hydroxylase (Section VI). Pyridine 2,4-dicarboxylate is only a very weak inhibitor of 2-oxoglutarate dehydrogenase (Majamaa et al., 1985), an enzyme that differs from the collagen hydroxylases in that its reaction mechanism does not involve any metal ion. This demonstrates the importance of chelation of the enzyme-bound Fe^{2+} for the binding of various compounds at the 2-oxoglutarate site of the collagen hydroxylases, and suggests that it should be possible to develop highly effective prolyl 4-hydroxylase inhibitors that show a high degree of specificity.

Systematic variation in the structure of pyridine 2,4-dicarboxylate demonstrates that if the chelating moiety is destroyed by omission or shifting of the C2 carboxyl group or substitution of the aromatic nitrogen, the inhibitory effect is substantially decreased (Majamaa et al., 1984). Omission of the other carboxyl group thought to be bound at subsite I, or shifting of this group to a position other than 4 or 5, also markedly reduces the inhibitory potential (Majamaa et al., 1984).

N-Oxalylglycine (Fig. 8) is also a potent prolyl 4-hydroxylase inhibitor, its apparent K_i varying between 0.5 and 8 μM depending on the conditions and the variable substrate used in its determination (Cunliffe et al., 1992; Baader et al., 1994). This compound differs from 2-oxoglutarate only by replacement of the methylene group at C3 by -NH-. Oxalylglycine inhibits competitively with respect to 2-oxoglutarate, but it cannot replace 2-oxoglutarate as a cosubstrate in spite of the presence of a 2-oxocarboxylic acid moiety in the molecule (Cunliffe et al., 1992; Baader et al., 1994). This supports the view that decarboxylation involves a nucleophilic attack by an electron pair in the dioxygen unit on the electron-deficient orbital of the C2 of 2-oxoglutarate (Section VI). The neighboring C3 atom in 2-oxoglutarate is sp^3 hybridized and cannot contribute electron density to the p_z orbital of C2 (Baader et al., 1994). In oxalyglycine the C3 is replaced by the sp^2 hybridized nitrogen atom. Its p_z orbital contains an electron pair that can contribute to the electron density of C2, so that a nucleophilic attack may no longer be possible (Baader et al., 1994). Oxalylalanine, which differs from oxalylglycine in that it carries a methyl group at the carbon atom corresponding to the C4 of 2-oxoglutarate, is also an effective inhibitor, with a K_i of 40 μM (Cunliffe et al., 1992; Baader et al., 1994).

Pyridine 2,5-dicarboxylate (Fig. 8) is likewise a competitive inhibitor with respect to 2-oxoglutarate (Majamaa et al., 1984). Its K_i for prolyl 3-hydroxylase and lysyl hydroxylase is distinctly higher than that of pyridine 2,4-dicarboxylate, but surprisingly, its K_i for prolyl 4-hydroxylase is lower than the K_i of pyridine 2,4-dicarboxylate, being 0.8 μM (Majamaa et al., 1984, 1985). The C5 carboxyl group can be replaced by acyl sulfonamides (Dowell and Hadley, 1992) and amides (Tucker and Thomas, 1992), with a retention or improvement in potency as compared with pyridine 2,5-dicarboxylate. Also, the C2 carboxyl group of pyridine 2-carboxylic acid can be successfully replaced by groups capable of forming a five-ring chelate with enzyme-bound Fe^{2+} (Dowell et al., 1993). Nevertheless, it has been pointed out that pyridine 2,5-dicarboxylate and 2-oxoglutarate may bind to prolyl 4-hydroxylase in different ways, and thus the data concerning modifications of the pyridine 2,5-dicarboxylate structure must be interpreted with caution (Cunliffe et al., 1992).

3,4-Dihydroxybenzoate also possesses functional groups required for binding at all the subsites of the 2-oxoglutarate binding site (Fig. 8) and is an effective inhibitor, with a K_i of about 5 μM for prolyl 4-hydroxylase (Majamaa et al., 1986). This compound and its derivatives nevertheless differ from the pyridine derivatives in their mode of inhibition with respect to ascorbate. 3,4-Dihydroxybenzoate is a competitive inhibitor with respect to both 2-oxoglutarate and ascorbate, whereas the pyridine derivatives are uncompetitive with respect to ascorbate (Majamaa et al., 1986). This agrees with the notion that the ascorbate binding site is partially identical to the binding site of 2-oxoglutarate (Section VI). The dihydroxybenzoate and pyridine inhibitors probably react with different enzyme forms, as determined by the oxidation state of the iron atom at the catalytic site (Majamaa et al., 1986).

Many of the prolyl 4-hydroxylase inhibitors show a low membrane permeability and are therefore ineffective in cultured cells and *in vivo*. Nevertheless, additional modifications of these structures can be used to overcome the problem. Lipophilic proinhibitor derivatives of these compounds do not in themselves inhibit the pure enzymes, but readily pass through cell membranes and become converted to the active inhibitor only inside the cell (Majamaa et al., 1987; Sasaki et al., 1987; Hanauske-Abel, 1991; Tschank et al., 1991; Baader et al., 1994). Such proinhibitors include ethylpyridine 2,4-dicarboxylate (Tschank et al., 1991), pyridine 2,4-dicarboxylic acid di(methoxyethyl)amide (HOE 077, Hoechst) (Hanauske-Abel, 1991), dimethyloxalylglycine (Baader et al., 1994), and ethyl 3,4-dihydroxybenzote (Majamaa et al., 1987; Sasaki et al., 1987). The HOE 077 proinhibitor has been shown to selectively inhibit collagen accumulation in the liver of rats in two models of hepatic fibrosis (Bickel et al., 1991; Böker et al., 1991), and it now seems possible that some of the prolyl 4-hydroxylase inhibitors could be developed into effective drugs for the treatment of patients with fibrotic disorders.

L-Mimosine (Fig. 8) is also a prolyl 4-hydroxylase inhibitor, the concentration required for 50% inhibition being about 120 μM (McCaffrey et al., 1995). In experiments with cultured human and rat vascular smooth muscle cells L-mimosine caused a reversible, dose-dependent inhibition of DNA synthesis within 24 h (McCaffrey et al., 1995), and concurrent inhibition was found in the hydroxylation of deoxyhypusine to hypusine, a rare amino acid required by the eukaryotic initiation factor eIF-5A and cell cycle transition. Prolyl 4-hydroxylase activity in vascular smooth muscle cells was inhibited by more than 80% with 400 μM L-mimosine, and the production of collagen by these cells was correspondingly markedly reduced. A single compound thus appears to cause specific inhibition of both deoxyhypusyl hydroxylase and prolyl 4-hydroxylase, leading to both antiproliferative and fibrosuppressive effects (McCaffrey et al., 1995).

A number of other compounds have also been shown to inhibit prolyl 4-hydroxylase, either in experiments with pure enzyme or cultured cells or *in vivo* (Kivirikko et al., 1992; Shigematsu et al., 1994), but their mode of action has not been elucidated in detail.

Many compounds have also been identified that serve as irreversible inactivators of collagen hydroxylases, probably by a suicide mechanism. As indicated in Section VI, all these compounds have been shown to become covalently bound to the catalytic site of the enzymes (Günzler et al., 1987, 1988a, b; Helaakoski, Myllylä, Günzler, Tripier, Henke, and Kivirikko, unpublished observations). The compounds can be divided into three groups depending on whether they react at the 2-oxoglutarate, ascorbate, or peptide binding sites.

The first group includes coumalic acid (Fig. 8), which appears to act as a 2-oxoglutarate analogue (Günzler et al., 1987). This inactivation can be prevented by the addition of 2-oxoglutarate or its competitive analogues before but not after the inactivation period (Günzler et al., 1987). Nevertheless, coumalic acid is not an effective inactivator, as a concentration of about 2 mM is required to inactivate prolyl 4-hydroxylase by 50% in 1 h, and an even higher concentration is required for a similar degree of inactivation of lysyl hydroxylase (Günzler et al., 1987). The high concentrations required are probably explained by the lack of a functional group able to bind to the Fe^{2+} at a catalytic site (Fig. 8).

The second group consists of the anthracyclines doxorubicin (Fig. 8) and daunorubicin (Günzler et al., 1988b). The inactivation caused by these compounds can be prevented by high concentrations of ascorbate or low concentrations of its competitive analogues, whereas 2-oxoglutarate and its analogues offer no protection. The anthracyclines thus probably react with the $Fe^{3+} \cdot O^-$ form of the enzyme in the uncoupled reaction cycles (Günzler et al., 1988b). Doxorubicin and daunorubicin inactivate prolyl 4-hydroxylase by 50% in 1 h at a concentration of 60 μM (Günzler et al., 1988b).

The third group of suicide inactivators consists of peptides in which the proline residue to be hydroxylated has been replaced by 5-oxaproline (Fig. 8), a proline analogue containing oxygen as part of the ring (Günzler et al., 1988a). The most effective of these peptides tested has the structure benzyloxycarbonyl-Phe-Oxaproline-Gly-benzylester and inactivates prolyl 4-hydroxylase by 50% in 1 h at a concentration of 0.8 μM (Günzler et al., 1988a). Replacement of the glycine by other amino acids, including β-alanine, completely eliminates the inactivating effect. Replacement of the phenylalanine by alanine, valine, or glutamate reduces its potency in this order (Günzler et al., 1988a). Replacement of either of the two aromatic blocking groups by an aliphatic residue likewise distinctly reduces the inactivating effect. Oxaproline-containing peptides also inactivate prolyl 4-hydroxylase in cultured human skin fibroblasts, but the concentration needed for a similar degree of inactivation is about one order of a magnitude higher than that required with the pure enzyme (Karvonen et al., 1990).

VIII. Detailed Structures of the Catalytically Important α Subunits of Prolyl 4-Hydroxylases

As indicated in Section VI, most, if not all, of the sequences contributing to the catalytic sites of the prolyl 4-hydroxylases are located in the α subunits. The α(I) subunit of the vertebrate prolyl 4-hydroxylases has now been cloned from humans (Helaakoski et al., 1989), rat (Hopkinson et al., 1994), mouse (Helaakoski et al., 1995), and chick (Bassuk et al., 1989), and the α(II) subunit from humans (Annunen et al., 1997) and mouse (Helaakoski et al., 1995). In addition, the α subunit has been cloned from C. elegans (Veijola et al., 1994, see also Section III). The final human, rat, and mouse α(I) subunits are 517 amino acid residues in length (Kuutti et al., 1975; Helaakoski et al., 1989; Hopkinson et al., 1994; Helaakoski et al., 1995), while the final chick α(I) subunit is one amino acid shorter (Bassuk et al., 1989). All these α subunits are synthesized in a form containing a signal peptide, the size of which is 17 amino acids in the human α(I) subunit (Helaakoski et al., 1989). The processed human and mouse α(II) subunits consist of 514 amino acids (Helaakoski et al., 1995; Annunen et al., 1997), while the processed C. elegans α subunit is longer, 542 amino acids (Veijola et al., 1994). This size difference is mainly due to a 32-amino acid C-terminal extension present in the C. elegans α subunit that is not found in the vertebrate α(I) and α(II) subunits (Fig. 9).

→

Figure 9.   Alignment of the amino acid residues of the mouse α(I) [Mα(I)], mouse α(II) [Mα(II)], and C. elegans α(C.e.α) subunits of prolyl 4-hydroxylase. An asterisk (*) = stop codon of translation. The caps (.) are introduced for maximal alignment of the polypeptides. White letters on a black background indicate identity and black letters on a gray background indicate similarity. Similar amino acids as in Figure 5. [Slightly modified from Helaakoski et al., 1995 with permission.]

```
Mα(I)      1   HPGFFTSIGQMTDLIHNEKDLVTSLKDYIKAEEDKLEQIK
Mα(II)     1   EFFTSIGHMTDLIYAEKDLVQSLKEYILVEEAKLAKIK
C.e.α      1   DLFTSIADMQNLLETERNIPKILDKYIHDEEERLVQLK

Mα(I)     41   KWAEKLDRLTSRATKDPEGFVGHPVNAFKLMKRLNTEWSELE
Mα(II)    39   SWASKMEALTSRSAADPEGYLAHPVNAYKLVKRLNTDWPALG
C.e.α     39   KLSEEYSKKNEISIENGLKDITNPINAFLLIKRKIFDWKEIE

Mα(I)     83   NLILKDMSDGFISNLTIQ..RQYFPNDEDQVGAAKALFRLQD
Mα(II)    81   DLVLQDASAGFVANLSVQ..RQFFPTDEDESGAARALMRLQD
C.e.α     81   SKMNANKAGNVVSSITDDSYGVRYPTADDLSGAAIGLLRLQD

Mα(I)    123   TYNLDTNNISKGNLPGVQHKSFLTAEDCFELGKVAYTEADYY
Mα(II)   121   TYKLDPDTISRGELPGTKYQAMLSVDDCFGLGRSAYNEGDYY
C.e.α    123   TYRLDTKDLADGKIYADQGNYTFSAKDCFEIARAAYNEHDFY

Mα(I)    165   HTELWMEQALTQLEEGELSTVDKVSVLDYLSYAVYQQGDLDK
Mα(II)   163   HTVLWMEQVLKQLDAGEEATVTKSLVLDYLSYAVFQLGDLHR
C.e.α    165   HTVMWMEEAQRRLGDEVEPTVEVEDILEYLAFALYKQNNLKH

Mα(I)    207   ALLLTKKLLELDPEHQRANGNLVYFEYIMSKEKDANKSASGD
Mα(II)   205   AVELTRRLLSLDPSHERAGGNLRYFERLLEEERGKSLS.NQT
C.e.α    207   ALKLTEELYKMNPTHPRAKGNVKWYEDLLEQEGVR...RSDM

Mα(I)    249   QSDQKTAPKKKGIAVDYLPERQKYEMLCRGEGIKMTPRRQKR
Mα(II)   246   DAGLATQENLYERPTDYLPERDVYESLCRGEGVKLTPRRQKK
C.e.α    246   RKNLPPIQNRRPDSVLGNTERTMYEALCRNE..VPVSRRHLR

Mα(I)    291   LFCRYHDGNRNPKFILAPAKQEDEWDKPRIIRFHDIISDAEI
Mα(II)   288   LFCRYHHGNRVPQLLIAPFKEEDEWDSPHIVRYYDVMSDEEI
C.e.α    286   LYCYYLAG..PSFLVYAPIKVEIKRFNPLAVLFKDVISDDEV

Mα(I)    333   EIVKYLAKPRLSRATVHDPETGKLTTAQYRVSKSAWLSGYED
Mα(II)   330   ERIKEIAKPKLARATVRDPKTGVLTVASYRVSKSSWLEEDDD
C.e.α    326   AAIQELAKPKLARATVHDSVTGKLVTATYRISKSAWLKEWEG

Mα(I)    375   PVVSRINMRIQDLTGLDVSTAEELQVANYGVGGQYEPHFDFA
Mα(II)   372   PVVARVNRRMQHITGLTVKTAELLQVANYGMGGQYEPHFDFS
C.e.α    368   DVVETVNKRIGYMTNLEMETAEELQIANYGIGGHYDPHFDHA

Mα(I)    417   RKDEPDAFRELGTGNRIATWLFYMSDVSAGGATVFPEVGASV
Mα(II)   414   RSDDEDAFKRLGTGNRVATFLNYMSDVEAGGATVFPDLGAAI
C.e.α    410   KKEESKSFESLGTGNRIATVLFYMSQPSHGGGTVFTEAKSTI

Mα(I)    459   WPKKGTAVFWYNLFASGEGDYSTRHAACPVLVGNKWVSNKWL
Mα(II)   456   WPKKGTAVFWYNLLRSGEGDYRTRHAACPVLVGCKWVSNKWF
C.e.α    452   LPTKNDALFWYNLYKQGDGNPDTRHAACPVLVGIKWVSNKWI

Mα(I)    501   HERGQEFRRPCTLSELE*
Mα(II)   498   HERGQEFLRPCGTTEVD*
C.e.α    494   HERGNEFRRPCGLKSSDYERFVGDLGYGPEPRNAPNVSPNLA

C.e.α    536   KDVWETL*
```

The overall amino acid sequence identity between the human and mouse α(I) subunits is 96% and that between the human and chick α(I) subunits is 87% (Helaakoski et al., 1989; Bassuk et al., 1989; Helaakoski et al., 1995). The degree of amino acid sequence identity between the mouse α(I) and α(II) subunits is 63%, and that between the mouse α(I) and C. elegans α subunits is 43%, which is about the same as between the mouse α(II) and C. elegans α subunits, 41% (Veijola et al., 1994; Helaakoski et al., 1995). This identity is not distributed equally, however, being highest within the C-terminal domain (Fig. 9). For example, the degree of amino acid sequence identity between the mouse α(I) and α(II) subunits for residues 394–511 (type I numbering) is 81%, while the degree for the corresponding region between the mouse α(I) subunit and the C. elegans α subunit is 69% and that between the mouse α(II) subunit and the C. elegans α subunit 62%. This region contains several highly conserved stretches of amino acids, such as a 19-amino acid sequence [residues 481–489 in the mouse α(I) subunit] in which the degree of identity between the mouse α(I) and α(II) subunits and the C. elegans α subunit is 95% (Fig. 9). The five critical residues at the cosubstrate binding site, His-412, Asp-414, His-483, Lys-493 and His-501 in the human α(I) subunit (Fig. 6) are all located in the C-terminal region. Binding studies with a radioactively labeled prolyl 4-hydroxylase suicide inactivator, a 5-oxaproline-containing peptide (Section VII), have likewise demonstrated that all the radioactivity covalently bound to the human type I enzyme tetramer can be traced to a single Staphylococcus aureus V8 protease peptide from the α subunit, extending from residue 398 to about 500 (Helaakoski, Myllylä, Günzler, Tripier, Henke, and Kivirikko, unpublished observations). It thus seems likely that many of the sequences contributing to the catalytic sites of the various prolyl 4-hydroxylases are located within the C-terminal region of their α subunits.

Two slightly different types of mRNA have been identified for the human (Helaakoski et al., 1989) and mouse (Helaakoski et al., 1995) α(I) subunits, differing over a stretch of 64 nucleotides. This is due to a mutually exclusive alternative splicing of sequences corresponding to two consecutive, homologous 71 bp exons 9 and 10 in the respective gene (Helaakoski et al., 1994). The differences in the nucleotide sequences of these two exons lead to differences in 10 out of the $23\frac{2}{3}$ amino acids coded by them. The amino acid sequences of the human and mouse α(II) subunits and the *C. elegans* α subunit clearly correspond to the type of α(I) subunit in which the alternatively spliced region is coded by exon 9, and therefore those sequences are shown in Fig. 9. Expression in insect cells has demonstrated that both types of α(I) subunit form an active prolyl 4-hydroxylase tetramer with the PDI/β polypeptide, there being no differences in catalytic properties between these two types of tetramer (Vuori et al., 1992b). The biological significance of the two alternatively spliced forms of the α(I) subunit, if any, thus remains unexplained.

The vertebrate α(I) subunits contain five cysteine residues (Fig. 9). Site-directed mutagenesis studies have suggested that the second and third cysteines form one intrachain disulfide bond and the fourth and fifth cysteines another, both these disulfide bonds being essential for tetramer assembly (John and Bulleid, 1994; Lamberg et al., 1995). Mutation of the first cysteine residue, the only cysteine that is not essential for tetramer assembly or stability, to serine had practically no effect on the catalytic activity of the tetramer (Lamberg et al., 1995). The α(II) subunit contains an additional cysteine residue between the fourth and fifth cysteines of the α(I) subunit (Helaakoski et al., 1995). Interestingly, this is located at a site where the best-conserved stretch of amino acids in interrupted in the mouse α(I) and *C. elegans* α subunits (Fig. 9).

The vertebrate α(I) and α(II) subunits each contain two potential attachment sites for asparagine-linked oligosaccharide units (Helaakoski et al., 1989, 1995; Bassuk et al., 1989; Hopkinson et al., 1994), while the *C. elegans* α subunit contains only one such site (Veijola et al., 1994). Site-directed mutagenesis studies on the human α(I) subunit have demonstrated that mutation of both glycosylated asparagines to glutamines causes no changes in tetramer assembly or in the specific activity of the resulting tetramer (Lamberg et al., 1995). As the extent of glycosylation of the two asparagines in the wild-type prolyl 4-hydroxylase is highly variable, the carbohydrate-free mutant tetramer may be more suitable than the wild-type enzyme for attempts at crystallization.

The *C. elegans* α subunit has features of both types of α subunit found in the vertebrate prolyl 4-hydroxylases (Fig. 9). The lack of the extra cysteine in the *C. elegans* α subunit is a feature that is especially characteristic of the α(I) subunit, whereas the lack of inhibition of the *C. elegans* enzyme by poly(L-proline) (Section VII) appears to be characteristic of the α(II) subunit (Helaakoski et al., 1995).

The human gene coding for the α(I) subunit is located in the q21.3–23.1 region of chromosome 10 (Pajunen et al., 1989). The transcribed part is more than 69 kb, and the gene consists of 16 exons (Helaakoski et al., 1994). The exons that encode only protein sequences vary from 54 to 240 bp, and the introns vary from 750 to more than 16,000 bp. The exon–intron boundaries of the human α(II) subunit gene are very similar to those of the α(I) subunit gene, but this gene has only one exon corresponding to No. 9 of the two alternatively spliced exons in the α(I) subunit gene (Nokelainen et al., 1996).

IX. Detailed Structure and Functions of the Multifunctional Protein Disulfide Isomerase/β Subunit Polypeptide

A. STRUCTURE

The PDI/β polypeptide has now been cloned from many animal species (e.g., Edman et al., 1985; Pihlajaniemi et al., 1987; Yamauchi et al., 1987; Gong et al., 1988; Kao et al., 1988; Parkkonen et al., 1988; Fliegel et al., 1990; Finken et al., 1994; Lucerno et al., 1994; Wilson et al., 1994; McKay et al., 1995; Veijola et al., 1996a) and also from yeast, microorganisms, and several plants (Noiva and Lennarz, 1992; Freedman et al., 1995). The final human PDI/β polypeptide is 491 amino acid residues in length and is synthesized in a form containing a signal peptide of 17 additional amino acids (Pihlajaniemi et al., 1987). The processed bovine PDI/β polypeptide likewise consists of 491 residues (Yamauchi et al., 1987), while the processed rabbit, rat, mouse, and chick polypeptides contain 489–493 residues (Edman et al., 1985; Yamauchi et al., 1987; Gong et al., 1988; Kao et al., 1988; Parkkonen et al., 1988; Fliegel et al., 1990). The processed PDI/β polypeptides from nonvertebrate species are slightly shorter, for examples, those from the nematodes *C. elegans* (Veijola et al., 1996a) and *Onchocera volvulus* (Wilson et al., 1994) are 477 residues and that from sea urchin 479 residues (Lucerno et al., 1994).

The PDI/β polypeptide appears to be a multidomain protein that has been suggested to consist of domains *a*, *e*, *b*, *b'*, *a'*, and *c* (Edman et al., 1985; Pihlajaniemi et al., 1987; Kivirikko et al., 1989, 1990, 1992; Noiva and Lennarz, 1992; Freedman et al., 1994). Recent structural characterization of the human domains *a* and *b* expressed in *Escherichia coli* has indicated, however, that these two domains are longer than thought previously, and the existence of domain *e* between these two is very unlikely (Kemmink et al., 1995, 1996; Darby et al., 1996). Domain *a'* is probably likewise longer than thought previously, whereas domain *c* is shorter (see below). A revised domain structure of the PDI/β polypeptide is shown in Fig. 10.

   Domains *a* and *a'* are similar to thioredoxin, a ubiquitous small protein that is involved in many cytoplasmic redox reactions (Buchanan et al., 1994; Holmgren, 1995; Martin, 1995). Both of these domains contain the sequence -Cys-Gly-His-Cys-, which have been shown by site-directed mutagenesis (Vuori et al., 1992a,c; LaMantia and Lennarz, 1993; Lyles and Gilbert, 1994), expression of the individual domains (Darby and Creighton, 1995) and chemical modification (Hawkins and Freedman, 1991) to represent two independently acting catalytic sites for disulfide isomerase activity. The corresponding sequence at the catalytic site of thioredoxin is -Cys-Gly-Pro-Cys- (Buchanan et al., 1994; Freedman et al., 1994; Holmgren,

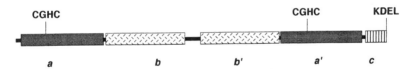

Figure 10.   Schematic representation of the inferred domain structure of the human PDI/β polypeptide. This structure has been revised from those shown previously (Kivirikko et al., 1992; Freedman et al., 1994) on the basis of recent nuclear magnetic resonance characterization of the human domains *a* and *b* expressed in *E. coli* (Kemmink et al., 1995, 1996), and the previously included domain *e* has been omitted (see text). Domain *a'* is also longer than thought previously, and domain *c* is correspondingly shorter. The two CGHC sequences indicate the two catalytic sites for PDI activity located in the two thioredoxin-like domains *a* and *a'*. Domains *b* and *b'* also have the thioredoxin fold while domain *c* is a highly acidic, putative Ca^{2+} binding region. Also shown is the C-terminal KDEL sequence, which is necessary for the retention of both the PDI/β polypeptide and prolyl 4-hydroxylase within the lumen of the endoplasmic reticulum.

1995; Martin, 1995). In fact, thioredoxin is able to catalyze protein disulfide isomerization but is only 2% as active as PDI in the classic disulfide isomerization assay, which is based on reactivation of incorrectly disulfide-bonded RNase (Hawkins et al., 1991; Freedman et al., 1994). Conversely, PDI can replace thioredoxin in the conventional assay for the latter but is far less active (Lundström and Holmgren, 1990). It has also been demonstrated that mutation of the proline to histidine at the catalytic site of thioredoxin alters the catalytic properties of the polypeptide and makes them more PDI-like (Lundström et al., 1992; Freedman et al., 1994). Nuclear magnetic resonance (NMR) studies of the human domain a expressed in $E.$ $coli$ have demonstrated that it has similar patterns of secondary structure and β sheet topology to thioredoxin (Kemmink et al., 1995, 1996). The highest degree of amino acid sequence identity between PDI/β polypeptides from various species is found within the two thioredoxin-like domains (Fig. 11). The sequences around the two -Cys-Gly-His-Cys- catalytic sites in these two domains are especially well conserved, a 23-amino acid sequence (residues 28–50, human numbering) in domain a and a 26-amino acid sequence (residues 369–394) in domain a' being 96 and 92% identical between the human and $C.$ $elegans$ polypeptides, respectively (Pihlajaniemi et al., 1987; Veijola et al., 1996a).

The region that was previously called domain e shows some similarity to a segment of the estrogen binding domain of the estrogen receptor (Tsibris et al., 1989). Nevertheless, NMR studies have demonstrated that amino acids that have been assigned to the N-terminal half of domain e belong to the stable folded structure of domain a and are part of the normal thioredoxin fold (Kemmink et al., 1995, 1996). Furthermore, these amino acids are detectably homologous to those in $E.$ $coli$ thioredoxin (Kemmink et al., 1995). Amino acids that have been assigned to the C-terminal half of domain e have correspondingly been found to belong to the stable structure of domain b (Darby et al., 1996). These findings cast doubt on the general significance of the amino acid sequence similarities to the estrogen receptor and indicate that the residues termed domain e do in fact not form a separate domain at all (Kemmink et al., 1995, 1996; Darby et al., 1996). Because domain a' is similar to domain a, it seems very likely that this domain is likewise longer than thought previously, and thus amino acids that have earlier been assigned to the N-termi-

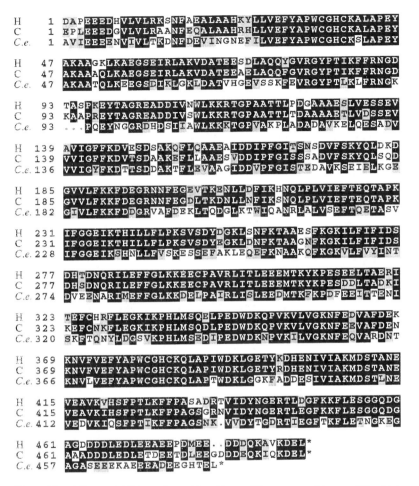

Figure 11.   Alignment of the amino acid residues of the processed human (H), chick (C), and *C. elegans* (*C. e.*) PDI/β polypeptides. Symbols and amino acid similarities as in Figure 9.

nal region of domain *c*, including sequences previously called *c*-1 (Koivunen et al., 1996), probably belong to domain *a'* rather than domain *c*.

Domains *b* and *b'* show a lower degree of amino acid sequence identity between species than domains *a* and *a'* (Fig. 11). Recent

NMR characterization of human domain *b* and preliminary analyses of domain *b'* have indicated that these two domains also have the thioredoxin-fold, even though they show no significant amino acid sequence similarity to any member of the thioredoxin family (Kemmink et al., 1997). The PDI polypeptide thus appears to be constructed of two catalytically active and two inactive thioredoxin modules (Kemmink et al., 1997). Domain *c* is rich in glutamate and aspartate residues and contains the -Lys-Asp-Glu-Leu motif or a functional variant of it (Pelham, 1990) as the C-terminal sequence, this being necessary for the retention of both the PDI/β polypeptide and the prolyl 4-hydroxylase tetramer within the lumen of the endoplasmic reticulum (Section IV).

The human gene coding for the PDI/β polypeptide is located in the q25 region of chromosome 17 (Pajunen et al., 1988, 1991; Popescu et al., 1988). The transcribed part is about 16.5 kb and consists of 11 exons (Tasanen et al., 1988). The chick gene is similar in its exon–intron organization, but the introns are considerably shorter than in the human gene, and therefore the chick gene is only about 9 kb (Nakazawa et al., 1988). The two thioredoxin-like domains *a* and *a'* are coded by exons numbers 1,2 and 8,9, respectively (Tasanen et al., 1988). The codons for the two -Cys-Gly-His-Cys- catalytic sites for disulfide isomerase activity are located 12 bp from the beginning of exons 2 and 9. The last three amino acids coded by exons 1 and 8 at the respective exon–intron junctions, and the first nine coded by exons 2 and 9, including a split codon for tyrosine, are identical (Tasanen et al., 1988). These findings suggest that evolution of this gene has involved exon shuffling and duplication of a two-exon unit, in which the internal exon–intron junctions have remained entirely conserved (Tasanen et al., 1988). The promoter region of the human PDI/β gene contains six CCAAT elements, five Sp1 binding sites and other regulatory elements, some of which are functionally redundant (Tasanen et al., 1992, 1993).

It has become evident in recent years that there is a family of PDI-like proteins (Freedman et al., 1994). They all contain two or three thioredoxin-like domains similar to domains *a* and *a'* of the PDI/β polypeptide, with each of these domains containing the -Cys-Gly-His-Cys- motif or a slightly modified variant. All these polypeptides also contain a functional variant of the endoplasmic reticulum retention signal (Pelham, 1990) -Lys-Asp-Glu-Leu. The members of this family include the polypeptides known as ERp60 (Bennett et al., 1988; Srivastava et al., 1991; Hirano et al., 1994, 1995; Mazzarella et al., 1994; Bourdi et al., 1995; Koivunen et al., 1996, 1997), PDIp (Desilva et al., 1996), a *C. elegans* PDI isoform (Veijola et al., 1996a), ERp72 (Mazzarella et al., 1990; Nguyen et al., 1993), P5 (Chaudhuri et al., 1992; Füllekrug et al., 1994), EUG1 (Tachibana and Stevens, 1992), and a human PDI-related polypeptide (Hayano and Kikuchi, 1995). ERp60 (Srivastava et al., 1991,1993; Hirano et al., 1995; Bourdi et al., 1995; Koivunen et al., 1996), PDIp (Desilva et al., 1996), the *C. elegans* PDI isoform (Veijola et al., 1996a), ERp72 (Mazzarella et al., 1990; Rupp et al., 1994; Lundström-Ljung et al., 1995); and P5 (Rupp et al., 1994; Lundström-Ljung et al., 1995) have been shown to have PDI activity, whereas EUG1 (Tachibana and Stevens, 1992) may have no such activity.

The ERp60 polypeptide (also known as ERp57, ERp61, and GRP58) has been cloned from human (Bourdi et al., 1995; Hirano et al., 1995; Koivunen et al., 1996, 1997), bovine (Hirano et al., 1994), rat (Bennett et al., 1988), mouse (Mazzarella et al., 1994), and *Drosophila* (Koivunen et al., 1996) sources. It was initially misidentified as a phosphatidyl inositol-dependent phospholipase C (Bennett et al., 1988), but has subsequently been shown to be devoid of any phospholipase activity (Martin et al., 1991; Srivastava et al., 1991, 1993; Mazzarella et al., 1994). The ERp60 polypeptide was recently proposed to function in combination with calnexin and calreticulin as a molecular chaperone of glycoprotein synthesis (Oliver et al., 1997). The size of the processed human ERp60 polypeptide is 481 amino acids (Bourdi et al., 1995; Hirano et al., 1995; Koivunen et al., 1996), that is, 10 residues shorter than the processed human PDI/β polypeptide. The overall amino acid sequence identity and similarity between the human ERp60 and PDI/β polypeptides are 29 and 56%, respectively (Koivunen et al., 1996). The ERp60 polypeptide has two thioredoxin-like domains in positions corresponding to the *a* and *a'* domains of the PDI/β polypeptide, and both these domains contain the -Cys-Gly-His-Cys- motif. The degree of amino acid sequence identity in domain *a* between the two human polypeptides is 49% and that in domain *a'* 54% (Koivunen et al., 1996). Significant similarity between the two polypeptides is also found in regions corresponding to domains *b* and *b'*, the only region that shows no similarity between the two polypeptides, except for the

C-terminal -Lys-Asp-Glu-Leu-like motif, being domain c (Koivunen et al., 1996).

The PDI isoform PDIp is highly expressed in human pancreas (Desilva et al., 1996). The size of the human PDIp polypeptide is 511 amino acids, but PDIp differs from the other members of the family of PDI-like proteins in that it may lack a signal peptide (Desilva et al., 1996). Nevertheless, it has a functional -Lys-Glu-Glu-Leu variant of the endoplasmic reticulum retention signal. The overall amino acid sequence identity and similarity between the human PDIp and PDI/β polypeptides are 46 and 66%, respectively, the highest degrees of identity being found in the two thioredoxin-like domains a and a' (Desilva et al., 1996). PDIp polypeptide is an active PDI, but a surprising feature is that its domain a' contains a -Cys-Thr-His-Gys- sequence instead of the -Cys-Gly-His-Cys- motif (Desilva et al., 1996).

The size of the processed *C. elegans* PDI isoform is 465 amino acids, that is, 12 amino acids shorter that the processed *C. elegans* PDI/β polypeptide (Veijola et al., 1996a). The overall amino acid sequence identity and similarity between the two *C. elegans* polypeptides are 45 and 73%, respectively, there being significant similarity between all the domains. The PDI isoform is clearly more closely related to the PDI/β polypeptide than to ERp60 and is an active PDI (Veijola et al., 1996a). Its domain a contains a -Cys-Val-His-Cys- sequence instead of the -Cys-Gly-His-Cys- motif (Veijola et al., 1996a).

The ERp72 polypeptide has been cloned from rat (Nguyen et al., 1993) and mouse (Mazzarella et al., 1990) cDNA libraries. In view of its Ca^{2+} binding activity, this polypeptide is also known as the endoplasmic reticulum luminal Ca^{2+} binding protein CaBP2 (Nguyen et al., 1993). The ERp72 polypeptide differs from most of the other PDI-like proteins in having three thioredoxin-like domains, each of them containing the -Cys-Gly-His-Cys- motif. The positions of the second and third thioredoxin-like domains correspond to those of the a and a' domains of the PDI/β polypeptide.

The P5 polypeptide has two thioredoxin-like domains, but their location is different from those in the PDI/β, ERp60, PDIp, and the *C. elegans* PDI isoform polypeptides, being directly adjacent to each other (Chaudhuri et al., 1992; Füllekrug et al., 1994). The P5 polypeptide is also known as the endoplasmic reticulum luminal Ca^{2+} binding protein CaBP1 (Füllekrug et al., 1994). EUG1, which has been cloned from yeast shows a 43% overall amino acid sequence identity to yeast PDI, but differs from all the other members of the family of PDI-like polypeptides in that its domains *a* and *a'* contain the sequences -Cys-Leu-His-Ser- and -Cys-Ile-His-Ser- in the place of the -Cys-Gly-His-Cys- motif, the more C-terminal cysteine in both catalytic site motifs thus being replaced by serine (Tachibana and Stevens, 1992). The EUG1 polypeptide appears to be inactive as a catalyst of protein disulfide formation, but may have some chaperone-like properties (Tachibana and Stevens, 1992).

The recently cloned human PDI-related polypeptide resembles ERp72 in having three -Cys-X-X-Cys-like motifs, but it differs markedly from ERp72 in being only 495 amino acids in size (Hayano and Kikuchi, 1995). Another surprising feature of this polypeptide is that the sequences of its three -Cys-X-X-Cys- like motifs are -Cys-Ser-Met-Cys-, -Cys-Gly-His-Cys-, and -Cys-Pro-His-Cys- (Hayano and Kikuchi, 1995). This polypeptide has not yet been expressed, and thus its probable PDI activity has not yet been demonstrated.

The possibility has been considered that some of the PDI-like polypeptides may be able to act as the β subunit of prolyl 4-hydroxylase. This aspect was studied by coexpressing the ERp60 polypeptide (Koivunen et al., 1996) or the *C. elegans* PDI isoform (Veijola et al., 1996a) in insect cells together with an α subunit of prolyl 4-hydroxylase. No association was found in these experiments, however, and no prolyl 4-hydroxylase activity was generated in the insect cells (Koivunen et al., 1996; Veijola et al., 1996a). It thus seems that the property of serving as the β subunit of prolyl 4-hydroxylases is specific to the PDI/β polypeptide and is not shared by other members of the family of PDI-like polypeptides.

B. FUNCTIONS

Molecular cloning and nucleotide sequencing of the β subunit of prolyl 4-hydroxylase indicated that this polypeptide is identical to PDI, and the polypeptide has been shown to have PDI activity even when present in the native prolyl 4-hydroxylase tetramer (Pihlajaniemi et al., 1987; Koivu et al., 1987; Parkkonen et al., 1988 , Section III). Subsequently this polypeptide has been reported to have several additional functions (Fig. 12). Two well-characterized ones are those of serving as the smaller subunit of the microsomal triglyceride transfer protein αβ dimer (Wetterau et al., 1990, 1991a) and of acting as a chaperone-like polypeptide that binds various peptides within the lumen of the endoplasmic reticulum (Noiva et al., 1991, 1993; Noiva and Lennarz, 1992) and probably catalyzes folding of many newly synthesized proteins (LaMantia and Lennarz, 1993; Cai et al., 1994; Otsu et al., 1994; Puig and Gilbert, 1994a,b; Puig et al., 1994; Rupp et al., 1994; Hayano et al., 1995; Song and Wang, 1995; Yao et al., 1997; Noiva et al., in press). Another well-characterized property is that of serving as a major cellular thyroid hormone-binding protein (Cheng et al., 1987; Yamauchi et al., 1987), but the physiological significance of this function is unknown, and this property may be related to the peptide binding function of the polypeptide. The PDI/β polypeptide is also a major phosphoprotein of the endoplasmic reticulum (Quéméneur et al., 1994) that undergoes ATP-dependent autophosphorylation (Guthapfel et al., 1996). This ATPase reaction is stimulated by the presence of denatured polypeptides (Guthapfel et al., 1996) and may thus be likewise related to the peptide binding function. Still further reported functions are as a calcium binding protein (Lebeche et al., 1994), a dehydroascorbate reductase (Wells et al., 1990), and a developmentally regulated retinal protein termed r-cognin (Krishna Rao et al., 1993).

Figure 12.   Schematic representation of some of the functions suggested for the multifunctional PDI/β polypeptide. The PDI/β polypeptide (open oval) is synthesized in an excess over the prolyl 4-hydroxylase α subunit (black oval) and enters a pool of the PDI/β polypeptide before being incorporated into the prolyl 4-hydroxylase tetramer. The α subunit is incorporated into the tetramer immediately after synthesis. The PDI/β polypeptide has PDI activity both when present as the homodimer and when present in the prolyl 4-hydroxylase tetramer. The PDI/β polypeptide also serves as the smaller subunit of the microsomal triglyceride transfer protein αβ dimer (the α subunit of which is shown by a shaded box), a chaperone-like polypeptide that nonspecifically binds various peptides within the lumen of the endoplasmic reticulum and probably catalyzes folding of many proteins, and a major cellular thyroid hormone (T_3) binding protein. An additional function shown is that of acting as a dehydroascorbate reductase. The wavy lines indicate newly synthesized polypeptides. [Modified from Kivirikko et al., 1989, 1990, 1992.]

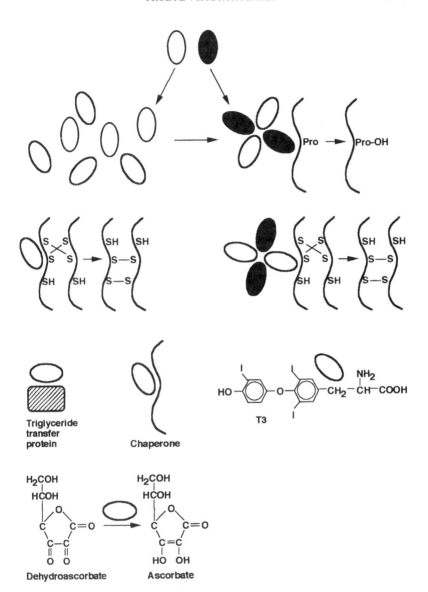

Triglyceride transfer protein

Chaperone

T3

Dehydroascorbate   Ascorbate

Early studies on prolyl 4-hydroxylase indicated that when the enzyme tetramer is dissociated by various means, the α subunit immediately forms insoluble aggregates that have no prolyl 4-hydroxylase activity (Kivirikko et al., 1992). Expression in insect cells has demonstrated that when an α subunit is expressed alone, without the PDI/β polypeptide, it forms insoluble aggregates with no prolyl 4-hydroxylase activity (Vuori et al., 1992b; Veijola et al., 1994; Helaakoski et al., 1995). The α subunit similarly forms aggregates when expressed in a cell-free translation/translocation system of dog pancreas microsomes depleted of PDI, BiP, and other soluble endoplasmic reticulum proteins (John et al., 1993). Thus one function of the PDI/β polypeptide appears to be to keep the α subunits in a nonaggregated conformation. Since this function resembles the role of certain chaperones with respect to other proteins (Vuori et al., 1992b; Freedman et al., 1994), attempts have been made to replace the PDI/β polypeptide with the chaperone BiP in insect cell expression experiments (Veijola et al., 1996b). The BiP chaperone was found to form both soluble and insoluble complexes with the α subunit, but the soluble complexes had no prolyl 4-hydroxylase activity (Veijola et al., 1996b). This indicates that the function of the PDI/β polypeptide in prolyl 4-hydroxylases is not only that of keeping the α subunits in solution but is more specific, probably that of keeping them in a catalytically active, nonaggregated conformation (Veijola et al., 1996b). Soluble α subunit–BiP complexes appear to be intermediates also under normal biosynthesis conditions of the enzyme tetramer, the α subunit probably interacting sequentially first with BiP and then with the PDI/β polypeptide (John and Bulleid, 1996). As discussed in Section IV, an additional function of the PDI/β polypeptide in prolyl 4-hydroxylases is to retain the enzyme within the lumen of the endoplasmic reticulum (Vuori et al., 1992c). Expression in insect cells with a double mutant PDI/β polypeptide in which both -Cys-Gly-His-Cys- catalytic sites had been inactivated by converting their sequences to -Ser-Gly-His-Cys- has demonstrated that the disulfide isomerase activity of the PDI/β polypeptide is not required for association of the α and PDI/β subunits into prolyl 4-hydroxylase, neither is it required for prolyl 4-hydroxylase activity (Vuori et al., 1992c).

The role of the PDI/β polypeptide in the microsomal triglyceride transfer protein dimer appears to be very similar to its role in the prolyl 4-hydroxylases. When the dimer is dissociated by chaotropic agents or low concentrations of guanidine hydrochloride, the free 97-kDa α subunit forms insoluble aggregates with no lipid transfer activity (Wetterau et al., 1991b), as it also does when expressed in insect cells without the PDI/β polypeptide (Ricci et al., 1995; Lamberg et al., 1996b). As in the case of the prolyl 4-hydroxylases, the disulfide isomerase activity of the PDI/β polypeptide is not required for triglyceride transfer protein dimer assembly or for lipid transfer activity of the dimer (Lamberg et al., 1996b).

The sites in the PDI/β polypeptide that nonspecifically bind various peptides have been shown to be distinct from the two catalytic sites for disulfide isomerase activity (LaMantia and Lennarz, 1993; Noiva et al., 1993). Photoaffinity labeling studies with a radioactive peptide probe have suggested the presence in PDI of only a single peptide binding site and have localized this site to a single 26-amino acid tryptic peptide in the C-terminal region of the rat PDI/β polypeptide (Noiva et al., 1993). The corresponding 27-amino acid sequence in the human PDI/β polypeptide extends from residue 452–478. Binding of the photoaffinity probe to this site has been shown to inhibit the chaperone activity of the PDI/β polypeptide (Puig et al., 1994), thus demonstrating that the site is indeed involved in the chaperone function. Inactivation of the two catalytic sites for disulfide isomerase activity has no effect on the chaperone activity of the PDI/β polypeptide in folding experiments with a protein containing no disulfide bonds (Quan et al., 1995), but the isomerase activity is required for chaperone activity in proteins having disulfide bonds (Puig et al., 1994; Yao et al., 1997), probably because the chaperone cannot assist the folding of polypeptides that have nonnative conformations stabilized by disulfide bonds.

The role of the PDI/β polypeptide in prolyl 4-hydroxylases and the microsomal triglyceride transfer protein, as described above, resembles that of some molecular chaperones such as Hsp-90 in other proteins (Vuori et al., 1992b; Freedman et al., 1994). To study whether the peptide binding site is also involved in prolyl 4-hydroxylase tetramer assembly, small deletions were introduced into the presumed peptide binding region of the human PDI/β polypeptide (Koivunen et al., submitted for publication). Deletion of the extreme N-terminal triplet, -Phe-Leu-Glu-, residues 452–454 in the 27-amino acid sequence (Fig. 11) totally eliminated prolyl 4-hydroxylase tetramer formation, whereas deletion of any other triplet in the sequence or deletion of the entire domain c had no inhibitory effect (Koivunen et al., submitted for publication). The critical region when studied by means of small deletions nevertheless extended further in the N-terminal direction and involved a total of 19 amino acids, residues 436–454 (Koivunen et al., submitted for publication). A critical region for the chaperone activity of the PDI/β polypeptide in folding experiments with a protein containing no disulfide bonds was found to overlap with this region and to involve a total of 12 amino acids, residues 446–457, while domain c again had no significant role (Koivunen et al., submitted for publication). Secondary structure predictions suggest that the critical region identified by means of small deletions for both the subunit function and the chaperone activity corresponds to the most C-terminal α helix of domain a' (Koivunen et al., submitted for publication). This region also overlaps with the previously identified peptide binding region. Nevertheless, this region is unlikely to be the only critical region for either the subunit function of the PDI/β polypeptide or its chaperone activity. The N-terminus of the PDI/β polypeptide is not critical for prolyl 4-hydroxylase assembly, as a polypeptide containing a six-histidine tag in its N-terminus forms an active vertebrate $[\alpha(I)]_2\beta_2$ and $[\alpha(II)]_2\beta_2$ tetramer and $C.$ $elegans$ αβ dimer as readily as the wild-type polypeptide (Annunen et al., 1997).

X. Detailed Structure of the Subunit of Lysyl Hydroxylase

Lysyl hydroxylase has now been cloned from human (Hautala et al., 1992a; Yeowell et al., 1992), rat (Armstrong and Last, 1995), and chick (Myllylä et al., 1991) sources, and very recently an isoenzyme termed lysyl hydroxylase 2 (Section III.B) was cloned from human sources (Valtavaara et al., 1997). The processed human lysyl hydroxylase 1 polypeptide is 709 amino acid residues in length and is synthesized in a form containing a signal peptide of 18 additional amino acids (Hautala et al., 1992a), while the processed rat (Armstrong and Last, 1995) and chick (Myllylä et al., 1991) polypeptides are both 710 residues in length (Fig. 13). The overall amino acid sequence identity between the human and rat lysyl hydroxylase 1 polypeptides is 92% and that between the human and chick polypeptides is 76% (Myllylä et al., 1991; Hautala et al., 1992a; Yeowell et al., 1992;

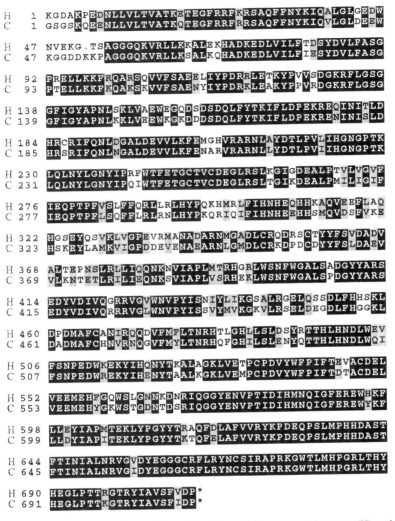

Figure 13. Alignment of the amino acid residues of the processed human (H) and chick (C) lysyl hydroxylase 1 (LH) polypeptides. Symbols and amino acid similarities as in Figure 9.

Armstrong and Last, 1995). The C-terminal region is especially well conserved, a 139-amino acid region, residues 570–709 (C-terminus of the polypeptide; human numbering), being 94% identical between the human and chick polypeptides and a 76-amino acid region, residues 621–697, 99% identical (Myllylä et al., 1991; Hautala et al., 1992a). The 76-amino acid region contains the three probable Fe^{2+} binding ligands, namely, His-638, Asp-640, and His-690 (Section VI). The overall amino acid sequence identity between the human lysyl hydroxylase 1 and 2 polypeptides is 75%, the degree of identity again being highest within the C-terminal domain, and all three probable Fe^{2+} binding ligands being conserved in lysyl hydroxylase 2 (Valtavaara et al., 1997). The lysyl hydroxylase polypeptides, like the α subunits of the prolyl 4-hydroxylases, are thus likely to contain functionally important sequences especially within their C-terminal region.

The human (Hautala et al., 1992a; Yeowell et al., 1992) and rat (Armstrong and Last, 1995) lysyl hydroxylase 1 polypeptides contain 10 cysteine residues, all except the extreme N-terminal one being conserved in the chick sequence, which contains 9 cysteines (Myllylä et al., 1991). The polypeptides from all three sources have four potential attachment sites for N-linked oligosaccharide units, the positions of the 2nd–4th sites of the human polypeptide being conserved in the chick polypeptide (Fig. 13). The C-terminus of the human and rat polypeptides has the sequence -Phe-Val-Asp-Pro, the valine being replaced by isoleucine in the chick polypeptide (Myllylä et al., 1991; Hautala et al., 1992a; Yeowell et al., 1992; Armstrong and Last, 1995). Thus the polypeptide has no retention signal typical of the luminal or transmembrane proteins of the endoplasmic reticulum (Section IV).

The human lysyl hydroxylase 1 gene is located in the p36.2–36.3 region of chromosome 1 (Hautala et al., 1992a). The transcribed part is about 40 kb, and the gene consists of 19 exons (Heikkinen et al., 1994). Introns 9 and 16 show extensive homology, resulting from five *Alu* sequences in intron 9 and eight in intron 16 (Heikkinen et al., 1994). The high homology and many short identical or complementary sequences in these introns generate many potential recombination sites, which appears to explain a large gene duplication. This duplication is the most common mutation in this gene found in patients with Ehlers–Danlos syndrome type VI (Section XI).

XI. Mutations in the Gene for Lysyl Hydroxylase in the Type VI Variant of the Ehlers–Danlos Syndrome

The Ehlers–Danlos syndrome is a heterogeneous group of disorders characterized by joint hypermobility, alterations in the skin, such as hyperextensibility, thinness and fragility, and other signs of connective tissue involvement. The syndrome has been divided into 10 subtypes, but even these are heterogeneous (for reviews, see Krane, 1984; Prockop and Kivirikko, 1984, 1995; Beighton, 1993; Kivirikko, 1993; Steinmann et al., 1993; Byers, 1995). The main clinical characteristics of this autosomal recessive subtype are muscular hypotonia, joint hypermobility and progressive kyphoscoliosis. The patients also show varying increases in tissue fragility, skin hyperextensibility, bruisability, deficient wound healing with poor scar formation, a potential for arterial rupture, and various ocular manifestations such as fragility of the globe, spontaneous retinal detachment, microcornea, and slightly blue sclerae (Krane, 1984; Prockop and Kivirikko, 1984, 1995; Wenstrup et al., 1989; Beighton, 1993; Kivirikko, 1993; Steinmann et al., 1993; Byers, 1995).

The biochemical abnormality in most but not all patients with Ehlers–Danlos syndrome type VI is a deficiency of lysyl hydroxylase activity (Krane et al., 1972; Pinnell et al., 1972; Krane, 1984; Prockop and Kivirikko, 1984, 1995; Wenstrup et al., 1989; Kivirikko et al., 1992; Beighton, 1993; Kivirikko, 1993; Steinmann et al., 1993; Byers, 1995). Molecular cloning of human lysyl hydroxylase 1 and its gene (Section X) has made it possible to characterize in detail mutations responsible for the disorder in various families. The two patients in one family, the third and fifth children of healthy parents, had a homozygous single base substitution in the lysyl hydroxylase gene which converted the codon for arginine at amino acid position 301 of the processed polypeptide to a translation termination codon (Hyland et al., 1992). The parents, who are first cousins, and two of the three healthy siblings of the patients were found to be heterozygous carriers of the same mutation (Hyland et al., 1992). One patient in another family is a compound heterozygote for two mutations (Ha et al., 1994). The mutation in one allele is a triple base deletion that results in the loss of one amino acid, while the mutation in the other allele is an amino acid substitution (Ha et al., 1994).

The most common mutation appears to be a large duplication

involving seven exons in the lysyl hydroxylase 1 gene (Hautala et al., 1993; Pousi et al., 1994; Heikkinen et al., 1997), caused by a homologous recombination of *Alu* sequences in introns 9 and 16 (Pousi et al., 1994). The prevalence of this mutant allele among 35 Ehlers–Danlos syndrome type VI families studied so far is 19% (Heikkinen et al., 1997). In one of these families an abnormal sequence in exon 17 of the gene, which leads to skipping of sequences coded by this exon during mRNA processing, is responsible for the nonfunctionality of the other allele in a compound heterozygous patient (Heikkinen et al., 1997).

An interesting finding in patients with Ehlers–Danlos syndrome type VI is that the deficiency of hydroxylysine in collagens varies between tissues (Krane, 1984; Kivirikko et al., 1992; Steinmann et al., 1993). The hydroxylysine content of skin collagen is usually about 5% of normal, whereas the few values that have so far been reported for other tissues have been higher than this (Steinmann et al., 1993). In agreement with this variation, no kidney function abnormalities have been reported in patients with Ehlers–Danlos syndrome type VI, even though the type IV collagens of basement membranes are particularly rich in hydroxylysine in healthy subjects. These findings are probably explained by the existence of tissue-specific lysyl hydroxylase isoenzymes (Section III.B).

XII. Regulation of Collagen Hydroxylase Activities

Prolyl 4-hydroxylase activity has been demonstrated in a number of cell types, both fibroblastic and nonfibroblastic, including cells of nonmesenchymal origin (Cardinale and Udenfriend, 1974; Prockop et al., 1976; Kivirikko and Myllylä, 1980; Kivirikko et al., 1992). A good correlation is usually found between the amount of prolyl 4-hydroxylase activity and the rate of collagen synthesis in cultured cells and *in vivo* in many physiological and pathological conditions (Cardinale and Udenfriend, 1974; Prockop et al., 1976; Kivirikko and Myllylä, 1980; Kivirikko et al., 1992). Nevertheless, an excess of prolyl 4-hydroxylase activity is clearly present in cells in some situations, while in others the activity is rate limiting for collagen production (Kivirikko et al., 1992; Lamberg et al., 1996a;

Myllyharju et al., 1997). A number of studies, which will be not discussed here in detail, are now available on changes in collagen hydroxylase activities in cells and tissues under various conditions (Cardinale and Udenfriend, 1974; Prockop et al., 1976; Kivirikko and Myllylä, 1980; Kivirikko et al., 1992).

The rates of synthesis of the α and PDI/β subunits of prolyl 4-hydroxylase are regulated differently. The PDI/β polypeptide is produced in a large excess over the α subunits and enters a pool of the PDI/β polypeptide (Fig. 12) before being incorporated into the enzyme tetramer (Kivirikko et al., 1992). In contrast, the α subunit appears to become incorporated into the tetramer immediately upon synthesis. Regulation of the amounts of active prolyl 4-hydroxylase therefore appears to occur mainly through regulation of the synthesis of the α subunit. The mechanisms by which this takes place are poorly understood, however.

Many studies have demonstrated that increases in prolyl 4-hydroxylase activity are associated with increases in levels of the mRNA for the α subunit, but no studies are yet available to indicate whether changes in the levels of mRNAs for the α(I) and α(II) subunits are similar. The levels of mRNA for the α subunit (the probes used probably hybridized to both types of α subunit mRNA) and the PDI/β polypeptide vary during the growth of chick embryo tendon fibroblasts, for example, changes in the level of the former being similar to those in the amounts of the mRNAs for the α1 and α2 chains of type I collagen (Bassuk and Berg, 1991). The addition of hydralazine to cultures of human skin fibroblasts rapidly increases the level of prolyl 4-hydroxylase α subunit mRNA about fourfold, and also the level of the PDI/β polypeptide mRNA, but more slowly (Yeowell et al., 1991). Differentiation of mouse F9 teratocarcinoma stem cells, which involves a marked increase in the synthesis rates and mRNA concentrations of basement membrane proteins such as type IV collagen, is also associated with a marked increase in the level of prolyl 4-hydroxylase activity (Roguska and Gudas, 1985) and an increase of up to 50-fold in the levels of mRNAs for the α and PDI/β subunits of prolyl 4-hydroxylase (Helaakoski et al., 1990).

The promoter regions of the genes for the α(I) and α(II) subunits of prolyl 4-hydroxylase contain many potential sites for the binding of various transcription factors, but the functional significance of these potential regulatory elements remains to be determined (Helaa-koski et al., 1994; Nokelainen et al., 1996). The 5' untranslated region of the human α(I) subunit mRNA has the potential for forming several loop structures that may participate in translational control (Helaakoski et al., 1994), but again no data are currently available on any translational control of this gene. An additional aspect of regulation of the α(I) subunit mRNA is the mutually exclusive alternative splicing of sequences coded by exons 9 and 10. Both types of mRNA have been found to be expressed in all tissues studied, but in some the type coding for exon 9 or 10 sequences is more abundant than the other type (Helaakoski et al., 1994).

Very recent studies have indicated that the level of association of the α and PDI/β subunits into the active enzyme tetramer is also dependent on the level of collagen synthesis (Vuorela et al., 1997). A highly unusual control system thus appears to exist in collagen synthesis in that production of a stable prolyl 4-hydroxylase tetramer requires expression of collagen polypeptide chains while the production of collagen molecules with stable triple helices requires expression of active prolyl 4-hydroxylase (Vuorela et al., 1997).

Lysyl hydroxylase and prolyl 3-hydroxylase activities also vary markedly between cell types, and in the same cell type in different experimental conditions (Kivirikko et al., 1992). A surprising aspect of the regulation of lysyl hydroxylase is that minoxidil, an antihypertensive piperidinopyrimidine drug, and many of its derivatives, specifically reduce this enzyme activity in cultured fibroblasts (Murad and Pinnell, 1987; Murad et al., 1992, 1994). A corresponding specific decrease is found in the amounts of lysyl hydroxylase protein (Hautala et al., 1992b) and mRNA (Yeowell et al., 1992; Hautala et al., 1992b), suggesting that the mechanism involves either decreased transcription of the lysyl hydroxylase gene or decreased stability of the mRNA. Hydralazine, which increases the concentration of prolyl 4-hydroxylase mRNAs (see above), also increases the level of lysyl hydroxylase mRNA (Yeowell et al., 1992).

Acknowledgments

We thank Ari-Pekka Kvist, for helping with the production of figures with sequence alignments and Auli Kinnunen and Marja-Leena Kivelä for secretarial help. The original research performed in our laboratory was supported in part by grants from the Health Sciences Council of the Academy of Finland.

References

Adams, E. and Frank, L., *Annu. Rev. Biochem.*, **49**, 1005–1061 (1980).

Ananthanarayanan, V. S., Saint-Jean, A., and Jiang, P., *Arch. Biochem. Biophys.*, **298**, 21–28 (1992).

Andrews, P. C., Hawke, D., Shively, J. E., and Dixon, J. E., *J. Biol. Chem.*, **259**, 15021–15024 (1984).

Annunen, P., Helaakoski, T., Myllyharju, J., Veijola, J., Pihlajaniemi, T., and Kivirikko, K. I., *J. Biol. Chem.*, **272**, 17342–17348 (1997).

Annunen, P., Autio-Harmainen, H., and Kivirikko, K. I., *J. Biol. Chem.*, **273** (1998), in press.

Armstrong, L. C. and Last, J. A., *Biochim. Biophys. Acta*, **1264**, 93–102 (1995).

Atreya, P. L. and Ananthanarayanan, V. S., *J. Biol. Chem.*, **266**, 2852–2858 (1991).

Baader, E., Tschank, G., Baringhaus, K.-H., Burghard, H., and Günzler, V., *Biochem. J.*, **300**, 525–530 (1994).

Bai, Y., Muragaki, Y., Obata, K.-I., Iwata, K., and Ooshima, A., *J. Biochem.*, **99**, 1563–1570 (1986).

Bassuk, J. A. and Berg, R. A., *Biochem. Biophys. Res. Commun.*, **1174**, 169–175 (1991).

Bassuk, J. A., Kao, W. W.-Y., Herzer, P., Kedersha, N. L., Seyer, J., DeMartino, J. A., Daugherty, B. L., Mark, G. E. III, and Berg, R. A., *Proc. Natl. Acad. Sci. USA*, **86**, 7382–7386 (1989).

Beighton, P., in *McKusick's Heritable Disorders of Connective Tissue*, Beighton, P., Ed., Mosby, St. Louis, Mo, 1993, pp. 189–251.

Bennett, C. F., Balcarek, J. M., Varrichio, A., and Crooke, S. T., *Nature (London)*, **334**, 268–270 (1988).

Berg, R. A. and Prockop, D. J., *Biochem. Biophys. Res. Commun.*, **52**, 115–119 (1973a).

Berg, R. A. and Prockop, D. J., *J. Biol. Chem.*, **248**, 1175–1182 (1973b).

Bickel, M., Baader, E., Brocks, D. G., Engelbart, K., Günzler, V., Schmidts, H. L., and Vogel, G. H., *J. Hepatol.* **13** (*Suppl. 3*), S26–S34 (1991).

Böker, K., Schwarting, G., Kaule, G., Günzler, V., and Schmidt, E., *J. Hepatol.* **13** (*Suppl. 3*), S35–S40 (1991).

Bourdi, M., Demady, D., Martin, J. L., Jabbour, S. K., Martin, B. M., George, J. W., and Pohl, L. R., *Arch. Biochem. Biophys.*, **323**, 397–403 (1995).

Brodsky, B., and Ramshaw, J. A. M., *Matrix Biol.* **15**, 545–554 (1997).

Buchanan, R. B., Schürmann, P., Decottignies, P., and Lozano, R. M., *Arch. Biochem. Biophys.*, **314**, 257–260 (1994).

Burgeson, R. E. and Nimni, M. E., *Clin. Orthop.*, **282**, 250–272 (1992).

Byers, P. H., in *The Metabolic and Molecular Basis of Inherited Diseases*, Scriver,

C. R., Beaudet, A. L., Sly, W. S., and Valle, D., Eds., McGraw-Hill, New York, 1995, pp. 4029–4077.

Cai, H., Wang, C.-C., and Tsou, C.-L., *J. Biol. Chem.*, **269**, 24550–24552 (1994).

Cardinale, G. J. and Udenfriend, S., *Adv. Enzymol.*, **41**, 245–300 (1974).

Carr, L. G., Skatrud, P. L., Scheetz, M. E., II, Queener, S. W., and Ingolia, T. D., *Gene*, **48**, 257–266 (1986).

Chaudhuri, M. M., Tonin, P. N., Lewis, W. H., and Srinivasan, P. R., *Biochem. J.*, **281**, 645–650 (1992).

Chauhan, U., Assad, R., and Peterkofsky, B., *Biochem. Biophys. Res. Commun.*, **131**, 277–283 (1985).

Cheng, S.-Y., Gong, Q.-H., Parkison, C., Robinson, E. A., Appella, E., Merlino, G. T., and Pastan, I., *J. Biol. Chem.*, **262**, 11221–11227 (1987).

Chopra, R. K. and Ananthanarayanan, V. S., *Proc. Natl. Acad. Sci. USA*, **79**, 7180–7184 (1982).

Colley, K. J. and Baenziger, J. U., *J. Biol. Chem.*, **262**, 10290–10295 (1987).

Cunliffe, C. J. and Franklin, T. J., *Biochem. J.*, **239**, 311–315 (1986).

Cunliffe, C. J., Franklin, T. J., Hales, N. J., and Hill, G. B., *J. Med. Chem.*, **35**, 2652–2658 (1992).

Darby, N. J. and Creighton, T. E., *Biochemistry*, **34**, 11725–11735 (1995).

Darby, N. J., Kemmink, J., and Creighton, T. E., *Biochemistry*, **35**, 10517–10528 (1996).

Davis, A. E., III and Lachmann, P. J., *Biochemistry*, **23**, 2139–2144 (1984).

de Jong, L. and Kemp, A., *Biochim. Biophys. Acta*, **709**, 142–145 (1982).

de Jong, L. and Kemp, A., *Biochim. Biophys. Acta*, **787**, 105–111 (1984).

de Jong, L., Albracht, S. P. J., and Kemp, A., *Biochim. Biophys. Acta*, **704**, 326–332 (1982).

de Jong, L., van der Kraan, I., and de Waal, A., *Biochem. Biophys. Acta*, **1079**, 103–111 (1991).

Desilva, M. G., Lu, J., Donadel, G., Modi, W. S., Xie, H., Notkins, A. L., and Lan, M. S., *DNA Cell Biol.*, **15**, 9–16 (1996).

de Waal, A. and de Jong, L, *Biochemistry*, **27**, 150–155 (1988).

de Waal, A., de Jong, L., Hartog, A. F., and Kemp, A., *Biochemistry*, **24**, 6493–6499 (1985).

de Waal, A., Hartog, A. F., and de Jong, L., *Biochim. Biophys. Acta*, **912**, 151–155 (1987).

de Waal, A., Hartog, A. F., and de Jong, L., *Biochim. Biophys. Acta*, **953**, 20–25 (1988).

Diekman, J. and Fischer, R. L., *EMBO J.*, **7**, 3315–3320 (1988).

Dowell, R. I. and Hadley, E. M., *J. Med. Chem.*, **35**, 800–804 (1992).

Dowell, R. I., Hales, N. H., and Tucker, H., *Eur. J. Med. Chem.*, **28**, 513–516 (1993).

Drickamer, K., Dordal, M. S., and Reynolds, L., *J. Biol. Chem.*, **261**, 6878–6887 (1986).

Edman, J. C., Ellis, L., Blacher, R. W., Roth, R. A., and Rutter, W. J., *Nature (London)* **317**, 267–270 (1985).

Elomaa, O., Kangas, M., Sahlberg, C., Tuukkanen, J., Sormunen, R., Liakka, A., Thesleff, I., Kraal, G., and Tryggvason, K., *Cell*, **80**, 603–609 (1995).

Fincher, G. B., Stone, B. A., and Clarke, A. E., *Annu. Rev. Plant. Physiol.*, **34**, 47–70 (1983).

Finken, M., Sobek, A., Symmons, P., and Kunz, W., *Mol. Biochem. Parasitol.*, **64**, 135–144 (1994).

Fliegel, L., Newton, E., Burns, K., and Michalak, M., *J. Biol. Chem.*, **265**, 15496–15502 (1990).

Floros, J., Steinbrink, R., Jacobs, K., Phelps, D., Kriz, R., Recny, M., Sultzman, L., Jones, S., Taeusch, H. W., Frank, H. A., and Fritsch, E. F., *J. Biol. Chem.*, **261**, 9029–9033 (1986).

Franklin, T. J., *Biochem. Pharmacol.*, **49**, 267–273 (1995).

Freedman, R. B., *Curr. Op. Struct. Biol.*, **5**, 85–91 (1995).

Freedman, R. B., Hirst, T. R., and Tuite, M. F., *Trends Biochem. Sci.*, **19**, 331–336 (1994).

Fujimoto, D. and Prockop, D. J., *J. Biol. Chem.*, **244**, 205–210 (1969).

Füllekrug, J., Sönniches, B., Wunsch, U., Arseven, K., Nguyen Van, P., Söling, H.-D., and Mieskes, G., *J. Cell Sci.*, **107**, 2719–2727 (1994).

Gautron, J.-P., Pattou, E., Bauer, K., and Kordon, C., *Neurochem. Int.*, **18**, 221–235 (1991).

Gong, Q.-H., Fukuda, T., Parkinson, C., and Cheng, S.-Y., *Nucleic Acids Res.*, **16**, 1203 (1988).

Günzler, V., Majamaa, K., Hanauske-Abel, H. M., and Kivirikko, K. I., *Biochim. Biophys. Acta*, **873**, 38–44 (1986).

Günzler, V., Hanauske-Abel, H. M., Myllylä, R., Mohr, J., and Kivirikko, K. I., *Biochem. J.*, **242**, 163–169 (1987).

Günzler, V., Brocks, D., Henke, S., Myllylä, R., Geiger, R., and Kivirikko, K. I., *J. Biol. Chem.*, **263**, 19498–19504 (1988a).

Günzler, V., Hanauske-Abel, H. M., Myllylä, R., Kaska, D., Hanauske, A., and Kivirikko, K. I., *Biochem. J.*, **251**, 365–372 (1988b).

Guthapfel, R., Guéguen, P., and Quéméneur, E., *J. Biol. Chem.*, **271**, 2663–2666 (1996).

Ha, V. T., Marshall, M. K., Elsas, L. J., Pinnell, S. R., and Yeowell, H. N., *J. Clin. Invest.*, **93**, 1716–1721 (1994).

Hanauske-Abel, H. M., *J. Hepatol.*, **13**, (*Suppl. 3*), S8–S16 (1991).

Hanauske-Abel, H. and Günzler, V., *J. Theor. Biol.*, **94**, 421–455 (1982).

Hautala, T., Byers, M. G., Eddy, R. L., Shows, T. B., Kivirikko, K. I., and Myllylä, R., *Genomics*, **13**, 62–69 (1992a).

Hautala, T., Heikkinen, J., Kivirikko, K. I., and Myllylä, R., *Biochem. J.*, **283**, 51–54 (1992b).

Hautala, T., Heikkinen, J., Kivirikko, K. I., and Myllylä, R., *Genomics*, **15**, 399–404 (1993).

Hawkins, H. C. and Freedman, R. B., *Biochem. J.*, **275**, 335–339 (1991).

Hawkins, H. C., Blackburn, E. C., and Freedman, R. B., *Biochem. J.*, **275**, 349–353 (1991).

Hayano, T. and Kikuchi, M., *FEBS Lett.*, **372**, 210–214 (1995).

Hayano, T., Hirose, M., and Kikuchi, M., *FEBS Lett.*, **377**, 505–511 (1995).

Heikkinen, J., Hautala, T., Kivirikko, K. I., and Myllylä, R., *Genomics*, **24**, 464–471 (1994).

Heikkinen, J., Toppinen, T., Yeowell, H., Krieg, T., Steinmann, B., Kivirikko, K. I., and Myllylä, R., *Am. J. Hum. Genet.*, **60**, 48–56 (1997).

Helaakoski, T., Vuori, K., Myllylä, R., Kivirikko, K. I., and Pihlajaniemi, T., *Proc. Natl. Acad. Sci. USA*, **86**, 4392–4396 (1989).

Helaakoski, T., Pajunen, L., Kivirikko, K. I., and Pihlajaniemi, T., *J. Biol. Chem.*, **265**, 11413–11416 (1990).

Helaakoski, T., Veijola, J., Vuori, K., Rehn, M., Chow, L. T., Taillon-Miller, P., Kivirikko, K. I., and Pihlajaniemi, T., *J. Biol. Chem.*, **269**, 27847–27854 (1994).

Helaakoski, T., Annunen, P., Vuori, K., MacNeil, I. A., Pihlajaniemi, T., and Kivirikko, K. I., *Proc. Natl. Acad. Sci. USA*, **92**, 4427–4431 (1995).

Hirano, N., Shibasaki, F., Kato, H., Sakai, R., Tanaka, T., Nishida, J., Yazaki, Y., Takenawa, T., and Hirai, H., *Biochem. Biophys. Res. Commun.*, **204**, 375–282 (1994).

Hirano, N., Shibasaki, F., Sakai, R., Tanaka, T., Nishida, J., Yazaki, Y., Takenawa, T., and Hirai, H., *Eur. J. Biochem.*, **234**, 336–342 (1995).

Holdsworth, M. J., Bird, C. R., Ray, J., Schuch, W., and Grierson, D., *Nucleic Acids Res.*, **15**, 731–739 (1987).

Holmgren, A., *Curr. Biol.*, **3**, 239–243 (1995).

Hopkinson, I., Smith, S. A., Donne, A., Gregory, H., Franklin, T. J., Grant, M. E., and Rosamund, J., *Gene*, **149**, 391–392 (1994).

Höyhtyä, M., Myllylä, R., Piuva, J., Kivirikko, K. I., and Tryggvason, K., *Eur. J. Biochem.*, **141**, 477–482 (1984).

Hulmes, D. J. S., *Essays Biochem.*, **27**, 49–67 (1992).

Hyland, J., Ala-Kokko, L., Royce, P., Steinmann, B., Kivirikko, K. I., and Myllylä, R., *Nature Genet.*, **2**, 228–231 (1992).

Ichijo, H., Hellman, U., Wennstedt, C., Gonez, L. J., Claesson-Welsh, L., Heldin, C.-H., and Miyazono, K., *J. Biol. Chem.*, **268**, 14505–14513 (1993).

Jensenius, J. C., Laursen, S. B., Zheng, Y., and Holmskov, U., *Biochem. Soc. Trans.*, **22**, 95–100 (1994).

Jiang, F., Peisach, J., Ming, L.-J., Que, L. J., Jr., and Chen, V. J., *Biochemistry*, **30**, 11437–11445 (1991).

Jiang, P. and Ananthanarayanan, V. S., *J. Biol. Chem.*, **266**, 22960–22967 (1991).

Jimenez, S. A., Harsch, M., and Rosenbloom, J., *Biochem. Biophys. Res. Commun.*, **52**, 106–114 (1973).

John, D. C. A. and Bulleid, N. J., *Biochemistry*, **33**, 14018–14025 (1994).

John, D. C. A. and Bulleid, N. J., *Biochem. J.*, **317**, 659–665 (1996).

John, D. C. A., Grant, M. E., and Bulleid, N. J., *EMBO J.*, **12**, 1587–1592 (1993).

Kao, W. W.-Y.,Nagazawa, M., Aida, T., Everson, W. V., Kao, C. W.-C., Seyer, J. M., and Hughes, S. H. *Connect. Tissue Res.*, **18**, 157–174 (1988).

Karvonen, K., Ala-Kokko, L., Pihlajaniemi, T., Helaakoski, T., Günzler, V., Henke, S., Kivirikko, K. I., and Savolainen, E.-R., *J. Biol. Chem.*, **265**, 8415–8419 (1990).

Kaska, D. D., Günzler, V., Kivirikko, K. I., and Myllylä, R., *Biochem. J.*, **241**, 483–490 (1987).

Kaska, D. D., Myllylä, R., Günzler, V., Gibor, A., and Kivirikko, K. I., *Biochem. J.*, **256**, 257–263 (1988).

Kedersha, N. L., Tkacz, J. S., and Berg, R. A., *Biochemistry*, **24**, 5952–5960 (1985a).

Kedersha, N. L., Tkacz, J. S., and Berg, R. A., *Biochemistry*, **24**, 5960–5967 (1985b).

Kellokumpu, S., Sormunen, R., Heikkinen, J., and Myllylä, R., *J. Biol. Chem.*, **269**, 30524–30529 (1994).

Kemmink, J., Darby, N. J., Dijkstra, K., Scheek, R. M., and Creighton, T. E., *Protein Sci.*, **4**, 2587–2593 (1995).

Kemmink, J., Darby, N. J., Dijkstra, K., Nilges, M., and Creighton, T. E., *Biochemistry*, **35**, 7684–7691 (1996).

Kemmink, J., Darby, N. J., Dijkstra, K., Nilges, M., and Creighton, T. E., *Current Biol.*, **7**, 239–245 (1997).

Kielty, C. M., Hopkinson, I., and Grant, M. E., in *Connective Tissue and Its Heritable Disorders. Molecular, Genetic and Medical Aspects*, Royce, P. M. and Steinmann, B., Eds., Wiley-Liss, New York, 1993, pp. 103–147.

Kikuchi, Y. and Tamiya, N., *Bull. Chem. Soc. Jpn.*, **55**, 1556–1560 (1982).

Kivirikko, K. I., *Ann. Med.* **25**, 113–126 (1993).

Kivirikko, K. I., in *Principles of Medical Biology, Vol. 3, Cellular Organelles and the Extracellular Matrix*, Bittar, E. E. and Bittar, N., Eds., JAI Press, Greenwich, CT, 1995, pp. 233–254.

Kivirikko, K. I. and Majamaa, K., in *Fibrosis*, Ciba Foundation Symposium 114, Evered, D. and Whelan, J., Eds., 1985, pp. 34–48.

Kivirikko, K. I., and Myllylä, R., *Int. Rev. Connect. Tissue Res.*, **8**, 23–72 (1979).

Kivirikko, K. I. and Myllylä, R., in *The Enzymology of Post-translational Modification of Proteins*, Freedman, R. B. and Hawkins, H. C., Eds., Academic, London, 1980, pp. 53–104.

Kivirikko, K. I. and Myllylä, R., *Methods Enzymol.*, **82**, 245–304 (1982).

Kivirikko, K. I. and Myllylä, R., *Ann. N.Y. Acad. Sci.*, **460**, 187–201 (1985).

Kivirikko, K. I. and Myllylä, R., *Methods Enzymol.*, **144**, 96–114 (1987).

Kivirikko, K. I. and Prockop, D. J., *J. Biol. Chem.*, **242**, 4007–4012 (1967a).

Kivirikko, K. I. and Prockop, D. J., *Proc. Natl. Acad. Sci. USA*, **57**, 782–789 (1967b).

Kivirikko, K. I. and Savolainen, E.-R., in *Liver Drugs: From Experimental Pharmacol to Therapeutic Application*, Testa, B. and Perrissound, D., Eds., CRC Press, Boca Raton, FL, (1988), pp. 193–222.

Kivirikko, K. I., Prockop, D. J., Lorenzi, G. P., and Blout, E. R., *J. Biol. Chem.*, **244**, 2755–2760 (1969).

Kivirikko, K. I., Shudo, K., Sakakibara, S., and Prockop, D. J., *Biochemistry*, **11**, 122–129 (1972).

Kivirikko, K. I., Ryhänen, L., Anttinen, H., Bornstein, P., and Prockop, D. J., *Biochemistry*, **12**, 4966–4971 (1973).

Kivirikko, K. I., Myllylä, R., and Pihlajaniemi, T., *FASEB J.*, **3**, 1609–1617 (1989).

Kivirikko, K. I., Helaakoski, T., Tasanen, K., Vuori, K., Myllylä, R., Parkkonen, T., and Pihlajaniemi, T., *Ann. N.Y. Acad. Sci.*, **580**, 132–142 (1990).

Kivirikko, K. I., Myllylä, R., and Pihlajaniemi, T., in *Post-Translational Modifications of Proteins*, Harding, J. J. and Crabbe, M. J. C., Eds., CRC Press, Boca Raton, FL, 1992, pp. 1–51.

Kodama, T., Freeman, M., Rohrer, L., Zabrecky, J., Matsudaira, P., and Krieger, M., *Nature (London)*, **343**, 531–535 (1990).

Koivu, J., Myllylä, R., Helaakoski, T., Pihlajaniemi, T., Tasanen, K., and Kivirikko, K. I., *J. Biol. Chem.*, **262**, 6447–6449 (1987).

Koivunen, P., Helaakoski, T., Annunen, P., Veijola, J., Räisänen, S., Pihlajaniemi, T., and Kivirikko, K. I., *Biochem. J.*, **316**, 599–605 (1996).

Koivunen, P., Horelli-Kuitunen, N., Helaakoski, T., Karvonen, P., Jaakkola, M., Palotie, A., and Kivirikko, K. I., *Genomics*, **42**, 397–404 (1997).

Koivunen, P., Karvonen, P., Helaakoski, T., Notbohm, H., and Kivirikko, K. I., submitted for publication.

Kovacevic, S. and Miller, J. R., *J. Bacteriol.*, **173**, 398–400 (1991).

Kovacevic, S., Weigel, B. J., Tobin, M. B., Ingolia, T. D., and Miller, J. R., *J. Bacteriol.*, **171**, 754–760 (1989).

Krane, S. M., in *Extracellular Matrix Biochemistry*, Piez, K. A. and Reddi, A. H., Eds., Elsevier, New York, 1984, pp. 413–463.

Krane, S. M., Pinnell, S. R., and Erbe, R. W., *Proc. Natl. Acad. Sci. USA*, **69**, 2899–2903 (1972).

Krishna Rao, A. S. M., and Hausman, R. E., *Proc. Natl. Acad. Sci. USA*, **90**, 2950–2954 (1993).

Krol, B. J., Murad, S., Walker, L. C., Marshall, M. K., Clark, W. L., Pinnell, S. R., and Yeowell, H. N., *J. Invest. Dermatol.*, **106**, 11–16 (1996).

Kuttan, R. and Radhakrishnan, A. N., *Adv. Enzymol.*, **37**, 273–347 (1973).

Kuutti, E.-R., Tuderman, L., and Kivirikko, K. I., *Eur. J. Biochem.*, **57**, 181–188 (1975).

LaMantia, M. and Lennarz, W. J., *Cell*, **74**, 899–908 (1993).

Lamberg, A., Pihlajaniemi, T., and Kivirikko, K. I., *J. Biol. Chem.*, **270**, 9926–9931 (1995).

Lamberg, A., Helaakoski, T., Myllyharju, J., Peltonen, S., Notbohm, H., Pihlajaniemi, T., and Kivirikko, K. I., *J. Biol. Chem.*, **271**, 11988–11995, (1996a).

Lamberg, A., Jauhiainen, M., Metso, J., Ehnholm, C., Shoulders, C., Scott, J, Pihlajaniemi, T., and Kivirikko, K. I., *Biochem. J.*, **315**, 533–536 (1996b).

Lamport, D. T. A., *Recent Adv. Phytochem.*, **11**, 79–115 (1977).

Lebeche, D., Lucero, H. A., and Kaminer, B., *Biochem. Biophys. Res. Commun.*, **202**, 556–561 (1994).

Lee, E., Némethy, G., Scheraga, H. A., and Ananthanarayanan, V. S., *Biopolymers*, **23**, 1193–1206 (1984).

Lu, J., Willis, A. C., and Reid, K. B. M., *Biochem. J.*, **284**, 795–802 (1992).

Lucero, H. A., Lebeche, D., and Kaminer, B., *J. Biol. Chem.*, **269**, 23112–23119 (1994).

Lundström, J. and Holmgren, A., *J. Biol. Chem.*, **265**, 9114–9120 (1990).

Lundström, J., Krause, G., and Holmgren, A., *J. Biol. Chem.*, **267**, 9047–9052 (1992).

Lundström-Ljung, J., Birnbach, U., Rupp, K., Söling, H.-D., and Holmgren, A., *FEBS Lett.*, **357**, 305–308 (1995).

Lyles, M. M. and Gilbert, H. F., *J. Biol. Chem.*, **269**, 30946–30952 (1994).

Maeda, H., Matsumura, Y., and Kato, H., *J. Biol. Chem.*, **263**, 16051–16054 (1988).

Maeda, K., Okubo, K., Shimomura, I., Funahashi, T., Matsuzawa, Y., and Matsubara, K., *Biochem. Biophys. Res. Commun.*, **221**, 286–289 (1996).

Majamaa, K., Hanauske-Abel, H. M., Günzler, V., and Kivirikko, K. I., *Eur. J. Biochem.*, **138**, 239–245 (1984).

Majamaa, K., Turpeenniemi-Hujanen, T. M., Latipää, P., Günzler, V., Hanauske-Abel, H. M., Hassinen, I. E., and Kivirikko, K. I., *Biochem. J.*, **229**, 127–133 (1985).

Majamaa, K., Günzler, V., Hanauske-Abel, H., Myllylä, R., and Kivirikko, K. I., *J. Biol. Chem.*, **261**, 7819–7823 (1986).

Majamaa, K., Sasaki, T., and Uitto, J., *J. Invest. Dermatol.*, **89**, 405–409 (1987).

Martin, J. L., *Curr. Biol.*, **3**, 245–250 (1995).

Martin, J. L., Pumford, N. R., LaRosa, A. C., Martin, B. M., Gonzaga, H. M. S., Beaven, M. A., and Pohl, L. R., *Biochem. Biophys. Res. Commun.*, **178**, 679–685 (1991).

Matsuda, J., Okabe, S., Hashimoto, T., and Yamada, Y., *J. Biol. Chem.*, **266**, 9460–9464 (1991).

Matsushita, M., Endo, Y., Taira, S., Sato, Y., Fujita, T., Ichikawa, N., Nakata, M., and Mizuochi, T., *J. Biol. Chem.*, **271**, 2448–2454 (1996).

Mayne, R. and Brewton, R. G., *Curr. Opin. Cell Biol.*, **5**, 883–890 (1993).

Mazzarella, R. A., Srinivasan, M., Haugejorden, S. M., and Green, M., *J. Biol. Chem.*, **265**, 1094–1101 (1990).

Mazzarella, R. A., Marcus, N., Haugejorden, S. M., Balcarek, J. M., Baldassare, J. J., Roy, B., Li, L., Lee, A. S., and Green, M., *Arch. Biochem. Biophys.*, **308**, 454–460 (1994).

McCaffrey, T. A., Pomerantz, K., Sanborn, T. A., Spokojny, A., Du, B., Park, M. H., Folk, J. E., Lamberg, A., Kivirikko, K. I., Falcone, D. J., Mehta, S., and Hanauske-Abel, H. M., *J. Clin. Invest.*, **95**, 446–455 (1995).

McGarvey, D. J., Yu, H., and Christoffersen, R. E., *Plant Mol. Biol.*, **15**, 165–167 (1990).

McKay, R. R., Zhum, L. Q., and Shortridge, R. D., *Insect Biochem. Mol. Biol.*, **25**, 647–654 (1995).

Ming, L.-J., Que, L. J., Jr., Kriauciunas, A., Frolik, C. A., and Chen, V. J., *Biochemistry*, **30**, 11653–11659 (1991).

Murad, S. and Pinnell, S. R., *J. Biol. Chem.*, **262**, 11973–11978 (1987).

Murad, S., Tennant, C., and Pinnell, S. R., *Arch. Biochem. Biophys.*, **292**, 234–238 (1992).

Murad, S., Walker, L. C., Tajima, S., and Pinnell, S., *Arch. Biochem. Biophys.*, **308**, 42–47 (1994).

Myllyharju, J., and Kivirikko, K. I., *EMBO J.*, **16**, 1173–1180 (1997).

Myllyharju, J., Lamberg, A., Notbohm, H., Fietzek, P. P., Pihlajaniemi, T., and Kivirikko, K. I., *J. Biol. Chem.*, **272**, 21824–21830 (1997).

Myllylä, R., Tuderman, L., and Kivirikko, K. I., *Eur. J. Biochem.*, **80**, 349–357 (1977).

Myllylä, R., Kuutti-Savolainen, E.-R., and Kivirikko, K. I., *Biochem. Biophys. Res. Commun.*, **83**, 441–448 (1978).

Myllylä, R., Schubotz, L. M., Weser, U., and Kivirikko, K. I., *Biochem. Biophys. Res. Commun.*, **89**, 98–102 (1979).

Myllylä, R., Majamaa, K., Günzler, V., Hanauske-Abel, H. M., and Kivirikko, K. I., *J. Biol. Chem.*, **259**, 5403–5405 (1984).

Myllylä, R., Pajunen, L., and Kivirikko, K. I., *Biochem. J.*, **253**, 489–496 (1988).

Myllylä, R., Kaska, D., and Kivirikko, K. I., *Biochem. J.*, **263**, 609–611 (1989).

Myllylä, R., Pihlajaniemi, T., Pajunen, L., Turpeenniemi-Hujanen, T., and Kivirikko, K. I., *J. Biol. Chem.*, **266**, 2805–2810 (1991).

Myllylä, R., Günzler, V., Kivirikko, K. I., and Kaska, D. D., *Biochem. J.*, **286**, 923–927 (1992).

Nakazawa, M., Aida, T., Everson, W. V., Gonda, M. A., Hughes, S. H., and Kao, W. W.-Y., *Gene*, **71**, 451–460 (1988).

Nguyen Van, P., Rupp, K., Lampen, A., and Söling, H.-D., *Eur. J. Biochem.*, **213**, 789–795 (1993).

Nokelainen, M., Nissi, R., Annunen, P., Helaakoski, T., Pihlajaniemi, T., and Kivirikko, K. I., *Abstr. XV FECTS Meeting*, Munich, F 16 (1996).

Noiva, R. and Lennarz, W. J., *J. Biol. Chem.*, **267**, 3553–3556 (1992).

Noiva, R., Kimura, H., Roos, J., and Lennarz, W. J., *J. Biol. Chem.*, **266**, 19645–19649 (1991).

Noiva, R., Freedman, R. B., and Lennarz, W. J., *J. Biol. Chem.*, **268**, 19210–19217 (1993).

Noiva, R., Onodera, S., Schwaller, M. D., and Irwin, V. A., in *Prolyl 4-Hydroxylase, Protein Disulfide Isomerase and Other Structurally Related Proteins*, Guzman, N. A., Ed., Marcel-Dekker, New York, in press.

Oliver, J. D., van der Wal, F. J., Bulleid, N. J., and High, S., *Science*, **275**, 86–88 (1997).

Otsu, M., Omura, F., Yoshimori, T., and Kikuchi, M., *J. Biol. Chem.*, **269**, 6874–6877 (1994).

Pajunen, L., Myllylä, R., Helaakoski, T., Pihlajaniemi, T., Tasanen, K., Höyhtyä, M., Tryggvason, K., Solomon, E., and Kivirikko, K. I., *Cytogenet. Cell. Genet.*, **47**, 37–41 (1988).

Pajunen, L., Jones, T. A., Helaakoski, T., Pihlajaniemi, T., Solomon, E., Sheer, D., and Kivirikko, K. I., *Am. J. Hum. Genet.*, **45**, 829–834.(1989).

Pajunen, L., Jones, T. A., Goddard, A., Sheer, D., Solomon, E., Pihlajaniemi, T., and Kivirikko, K. I., *Cytogenet. Cell Genet.*, **56**, 165–168 (1991).

Parkkonen, T., Kivirikko, K. I., and Pihlajaniemi, T., *Biochem. J.*, **256**, 1005–1011 (1988).

Pelham, H. R. B., *Trends Biochem. Sci.*, **15**, 483–486 (1990).

Pelicci, G., Lanfrancone, L., Grignani, F., McGlade, J., Cavallo, F., Forni, G., Nicoletti, I., Grignani, F., Pawson, T., and Pelicci, P. G., *Cell*, **70**, 93–104 (1992).

Peterkofsky, B., Tschank, G., and Luedke, C., *Arch. Biochem. Biophys.*, **254**, 282–289 (1987).

Pihlajaniemi, T. and Rehn, M., *Progr. Nucl. Acid. Res. Mol. Biol.*, **50**, 225–262 (1995).

Pihlajaniemi, T., Helaakoski, T., Tasanen, K., Myllylä, R., Huhtala, M.-L., Koivu, J., and Kivirikko, K. I., *EMBO J.*, **6**, 643–649 (1987).

Pinnell, S. R., Krane, S. M. Kenzora, J. E., and Glimcher, M. J., *N. Engl. J. Med.*, **286**, 1013–1020 (1972).

Pirskanen, A., Kaimio, A.-M., Myllylä, R., and Kivirikko, K. I., *J. Biol. Chem.*, **271**, 9398–9402 (1996).

Popescu, N. C., Cheng, S-Y., and Pastan, I., *Am. J. Hum. Genet.*, **42**, 560–564 (1988).

Pousi, B., Hautala, T., Heikkinen, J., Pajunen, L., Kivirikko, K. I., and Myllylä, R., *Am. J. Hum. Genet.*, **55**, 899–906 (1994).

Prescott, A. G., *J. Exp. Bot.*, **44**, 849–861 (1993).

Prockop, D. J., Berg, R. A., Kivirikko, K. I., and Uitto J., in *Biochemistry of Collagen*, Ramachandran, G. N. and Reddi, A. H., Eds., Plenum, New York, 1976, pp. 164–273.

Prockop, D. J. and Kivirikko, K. I., *J. Biol. Chem.*, **244**, 4838–4842 (1969).

Prockop, D. J. and Kivirikko, K. I., *N. Engl. J. Med.*, **311**, 376–386 (1984).

Prockop, D. J. and Kivirikko, K. I., *Annu. Rev. Biochem.*, **64**, 403–434 (1995).

Puig, A. and Gilbert, H. F., *J. Biol. Chem.*, **269**, 7764–7771 (1994a).

Puig, A. and Gilbert, H. F., *J. Biol. Chem.*, **269**, 25889–25896 (1994b).

Puig, A., Lyles, M. M., Noiva, R., and Gilbert, H. F., *J. Biol. Chem.*, **269**, 19128–19135 (1994).

Puistola, U., Turpeenniemi-Hujanen, T. M., Myllylä, R., and Kivirikko, K. I., *Biochim. Biophys. Acta,* **611**, 40–50 (1980a).

Puistola, U., Turpeenniemi-Hujanen, T. M. Myllylä, R., and Kivirikko, K. I., *Biochim. Biophys. Acta,* **611**, 51–60 (1980b).

Quan, H., Fan, G. B., and Wang, C. C., *J. Biol. Chem.*, **270**, 17078–17080 (1995).

Quéméneur, E., Guthapfel, R., and Guéguen, P., *J. Biol. Chem.*, **269**, 5485–5488 (1994).

Randall, C. R., Zang, Y., True, A. E., Que, L. J., Jr., Charnock, J. M., Garner, C. D., Fujishima, Y., Schofield, C. J., and Baldwin, J. E., *Biochemistry*, **32**, 6664–6673 (1993).

Rao, N. V. and Adams, S., *J. Biol. Chem.*, **253**, 6327–6330 (1978).

Reid, K. B. M., *Biochem. J.*, **179**, 367–371 (1979).

Reid, K. B. M., *Biochem. Soc. Transact.*, **21**, 464–468 (1993).

Rhoads, R. E. and Udenfriend, S., *Proc. Natl. Acad. Sci. USA*, **60**, 1473–1478 (1968).

Ricci, B., Sharp, D., Orourke, E., Kienzle, B., Blinderman, L., Gordon, D., Smithmonroy, C., Robinson, G., Gregg, R. E., Rader, D. J., and Wetterau, J. R., *J. Biol. Chem.*, **270**, 14281–14285 (1995).

Risteli, J., Tryggvason, K., and Kivirikko, K. I., *Eur. J. Biochem.*, **73**, 485–492 (1977).

Roach, P. L., Clifton, I. J., Fülöp, V., Harlos, K., Barton, G. J., Hajdu, J., Andersson, I., Schofield, C. J., and Baldwin, J. E., *Nature London*, **375**, 700–704 (1995).

Roguska, M. A. and Gudas, L. J., *J. Biol. Chem.*, **260**, 13893–13896 (1985).

Rohrer, L., Freeman, M., Kodama, T., Penman, M., and Krieger, M., *Nature (London)*, **343**, 570–572 (1990).

Rosenberry, T. L., Barnett, P., and Mays, C., *Methods Enzymol.*, **82**, 325–339 (1982).

Rosenbloom, J., in *Connective Tissue and Its Heritable Disorders. Molecular, Genetic and Medical Aspects*, Royce, P. M. and Steinmann, B., Eds., Wiley-Liss, New York, 1993, pp. 167–188.

Royce, P. M. and Barnes, M. J., *Biochem. J.*, **230**, 475–480 (1985).

Rupp, K., Birnbach, U., Lundström, J., Nguyen Van, P., and Söling, H.-D., *J. Biol. Chem.*, **269**, 2501–2507 (1994).

Ryhänen, L. and Kivirikko, K. I., *Biochim. Biophys. Acta*, **343**, 129–137 (1974).

Samson, S. M., Dotzlaf, J. E., Slisz, M. L., Becker, G. W., Van Frank, R. M., Veal, L. E., Yeh, W.-K., Miller, J. R., Queener, S. W., and Ingolia, T. D., *Biotechnology*, **5**, 1207–1211 (1987).

Sasaguri, M., Ikeda, M., Ideishi, M., and Arakawa, K., *Biochem. Biophys. Res. Commun.*, **150**, 511–516 (1988).

Sasaki, T., Majamaa, K., and Uitto, J., *J. Biol. Chem.*, **262**, 9397–9403 (1987).

Shigematsu, T., Tajima, S., Nishikawa, T., Murad, S., Pinnell, S. R., and Nishioka, I., *Biochim. Biophys. Acta.*, **1200**, 79–83 (1994).

Showalter, A. M. and Varner, J. E., in *The Biochemistry of Plants: A Comprehensive Treatise*, Vol. 15, Marcus, A., Ed., Academic, New York, 1989, pp. 485–520.

Song, J.-L. and Wang, C.-C., *Eur. J. Biochem.*, **231**, 312–316 (1995).

Spiess, J. and Noe, B. D., *Proc. Natl. Acad. Sci. USA*, **82**, 277–281 (1985).

Srivastava, S. P., Chen, N. Q., Liu, Y. X., and Holtzman, J. L., *J. Biol. Chem.*, **266**, 20337–20344 (1991).

Srivastava, S. P., Fuchs, J. A., and Holtzman, J. L., *Biochem. Biophys. Res. Commun.*, **193**, 971–978 (1993).

Steinmann, B., Royce, P. M., and Superti-Furga, A., in *Connective Tissue and Its*

Heritable Disorders. Molecular, Genetic and Medical Aspects, Royce, P. M. and Steinmann, B., Eds., Wiley-Liss, New York, 1993, pp. 351–407.

Stetten, M. R., *J. Biol. Chem.*, **181**, 31–37 (1949).

Stetten, M. R. and Schoenheimer, R., *J. Biol. Chem.*, **153**, 113–132 (1944).

Tachibana, C. and Stevens, T. H., *Mol. Cell. Biol.*, **12**, 4601–4611 (1992).

Tasanen, K., Parkkonen, T., Chow, L. T., Kivirikko, K. I., and Pihlajaniemi, T., *J. Biol. Chem.*, **263**, 16218–16224 (1988).

Tasanen, K., Oikarinen, J., Kivirikko, K. I., and Pihlajaniemi, T., *J. Biol. Chem.*, **267**, 11513–11519 (1992).

Tasanen, K., Oikarinen, J., Kivirikko, K. I., and Pihlajaniemi, T., *Biochem. J.*, **292**, 41–45 (1993).

Thomas, D., Patterson, S. D., and Bradshaw, R. A., *J. Biol. Chem.*, **270**, 28924–28931 (1995).

Tryggvason, K., Risteli, J., and Kivirikko, K. I., *Biochem. Biophys. Res. Commun.*, **76**, 275–281 (1977).

Tryggvason, K., Majamaa, K., Risteli, J., and Kivirikko, K. I., *Biochem. J.*, **183**, 303–307 (1979).

Tschank, G., Hanauske-Abel, H. M., and Peterkofsky, B., *Arch. Biochem. Biophys.*, **261**, 312–323 (1988).

Tschank, G., Brocks, D. G., Engelbart, K., Mohr, J., Baader, E., Günzler, V., and Hanauske-Abel, H. M., *Biochem. J.*, **275**, 469–476 (1991).

Tschank, G., Sanders, J., Baringhaus, K.-H., Dallacker, F., Kivirikko, K. I., and Günzler, V., *Biochem. J.*, **300**, 75–79 (1994).

Tsibris, J. C. M., Hunt, L. T., Ballejo, G., Barker, W. C., Toney, L. J., and Spellacy, W. N., *J. Biol. Chem.*, **264**, 13967–13970 (1989).

Tucker, H. and Thomas, D. F., *J. Med. Chem.*, **35**, 804–807 (1992).

Tuderman, L., Kuutti, E.-R., and Kivirikko, K. I., *Eur. J. Biochem.*, **52**, 9–16 (1975).

Tuderman, L., Myllylä, R., and Kivirikko, K. I., *Eur. J. Biochem.*, **80**, 341–348 (1977).

Turpeenniemi-Hujanen, T. M., Puistola, U., and Kivirikko, K. I., *Biochem. J.*, **189**, 247–253 (1980).

Turpeenniemi-Hujanen, T. M., Puistola, U., and Kivirikko, K. I., *Coll. Rel. Res.*, **1**, 355–366 (1981).

Valtavaara, M., Papponen, H., Pirttilä, A.-M., Hiltunen, K., Helander, H., and Myllylä, R., *J. Biol. Chem.*, **272**, 6831–6834 (1997).

van der Rest, M. and Bruckner, P., *Curr. Opin. Struct. Biol.*, **3**, 340–346 (1993).

van der Rest, M. and Garrone, R., *FASEB J.*, **5**, 2814–2823 (1991).

Veijola, J., Koivunen, P., Annunen, P., Pihlajaniemi, T., and Kivirikko, K. I., *J. Biol. Chem.*, **269**, 26746–26753 (1994).

Veijola, J., Annunen, P., Koivunen, P., Page, A., Pihlajaniemi, T., and Kivirikko, K. I., *Biochem. J.*, **317**, 721–729 (1996a).

Veijola, J., Pihlajaniemi, T., and Kivirikko, K. I., *Biochem. J.*, **315**, 613–618 (1996b).

Vuorela, A., Myllyharju, J., Nissi, R., Pihlajaniemi, T., and Kivirikko, K. I., *EMBO J.* **16**, 6702–6712 (1997).

Vuori, K., Myllylä, R., Pihlajaniemi, T., and Kivirikko, K. I., *J. Biol. Chem.*, **267**, 7211–7214 (1992a).

Vuori, K., Pihlajaniemi, T., Marttila, M., and Kivirikko, K. I., *Proc. Natl. Acad. Sci. USA*, **89**, 7467–7470 (1992b).

Vuori, K., Pihlajaniemi, T., Myllylä, R., and Kivirikko, K. I., *EMBO J.*, **11**, 4213–4217 (1992c).

Wells, W. W., Xu, D. P., Yang, Y., and Rocque, P. A., *J. Biol. Chem.*, **265**, 15361–15364.

Wenstrup, R. J., Murad, S., and Pinnell, S. R., *J. Pediatr.*, **115**, 405–409 (1989).

Wetterau, J. R., Combs, K. A., Spinner, S. N., and Joiner, B. J., *J. Biol. Chem.*, **265**, 9800–9807 (1990).

Wetterau, J. R., Aggerbeck, L. P., Laplaud, P. M., and McLean, L. R., *Biochemistry*, **30**, 4406–4412 (1991a).

Wetterau, J. R., Combs, K. A., McLean, L. R., Spinner, S. N., and Aggerbeck, L. P., *Biochemistry*, **30**, 9728–9735 (1991b).

Wijffels, G. L., Panaccio, M., Salvatore, L., Wilson, L., Walker, I. D., and Spithill, T. W., *Biochem. J.*, **299**, 781–790 (1994).

Wilson, W. R., Tuan, R. S., Shepley, K. J., Freedman, R. B., Greene, B. M., and Awadzi, K., *Mol. Biochem. Parasitol.*, **68**, 103–117 (1994).

Yamauchi, K., Yamamoto, T., Hayashi, H., Koya, S., Takikawa, H., Toyoshima, K., and Horiuchi, R., *Biochem. Biophys. Res. Commun.*, **146**, 1485–1492 (1987).

Yao, Y., Zhow, Y.-C., and Wang, C.-C., *EMBO J.*, **16**, 651–658 (1997).

Yeowell, H. N., Murad, S., and Pinnell, S. R., *Arch. Biochem. Biophys.*, **289**, 399–404 (1991).

Yeowell, H. N., Ha, V., Walker, L. C., Murad, S., and Pinnell, S. R., *J. Invest. Dermatol.*, **99**, 864–869 (1992).

SUBJECT INDEX

433